Surface Engineering

Surface engineering is considered an important aspect in the reduction of friction and wear. This reference text discusses a wide range of surface engineering technologies along with applications in a comprehensive manner.

The book describes various methods in surface engineering technology with a thorough explanation of various aspects of each process that comes under this domain. Apart from an enhanced explanation of the process and its attributes, this book also gives insight into the types of materials, applications, and optimization of surface engineering techniques. It discusses important topics including surface engineering of the functionality of graded materials, materials characterization, processing of biomaterials, design, surface modification technologies and process control, smart manufacturing, artificial intelligence, and machine learning applications.

The book

- discusses computational and simulation analyses for better selection of process parameters.
- covers optimizations of processes with state-of-the-art technologies.
- discusses applications of surface engineering in medical, agricultural, architecture engineering, and allied sectors.
- covers processing techniques of biomaterials in surface engineering.

The text is useful for senior undergraduate, graduate students, and academic researchers working in diverse areas such as industrial and production engineering, mechanical engineering, materials science, and manufacturing science. It covers a hybrid process for surface modification, modeling techniques, and issues in surface engineering.

Surface Engineering
Methods and Applications

Edited by
R. S. Walia
Qasim Murtaza
Shailesh Mani Pandey
Ankit Tyagi

CRC Press
Taylor & Francis Group
Boca Raton London New York

CRC Press is an imprint of the
Taylor & Francis Group, an **informa** business

First edition published 2023
by CRC Press
6000 Broken Sound Parkway NW, Suite 300, Boca Raton, FL 33487-2742

and by CRC Press
4 Park Square, Milton Park, Abingdon, Oxon, OX14 4RN

CRC Press is an imprint of Taylor & Francis Group, LLC

© 2023 selection and editorial matter, R. S. Walia, Qasim Murtaza, S. M. Pandey and Ankit Tyagi individual chapters, the contributors

Library of Congress Cataloging-in-Publication Data
Names: Walia, R. S., editor.
Title: Surface engineering : methods and applications / edited by R.S.
   Walia, Qasim Murtaza, S. M Pandey, and Ankit Tyagi.
Other titles: Surface engineering (CRC Press)
Description: First edition. l Boca Raton : CRC Press, [2023] l Includes
   bibliographical references and index.
Identifiers: LCCN 2022026128 (print) l LCCN 2022026129 (ebook) l ISBN
   9781032055015 (hbk) l ISBN 9781032333700 (pbk) l ISBN 9781003319375
   (ebk)
Subjects: LCSH: Surfaces (Technology)
Classification: LCC TA418.7 .S848 2023 (print) l LCC TA418.7 (ebook) l
   DDC 620/.44--dc23/eng/20221011
LC record available at https://lccn.loc.gov/2022026128
LC ebook record available at https://lccn.loc.gov/2022026129

ISBN: 978-1-032-05501-5 (hbk)
ISBN: 978-1-032-33370-0 (pbk)
ISBN: 978-1-003-31937-5 (ebk)

DOI: 10.1201/9781003319375

Typeset in Sabon
by SPi Technologies India Pvt Ltd (Straive)

Contents

Preface

Surface engineering has been regarded as one of the most important approaches to the reduction of friction and wear. The product requires surface engineering for excellent mechanical, chemical, and tribological properties and higher performance. With advancements in surface engineering processes, the product performance and efficiency have increased and are ever growing. From traditional to advance engineering, the processes have evolved immensely over time. One domain of advanced surface engineering processes is the involvement of biomaterials, smart materials, and hybrid materials in the processes.

This book details out a wide range of surface engineering technologies along with applications in a comprehensive manner. It elaborates on various technologies invented and employed for surface engineering of products with different materials and modifications. The primary aim of this book is to give the readers a well-informed layout of the surface engineering technology that ranges from nanomaterials to hybrid and smart materials, Industry 4.0, smart manufacturing, and technologies involved in the industry-specific applications each possesses. It also talks about various aspects of surface engineering that are different from process-specific details. Some of these aspects are computational and simulation analysis, the use of artificial intelligence and machine learning, design modification, process control, and their characteristics for engineering compatibility. The authors have very distinctively brought together current research archives from various sources and have come up with a platform that has surface engineering technology together with latest research technologies used in these processes for the readers to benefit from. The readers perceive this platform in a manner which makes it easy and better for them to understand and gain from. This easy manner of understanding provides the readers with an opportunity to have a deep understanding of the methods, materials, applications, and optimization of surface engineering technology.

The book has a wide scope of reaching out to a diverse audience including but not limited to production or manufacturing engineers, researchers, and academicians. The audience will be guided by this book to be able to make appropriate process selections for different types of materials used for surface engineering. The researcher scope of this book is that it will help the research fraternity in narrowing down steps for further research in similar domain. The book very distinctively provides scope of further enhancements to the surface engineering technology by providing explanations on latest advancements in the manufacturing processes. With industries advancing in the Fourth Industrial Revolution, these processes are needed to be revamped with new materials and techniques to keep their pace with industrial production standards. A section of the book is dedicated to the application and optimization of surface engineering techniques.

Editors

R. S. Walia is a Professor and Head of the Department of Production & Industrial Engineering, Punjab Engineering College, Chandigarh, India. He earned his Ph.D. (2006) in the area of Centrifugal Force Assisted Abrasive Flow Machining Process from IIT Roorkee, India. He has also worked as Professor at Delhi Technological University, Delhi, India. He has visited various academic institutions and industries in several countries for academic pursuits.

His research interests include Advanced Manufacturing Processes, Industrial Engineering, Work System Design and Ergonomics, Operation Research, Industrial Quality Control, Precision Manufacturing, and Advanced Manufacturing Processes such as metal coatings, free-standing metal coating components, hybrid stir casting & including tribological assessment, super-finishing processes, and re-manufacturing. He has guided 45 M.Tech theses and 14 Ph.D.s. He has published more than 100 research papers in reputed journals and conference proceedings and recieved best paper awards from Elsevier, Emerald, and Springer publication houses. He received the Commendable Research Award in recognition for his research during the year 2017, DTU, Delhi, and the Best Performance Award International Teams for NASA Moon Buggy Race at US Space and Rocket Centre, USA, April 2011. He was part of interview boards as expert member in public sector companies and government service commissions and also worked in the examination/research boards of various institutions and was involved in multiple institutional activities related to policy planning curriculum design, admission, evaluation & examination, and other academic processes.

Qasim Murtaza is a Professor in the Department of Mechanical Engineering, Delhi Technological University, (Erstwhile Delhi College of Engineering) Delhi, India. He earned his Ph.D. (2006) in the area of Manufacturing Process, Dublin City University, Ireland. He has also worked as Research Associate at Metropolitan Manchester University, Manchester, UK. He has visited various academic institutions and industries in several countries for academic pursuits. His research interests include Precision

Manufacturing and Advanced Manufacturing Processes such as metal coatings, free-standing metal coating components, hybrid stir casting & including tribological assessment, super-finishing processes, and re-manufacturing. He has guided 33 M.Tech theses and 7 Ph.D.s. He has published more than 97 research papers in reputed journals and conference proceedings and has received best paper awards from Elsevier, Emerald, and Springer publication houses.

He was part of interview boards as expert member in public sector companies and government service commissions and also worked in the examination/research boards of various institutions and was involved in multiple institutional activities related to policy planning, curriculum design, admission, evaluation & examination, and other academic processes.

Shailesh Mani Pandey is an Assistant Professor in the Department of Mechanical Engineering, National Institute of Technology (NIT) Patna, Patna, India. He earned his Ph.D. (2017) in the area of Manufacturing Process in Advance Coating, Delhi Technological University, Delhi, India. He has also worked as Assistant Professor at Delhi Technological University. His areas of research include Surface Modification, Coating, Tribology, and Material Processing; Composite Materials; Metal Matrix Nanocomposites; Nano-Coatings; Wear; Deformation and Corrosion. He also has a good command over teaching subjects like Welding Technology, Casting, Wear Mechanics, Material Science, Non-Destructive Testing, Metrology, and Composite Materials. Dr. Pandey has over 2 years of Industry Experience at Shriram Pistons and Rings Limited (SPRL) Ghaziabad in the field of Production Planning and Control and 8.5 plus years of teaching and research in various reputed organizations such as Delhi Technological University and NIT Patna. During this period, he has published 35 research papers in peer-reviewed international journals having good impact factors and 28 research papers in reputed international and national conferences in India as well as in abroad. Apart from this, recently, he has also published 1 chapter in books published by Springer. Nine students have completed their B.Tech projects and 6 students have completed their M.Tech dissertations under his guidance.

Ankit Tyagi is an Assistant Professor in the Department of Mechanical Engineering, SGT University, Haryana, India. He is pursuing his Ph.D. in the area of Manufacturing Process in Advance Coating, Delhi Technological University, Delhi, India. He has visited various academic institutions and industries in several countries for academic pursuits. His areas of research include Surface Modification, Coating, Tribology, and Material Processing; Composite Materials; Metal Matrix Nanocomposites; Nano-Coatings; Wear; Deformation and Corrosion. He also has a good command over teaching subjects like Welding Technology, Casting, Wear Mechanics, Material Science, Non-Destructive Testing, Metrology, and Composite Materials. He has published more than 30 research papers in reputed journals and conference proceedings.

Contributors

N. Adithya
Chennai Institute of Technology
Chennai, Tamil Nadu, India

Bijaya Bikram Samal
Advanced Technology Development
    Centre
Indian Institute of Technology
    Kharagpur
Kharagpur, West Bengal, India

Vasavi Boggarapu
National Institute of Technology
    Warangal
Warangal, Telangana, India

Shubhangi Chourasia
Global Institute of Technology and
    Management
Gurugram, Haryana, India
Delhi Technological University
Delhi, India

R. Deepak Suresh Kumar
Chennai Institute of Technology
Chennai, Tamil Nadu, India

T. Deepika
Amrita Vishwa Vidyapeetham
Coimbatore, Tamil Nadu, India

Pradnya D. Desai
Institute of Chemical Technology
Mumbai, Maharashtra, India

Abdul Faheem
University Polytechnic, AMU
Aligarh, Uttar Pradesh, India

S. Ganeshkumar
Sri Eshwar College of
    Engineering
Coimbatore, Tamil Nadu, India

Arka Ghosh
National Institute of Technology
    Rourkela
Rourkela, Odisha, India

Kalpana Gupta
Delhi Technological University
Delhi, India

Faisal Hasan
Zakir Husain College
    of Engineering and
    Technology, Aligarh Muslim
    University
Aligarh, Uttar Pradesh, India

Syed Ismail
National Institute of Technology
  Warangal
Warangal, Telangana, India

Justin Joseph
Chennai Institute of Technology
Chennai, Tamil Nadu, India

Ajith G. Joshi
Canara Engineering College
Bantwal, Karnataka, India

B. Krishna Prabhu
Canara Engineering College
Bantwal, Karnataka, India

S. M. Kulkarni
National Institute of Technology
  Karnataka
Surathkal, Karnataka, India

Abhishek Kumar
Indian Institute of Technology
  Kharagpur
Kharagpur, West Bengal, India

Satyajeet Kumar
National Institute of Technology
  Patna
Patna, Bihar, India

Chandan Kumar Biswas
National Institute of Technology
  Rourkela
Rourkela, Odisha, India

Ashish Kumar Nath
LASER Lab
Indian Institute of Technology
  Kharagpur
Kharagpur, West Bengal, India

Manab Mallik
National Institute of Technology
  Durgapur
Durgapur, West Bengal, India

M. Manjaiah
National Institute of Technology
  Warangal
Warangal, Telangana, India

Monalisha Mohanta
National Institute of Technology
  Rourkela
Rourkela, Odisha, India

Aarti P. More
Institute of Chemical
  Technology
Mumbai, Maharashtra, India

Qasim Murtaza
Delhi Technological University
Delhi, India

Syed Nasimul Alam
National Institute of Technology
  Rourkela
Rourkela, Odisha, India

V. Naveenprabhu
Sri Eshwar College of Engineering
Coimbatore, Tamil Nadu, India

Sikta Panda
National Institute of Technology
  Rourkela
Rourkela, Odisha, India

Shailesh Mani Pandey
National Institute of Technology
  Patna
Patna, Bihar, India

Subhankar Paul
National Institute of Technology
  Rourkela
Rourkela, Odisha, India

Mainak Saha
Indian Institute of Technology
  Madras
Chennai, Tamil Nadu, India

K. Sathish
Sri Eshwar College of Engineering
Coimbatore, Tamil Nadu, India

Pankaj Shrivastava
National Institute of Technology
  Rourkela
Rourkela, Odisha, India

Cheruvu Siva Kumar
Direct Digital Manufacturing Lab
Indian Institute of Technology
  Kharagpur
Kharagpur, West Bengal, India

R. Suresh
M S Ramaiah University of Applied
  Sciences
Bangalore, Karnataka, India

R. Sureshkumar
Sri Eshwar College of Engineering
Coimbatore, Tamil Nadu, India

A. Thirugnanam
National Institute of Technology
  Rourkela
Rourkela, Odisha, India

Ankit Tyagi
SGT University
Gurugram, Haryana, India

**Peddakondigalla Venkateswara
  Babu**
Vasireddy Venkatadri Institute of
  Technology
Guntur, Andhra Pradesh, India

K. B. Vigneshwara
Chennai Institute of
  Technology
Chennai, Tamil Nadu, India

R. Vivek
Sri Eshwar College of
  Engineering
Coimbatore, Tamil Nadu, India

R. S. Walia
Punjab Engineering College
Chandigarh, Punjab, India

K. T. Yugendheran
Chennai Institute of
  Technology
Chennai, Tamil Nadu, India

Chapter 1

Surface engineering
Redefining surface in global perspective

Shubhangi Chourasia
Global Institute of Technology and Management, Gurugram, Haryana, India
Delhi Technological University, Delhi, India

Shailesh Mani Pandey
National Institute of Technology Patna, Patna, Bihar, India

Kalpana Gupta
Delhi Technological University, Delhi, India

Abdul Faheem
University Polytechnic, AMU, Aligarh, Uttar Pradesh, India

Qasim Murtaza
Delhi Technological University, Delhi, India

R. S. Walia
Punjab Engineering College, Chandigarh, Punjab, India

CONTENTS

DOI: 10.1201/9781003319375-1

1.1 INTRODUCTION

The wear and tear in the moving components are typical problems from many years that generally arise during operating conditions, leading to a steady decrease in working efficiency, regular performance, and several work-dependent parameters of components. At this time, damaged components are required to be changed or replaced with new parts, which results in a momentary shutdown of the machine or the complete work process [1]. Constantly, industries are working to enhance the efficiency, component life, and effectiveness of high-tech technological processes through the concept of surface engineering. Thus, industries continuously develop new materials, coatings, new production methods, and advanced structures that could work successfully on every operating condition to deliver and confirm the maximum resilient output with minimum capital cost. Surface engineering is the vast boundless area envisioned to modify the materials' properties, enhancing the life of components, durability, resilience, and operational functioning of engineering components. Surface engineering modifies or enhances the surface properties such as wettability, visual, optical, corrosion resistance, wear resistance, tear resistance, or tribological properties autonomously. Surface engineering involves varying the properties of surface and surface phases to reduce surface deprivation over the period and can be accomplished by making a robust surface against the environment by employing many surface engineering technologies. The objective of surface engineering is to enhance the wear resistance behavior of the components and transform the frictional behavior with and without changing the composition like surface hardening, carburizing treatments, and coatings. The coating is the surface engineering technique, where surface engineered layers are smeared onto the substrate materials to improve the tribological properties of materials; the range of methods are PVD (physical vapor deposition), CVD (chemical vapor deposition), HVOF (high-velocity oxy-fuel), and thermal spray. Surface engineering applications are the best way to fight against corrosion, wear, erosion, fracture, etc., because gradual deterioration of metallic surfaces and components due to corrosion, wear, and erosion leads to forfeiture in cost, plant efficiency, and leaves shutdown impacts. It can be seen in many conducted studies by ASM [2] that the U.S. economy spends 300 billion dollars per year to save metal from corrosion, which is 4.2% of the total national product income. The collective impacts of corrosion and wear could cause a hazardous effect on components, leading to material loss, shortening of material life, and much more than a single process affecting the individuals. Studies have observed that corrosion can solely occur (without mechanical wear), but the reverse is not possible. When corrosion and wear combine, it causes adverse damage to numerous industries such as pulp, mining, paper manufacturing, chemical processing, and energy production.

Table 1.1 Synergistic influences and mechanism of corrosion and wear in a damp and aqueous environment [3]

Property	Influences and mechanism
Abrasion	• Removes metal elements, shielding oxidized metal and diverged/polarized metal • Formation of microscopic size grooves and dents • Strain-hardened surfaces are removed • Plastic deformation occurs due to high stress
Corrosion	• Micro-cracking • Hydraulic splitting due to micro-cracks at pits • Rough surface • Attacks on grain boundaries cause to weakening of the corresponding metals
Influences of abrasion and cracking	• Plastic deformation creates some elements which are more vulnerable to corrosion • Brittle elements and ductile elements are cracked with hydraulic splitting • Splitting, cavitation, and jet erosion occur in metal due to pressurized mill water

Table 1.1 depicts the synergistic influences and mechanism of corrosion and wear, which causes ruining of materials in a damp and aqueous environment [4].

1.2 JOURNEY OF SURFACE ENGINEERING

The surface engineering voyage started in ancient times when primeval China- and Greece-origin people were practicing tempering, case hardening, and heat treatment processes by employing organic materials in solid form. The new journey of surface engineering turned on in the 20th century with the dawn of extensive research into the field of modern surface technologies taking place. Ion beam, plasma spray, and laser techniques were used to develop the composite materials [5]. Surface engineering started around 20 years ago; in 1983, the first surface research engineering institute was entrenched in the Birmingham Research Institute, United Kingdom. The first international journal of surface engineering was dispensed in late 1985. China came up with its first journal of China's surface engineering in 1987. Nowadays, research in surface engineering is spreading its feet very speedily, in the area of electromagnetic energy, thermal spray techniques, laser melting techniques, nano-surface engineering, etc., which have been undergoing rigorous studies. In the last 20 years, surface engineering has been categorized under three developing phases: thermal spraying, PVD, and CVD (laser method, ion implantation, and many surface treatment

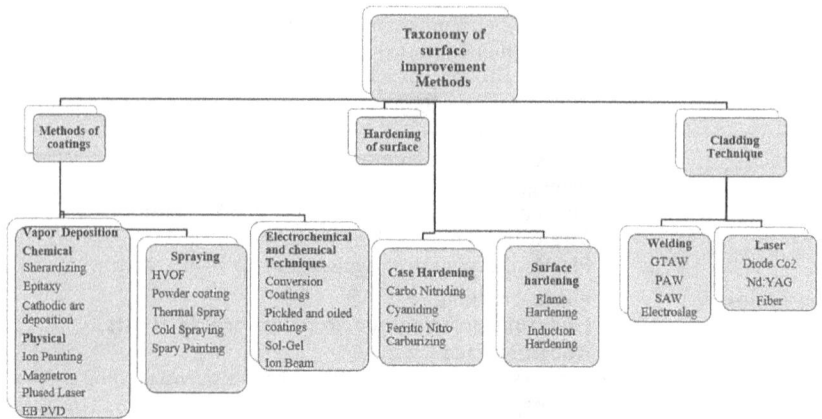

Figure 1.1 Surface improvement engineering methods.

processes). The second category is divided into various surface-coating processes. The third is the use of nanotechnology in nano-based materials. Figure 1.1 shows surface improvement engineering methods.

1.3 ENGINEERING PROPERTIES OF SURFACE ENGINEERING

The purpose of surface engineering is to provide optimum modification to the component surface, microstructure/compositions, and grain structure by improving the surface-reliant properties such as wear, erosion, corrosion, adhesive, abrasive, and oxidation [6, 7]. In surface modification, generally, conventional and advanced processing techniques are applied to modify surfaces, microstructure or compositional, and simultaneously for any section of components. Figure 1.2 presents the primary taxonomy of surface engineering enactment manufacturing/metallurgical industries. Wear is one of the surface engineering properties, which is defined as the gradual loss of materials from the working surface of components when they are in motion [8]. In another way, when tribological interaction occurs between the surfaces of the components with the atmosphere, it tends to result in the loss of material from the component surface. The progression of material loss from the surface of components is called wear. The surface roughness, macrostructure, and microstructure play a vital role in the wear process. Abrasive wear is another form of wear that occurs when a hard/solid surface slides over a soft surface, beginning material loss [9]. Erosive wear is because of solid particles in water or fluid sliding or rolling over the component's surface [10]. In most cases, the occurrence of erosion tends to increase in the corrosive environment. The loss of critical properties and material in the aqueous environment because of chemical and electrochemical reactions that respond to corrosion. Oxidation wear occurs when the material is placed in an oxidizing environment (rich in oxygen) and loss of material occurs.

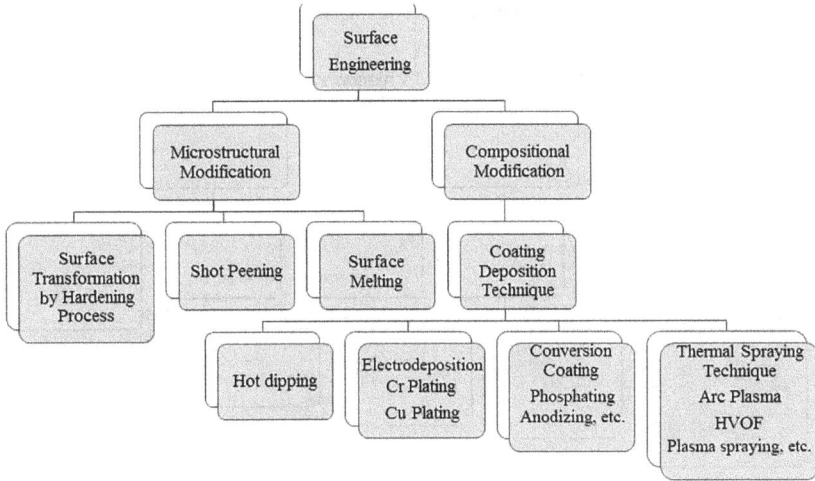

Figure 1.2 Primary taxonomy of surface engineering enactment manufacturing/metallurgical industries.

1.4 TYPES OF SURFACE ENGINEERING

Surface engineering is divided into three categories of processes as shown in Figure 1.3. In the first category, the surface of components is transformed without altering its compositional properties such processes are hardening and melting surface. Hardening is a process to produce hard surface material (e.g., carbon steel). In the hardening process, selective material is heated to the austenite stage, then allowed to be quenched, producing martensite that is successively followed by tempering. The surface hardening mostly depends on the amount of heat provided to the surface layer during the hardening process, the time of heating, and heat loss when the cooling occurs through the conduction method. Standard techniques used for surface hardening are beam hardening, flame hardening, induction hardening, and laser hardening. Surface melting refers to the liquid to solid phase transformation of material, which is heated up to its melting stage and then allowed to solidify. In the second category, components' surface transformation is transformed by altering their compositional properties (microstructure and mechanical properties). Such processes include carburizing (case hardening), carbonitriding, boronizing, nitriding, and chromizing. Carburizing is the process that comprises the diffusion of carbon particles onto the surface of the substrate (steel). At the same time, it became the austenitic and martensitic structure, which formed after the quenching in the carburizing process, which is responsible for providing hardness and grounds the reason for distortion of materials due to transformation in lattice structure and volume. The third category consists of techniques where the materials are smeared onto the surface of components denoted as coating techniques, paints, electroplating, PVD, CVD, and weld hard facing.

Classification of Surface Engineering Methods		
Methods of coatings	**Hardening of Surface**	**Cladding Technique**

Vapour Deposition	**Spraying**					
Chemical	HVOF					
Sheradizing	Powder coating	**Electrochemical and chemical Techniques**	**Case Hardening**		**Welding**	**Laser**
Epittaxy	Thermal Spray	Conversion Coatings	Carbo Nitriding	**Surface hardening**	GTAW	Diode
Cathodic arc deposition	Cold Spraying	Pickled and oiled coatings	Cyaniding	Flame Hardening	PAW	Co2
Physical	Spary Painting	Sol-Gel	Ferritic	Induction Hardening	SAw Electrosl ag	Nd:YAG
Ion Painting		Ion Beam	Nitro Carburizing			Fiber
EB PVD						

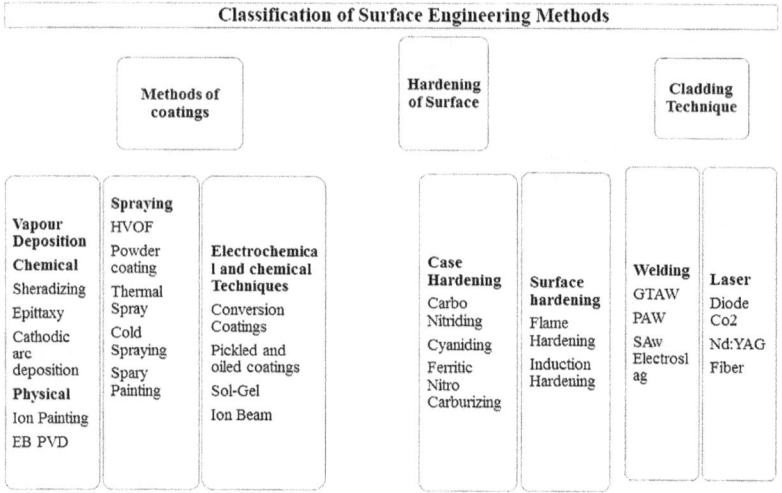

Figure 1.3 Types of surface engineering.

The coating is the surface engineering technique, where surface engineered layers are smeared onto the substrate materials for improving the tribological properties (corrosion, wear resistance, reduced COF, improved thermal insulation, mechanical properties, and aesthetic look). The above matter of content can be improved by providing surface treatment processes such as metallurgical, chemical, and mechanical treatment, as shown in Table 1.2.

Table 1.2 Surface engineering techniques and benefits

Surface treatments	Techniques	Benefits
Modification in surface metallurgy	Surface hardening	Hard martensitic surface developed results of this attained improved wear resistance
	Laser melting process	Grain refinement results of this attained improved wear resistance
	Shot peening process	Inducement of compressive stress on the bare metal surface results in improved fatigue strength
Modification in surface chemistry	The coating is based on phosphate chemical	They are used for steel for increasing adhesion and corrosion resistance
	Steam treating	Used in ferrous powder metallurgy process to enhance wear resistance
	Carbonitriding	Improvement in wear resistance properties of steel
	Ion Implantation	Wear resistance and friction improved
	Laser alloying	Wear resistance improved
	Boriding	Surface fatigue, wear resistance, and oxidation wear can be improved

(Continued)

Table 1.2 (Continued) Surface engineering techniques and benefits

Surface treatments	Techniques	Benefits
Coating or addition of layers	Zinc–nickel alloy	Enhanced corrosion resistance in steels
	Terne coatings	It has enhanced corrosion resistance in steel
	Hot-dip aluminizing	It has enhanced corrosion resistance in steel
	Slip ceramic coatings	Enhanced wear and heat resistance in materials
	PVD (physical vapor deposition)	Enhanced wear resistance and optical properties
	CVD (chemical vapor deposition)	Improved corrosion and erosion resistance and used in semiconductors
	Thermal spray method	Enhanced wear resistance and optical properties (ceramics, MCrAlY, thermal barrier coatings)
	Cladding	Improved corrosion resistance properties
	Mechanical platting	Improved corrosion resistance properties
	Weld edge	Improved wear resistance in the hard-facing alloy, corrosion resistance in nickel base alloy, and S.S. alloy

1.5 SIMULATION IN SURFACE ENGINEERING

Surface engineering has played a significant role in battling material-ruining difficulties with the swift developments in surface engineering technologies to meet the demand of industries. Presently, lots of diverse options are available for surface modeling methods, which could offer various opportunities for improving and enhancing the performance of materials and components in adverse/operating conditions. The appropriate design of the surface engineering system is essential for obtaining excellent performance, where focus is required to carry out a surface engineering design system. Surface engineering chiefly involves substrate material, grouped surface, and subsurface materials' properties, purposes, and structures. These properties affect the system's output during working conditions. Designing a surface engineering system involves the principle of simulation, the principle of simulating the properties and nature of simulation, research and development, and operations. Figure 1.4 shows the concept of simulation in surface engineering. All these design principles are interrelated and employed to develop the engineering design framework for surface engineering [11]. Simulation is the process where virtual models

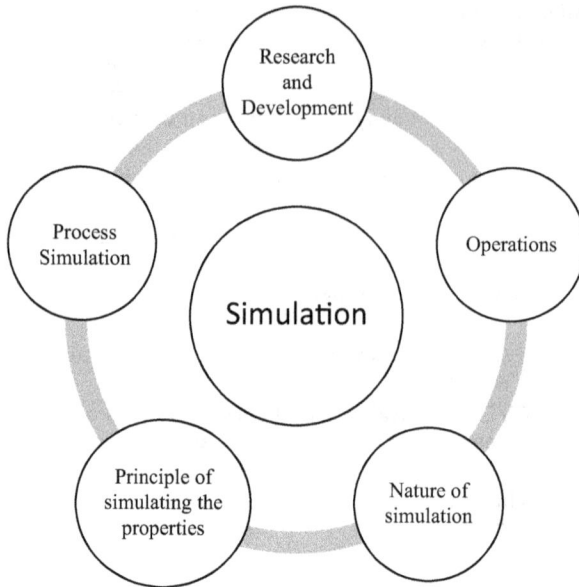

Figure 1.4 Concept of simulation in surface engineering.

or virtual experiments are to be served for understanding the nature of the designed system and evaluating the present method that is used for precise and accurate explanation of physical/chemical/biological properties of a system with the use of mathematical models/differential equations/algebraic equations [13]. Process simulation is a technique used for designing, developing, and analyzing to optimize several technical practices. However, process simulation is a valuable tool for designers to design specific designs of various components for optimizing and controlling the process of machine components. Process simulation provides critical technical data and various basic changes such as chemical, physical, mechanical, biological, and operations, and optimizes several technical operations. Twenty years ago, system designers made erudite models for one of the surface engineering processes, that is, carburizing, which was based on old diffusion laws [12]. Such erudite models have been magnificently integrated with industries, for controlling the process and merging with industries to develop new designs such as gear design and bearing design [14].

In the present scenario, rapid progress has been seen in research and development, where the statistical design of experiments is archaic, and the use of process simulation can only minimize and aid the experiment-based research cannot remove it. The experiment design model can be verified, but the reverse of it is not possible. A simulation-model-based software package can significantly profit experimental research; for example, in the thermo-dynamic field, the feasibility of feasible study is employed to increase the simple vapor–liquid experiments in the lab in the initial processes. Contrariwise, when data is not available, simple vapor–liquid experiments can be calibrated using thermodynamics design models with the amalgamation of the simulator. Simulation provides novel answers to solutions that are tough to calculate by conducting experiments. For example, the simulation process can solve the prototype models of computational fluid dynamics for process integration [13]. Simulating the tribological and mechanical properties of surface engineering is possibly challenging when using simulation models. Because in reality, different materials induced different properties (chemical and mechanical surface roughness, structure gradient, hardness, and tribological) in different environmental conditions to the substrate or component. Without knowing the exact relationships among the various properties of materials and their linked factors, it's dreadful to simulate the surface engineering properties of composite/ceramics/metal/non-metals materials [11]. That's why studying the relationships among the various properties of materials is essential before making its simulation models.

1.6 SUMMARY

In this chapter, a brief review of innovative development in surface engineering, various methods, the journey of surface engineering, surface engineering properties, surface engineering techniques and their benefits, surface modeling design principle, and applications of surface engineering is presented. As a common conclusion, surface engineering has a large caliber and can significantly influence the R&D, simulation, and manufacturing sectors. From this perspective, more focus is required on R&D, which focuses on the novel and supple concepts of surface modifications to structure to attain optimized, cost-effective solutions and effective performance of non-standard surface engineering methods and tools.

REFERENCES

1. Bunting, P. Degradation and Surface Engineering, https://shusuperblog. wordpress.com/authors/
2. Economic Effects of Metallic Corrosion in the United States, *Battelle Columbus Laboratories and the National Institute of Standards and Technology*, 1978 and Battelle updates in 1995.
3. Dunn, D. J. 1985. Metal Removal Mechanisms Comprising Wear in Mineral Processing. *Wear of Materials*, K. C. Ludema, Ed., American Society of Mechanical Engineers, pp. 501–508.
4. Introduction to Surface Engineering for Corrosion and Wear Resistance, © *2001 ASM International. All Rights Reserved. Surface Engineering for Corrosion and Wear Resistance (#06835G).*

5. Bell, T. 1990. Surface Engineering: Past, Present and Future. *Surface Engineering* 6, no. 1.

6. Majumdar, J. D., and Manna, I. 2015. Laser Surface Engineering of Titanium and Its Alloys for Improved Wear, Corrosion and High-Temperature Oxidation Resistance. *Laser Surface Engineering.* http://dx.doi.org/10.1016/B978-1-78242-074-3.00021-0

7. Majumdar, J. D., and Manna, I. 2001. Laser Material Processing. *Int. Mater. Rev.* 56: 341–388.

8. Budinski, K.G. (Ed.) 1988. *Surface Engineering for Wear Resistance.* New Jersey: Prentice Hall.

9. Bhushan, B. 1999. *Principles and Application of Tribology.* New York: A Wiley-Interscience Publication.

10. Budinski, K. G. (Ed.) 1988. *Surface Engineering for Wear Resistance.* New Jersey: Prentice Hall.

11. Bell, T., Mao, K., and Sun, Y. 1998. Surface Engineering Design: Modelling Surface Engineering Systems for Improved Tribological Performance, *Surface and Coatings Technology* 108–109: 360–368.

12. Pavlossoglou, J. 1977. *Hart. Techn. Mitt.* 32: 2.

13. Dimian, A. C. 2004. Introduction in Process Simulation. *Computer Aided Chemical Engineering Integrated Design and Simulation of Chemical Process,* vol. 35, pp. 35–71, doi: 10.1016/B978-0-444-62700-1.00002-4.

14. Sharm, V. K., Walter, G. H., and Breen, D. H. 1983. *J. Heat Treating* 3: 20.

Chapter 2

Tribological hybrid materials

Aarti P. More and Pradnya D. Desai

Institute of Chemical Technology, Mumbai, Maharashtra, India

CONTENTS

DOI: 10.1201/9781003319375-2

2.1 INTRODUCTION

Wear is the leading cause of material waste and low mechanical performance, and any decrease in wear can save a lot of money [1]. Friction is a significant source of wear and energy waste, and better friction control can save money. One-third of the world's current power sources are expected to be required to overcome friction in some way [2]. The tribology is related to friction and wear. The goal of tribology study is to reduce and eliminate losses caused by friction and wear [3]. According to some estimates, tribology-related losses in the United States account for about 4% of the country's GDP, with friction accounting for roughly a third of the world's current energy supplies. Improved tribological methods can save an industrial country about 1% of its gross domestic product [4]. The study of tribology [5] focuses on improving efficiency and performance while reducing breakdowns and saving money. The scope and depth of research in the discipline of tribology have drastically increased in recent years [1]. In a variety of uses, including solar panels, tidal turbines, and wind turbines, tribological concerns are crucial. Many of these energy-generation systems are plagued by tribological problems [6]. Manufacturing companies throughout the world are constantly seeking ways to improve the efficiency and efficacy of their technological processes, machinery, and vehicles. As a result, working conditions for tool and machine parts are becoming increasingly challenging. The industry is always on the lookout for new structural and tool materials, as well as manufacturing processes, that can ensure maximum durability under specified working circumstances while keeping unit costs low. One of the most significant and exciting breakthroughs in material engineering and production processes is the design and introduction of hybrid materials to the industry [7].

Many engineering materials have been adapted to create various tribological hybrid components, including steel, copper, cast iron, ceramics, special alloys, graphene, carbon nanotubes, and polymeric materials such as epoxy, polyether ether ketone, and so on [8]. This chapter covers the basics of tribology, as well as a detailed examination of hybrid materials for common tribological components, their subclassifications, applications in diverse industries, and a look at their synthesis and evaluation. This chapter addresses the fundamentals of tribology, as well as the various types of materials that were researched for tribological applications and their classification. The study's originality is a thorough evaluation of hybrid materials as tribological components, which covers every element of hybrid material. Organic and inorganic components make up the majority of hybrid materials. Due to their low density and high mechanical properties, hybrid materials have become the most powerful and outstanding materials in recent decades. The volume and variety of composite material applications are continually increasing, resulting in the creation of new markets.

The classification of hybrid composites as well as tribological hybrid materials is discussed in depth here. Metallic-based, ceramic-based, polymeric-based, and fibre-reinforced composites are among the hybrid composites thoroughly explored in this chapter. The subcategories in each category are explored, such as metallic-based aluminium, copper, molybdenum, and magnesium, and ceramic-based carbide, oxide, and nitride-based hybrid materials. Epoxy, polyetheretherketone (PEEK), polyimide, and poly(methyl methacrylate) (PMMA) are among the polymeric materials covered extensively, whereas fibre-reinforced composites include a variety of natural and synthetic fibre-based composites. To our knowledge, no literature goes into such detail about hybrid materials for tribological applications. The chapter provides an outline of the mechanics and applicability of hybrid materials, as well as a list of modelling methodologies used in various investigations. As a result, the chapter provides in-depth coverage of tribological hybrid materials.

2.1.1 Concept of tribology

In 1967, the Organization for Economic Cooperation and Development designated tribology as a new study discipline. Tribology means "rubbing science" and is derived from the Greek word *tribos*, which means "rubbing". Tribology is the investigation of lubrication, friction, and wear and is the science of joining faces in relative movement [9]. It is the process of applying operational analysis to large-scale economic issues such as dependability, service, and wear of technical equipment, which can range from spacecraft to domestic appliances [10].

Modern technologies that use gliding and tumbling surfaces require a thorough understanding of tribology. Railroads, autos, aircraft, and the production process of machine components are just a few examples of industrial uses that require relative motion. Bearings, seals, gears, and metal cutting are some of the tribological mechanical parts employed in these applications [11]. Micro/nanoelectromechanical systems, magnetic storage devices (NEMS/NEMS), and medicinal and cosmetic care items are among the other applications [4]. Not only in major industries but also in our daily life, tribology is critical. For example, writing is a tribological process. It is necessary to have a regulated transfer of lead or ink to the paper as well as good adhesion to the paper. Body joints are lubricated to reduce friction and wear so that they avoid osteoarthritis, joint replacement, etc. [12].

2.2 TRIBOLOGY AND SURFACE ENGINEERING

"Tribology and Surface Engineering" is a comprehensive tool for learning about the origins and effects of surface wear, as well as solutions for enhancing the tribological qualities of working surfaces [13–17].

Substantial improvements have already been accomplished in the development of surface engineering methods over the previous few decades, resulting in produced surfaces with thicknesses ranging from nanometres to millimetres [18]. The growing demand for combination characteristics in modern machines has driven the Creation of multi-modal engineering processes [19–22]. Many surface engineering systems are already accessible to tackle various equipment degradation issues and to satisfy the demands of modern mechanical system designs functioning in more severe environments [25], including single, double, and multilayer systems [23, 24].

Hybrid composites, which are made by adding a reinforcing phase of two sorts to the matrix, are an intriguing answer to this challenge. According to current findings, the use of heterophase reinforcement is an option that enables the variety and growth of friction coupling tribological properties to large extent. The proper selection of matrix and reinforcing phases enables friction coupling element wear to be reduced and the friction coefficient to be stabilised [26–33].

2.2.1 Materials used in tribological applications

Various materials are explored for tribological applications and their classification is given in Figure 2.1 [34].

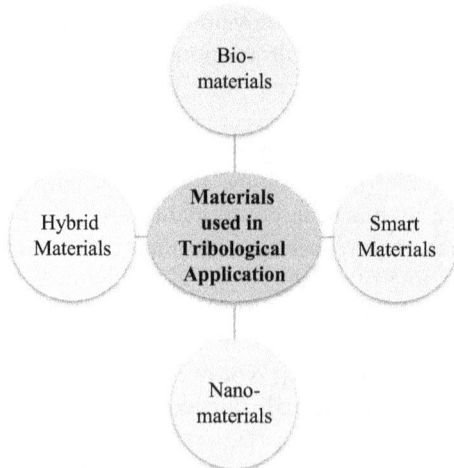

Figure 2.1 Classification of materials used in tribological applications.

2.2.1.1 Biomaterials

Artificial biomaterials, such as alloys, ceramics, and polymers, have tribological properties that are exploited in engineering, and some of them play a critical role in biomedicine [35]. Natural materials like wood rubber and glue, as well as tissues from biological entities and manmade materials including iron, glass, gold, and zinc, were used as biomaterials. Metals, polymers, ceramics, and composites are the most prevalent materials utilised in biomedical applications [36].

Metals, polymers, ceramics, and composites are the most commonly utilised biomaterials. Stainless steel, gold, tantalum, titanium alloys, and Co-Cr alloys are all metals. Alumina, zirconia, carbon, titania, bioglass, hydroxyapatite (HA), and different composites are all employed as biomaterials. Biomaterials include CF (Carbon fibre)/UHMWPE (ultra-high molecular weight polyethylene) [37], CF/PTFE (polytetrafluoroethylene) [38], Aluminum oxide (Al$_2$O$_3$)/PTFE [39] as well wide variety of polymers such as polyurethane[40], PTFE [41], polymethylmethacrylate [42], polyethylene terephthalate (PET) [43], silicone rubber (SR) [44], and PEEK [45] are used as biomaterials.

2.2.1.2 Nanomaterials

One of the most extensively utilised biomaterials is high-performance polymer-based composites epoxy resin (EP). In general, the production of a transfer film, as well as improvements in mechanical and thermal properties, increase the tribological properties of various polymer-based composites [35]. For usage in acetabular cup prostheses, polymer composites comprising UHMWPE and Al–Cu–Fe quasicrystals were studied [46]. The use of alumina-filled polymer composites to increase the mechanical strength of polymers has been investigated for industrial [47] and orthopaedic [48] applications.

Traditional ion implantation and plasma immersion ion are two novel surface modification approaches for improving biomaterial biocompatibility. These approaches, which attempt to create a hydrophobic/hydrophilic combination, should provide adequate lubrication and a low friction coefficient across the field [49]. Applications of biomaterials have been explored in bioengineering in the biomedical area. Graphene-based biomaterials commit great promise as antibacterial agents, biosensors, photothermal therapies (PTT), bioimaging tools, and stem cell and tissue engineering components [50].

Nanomaterials differ from bulk materials in that they have an effective surface area and are exceedingly small in size. The choice of nanoparticles is a critical step in improving tribological performance and boosting characteristics [51]. Nanomaterials have become a popular research area in recent years and are used as lubricant additives [52–69].

Nanoparticles (NPs) have several features, including their nanoscale size, which is suitable for optimal packing of the friction contact, allowing the integration of several properties such as extreme pressure (Ex P), friction modifiers (FM) additives, and anti-wear (AW) [70].

By producing a protective layer on materials and creating a tumbling motion between friction surfaces, nanomaterials added to lubricating oils can greatly increase tribological properties [71]. Nanoparticle additives were utilised by the researchers to minimise friction in this interaction. The authors experimented with Al_2O_3 [72], copper oxide (CuO) [73–75], zinc oxide (ZnO) [76, 77], titanium dioxide (TiO_2) [78], Tungsten disulphide (WS_2) [79], cerium oxide (CeO_2) [80], ZnO–multiwalled carbon nanotubes (MWCNTs) [81, 82], graphene [83, 84], and others cellulose nanocrystals [85], carbon dots [86], carbon nano-horn [87] nanoparticles additives and found promising anti-wear and anti-frictional properties.

Two-dimensional (2D) nanomaterials have been thoroughly investigated as potential lubricant additives [88]. 2D materials like molybdenum disulfide (MoS_2), hexagonal boron nitride (h-BN), graphene, and its derivatives have gotten a lot of attention because of their appealing chemical and physical properties [89–107]. Pure metal nanoparticles and their combinations, metal oxides, layered silicates, sulphides, and carbon nanomaterials (CNMs) are increasingly used in tribological applications. Applications of nanomaterials are in gas sensors, saturated absorbers, as well as lubricant additives (nanolubricant); other important applications are aviation and space equipment, chemical industry, optics, solar hydrogen, pharmaceuticals and cosmetic industry, in paints, the field of green chemistry, etc. [108].

2.2.1.3 Smart materials

The response of Smart materials can be controlled in a reversible manner as well as we can design the change in some of their properties as per external stimuli such as mechanical stress or temperature [109]. A smart fluid can get excited due to magnetic fields, and such fluids can be named magnetorheological (MR) fluids. The combination of base oil with additives in presence of ferromagnetic particles can make up such fluids.

MR (Magneto-rheological) fluid is a type of classic smart material composed of stable suspensions of small magnetically polarisable particles such as carbonyl iron suspended in a carrier liquid such as silicone oil or mineral oil [110]. A smart fluid is an external magnetic field-excited MR fluid [111].

Designing and producing materials with low coefficients of friction (CoFs) and low wear rates (WR) throughout a wide range of working situations is one of the major difficulties for modern industrial tribological systems [112, 113]. Solid lubricants (SLs) are the only practical option for reducing friction in many extreme settings, particularly those with temperatures above 350°C because liquid lubricants deteriorate quickly in these temperatures [114]. TMDCs (transition metal dichalcogenide such as MoS_2), graphite, and silicon nitride (Si_3N_4) are the most often utilised SLs in the industry [115–117]. These materials are effective lubricants due to their low CoF and simpler lamellar structure. Other SLs include metal oxides (PbO, MoO_3, B_2O_3, and NiO), noble metals (Pt, Au, Cu, and Ag), and inorganic fluorides (LiF, BaF_2, and CaF_2) [118].

Apart from the above-mentioned materials, the "Tribological Hybrid Material" is another aspect that will be discussed in depth here.

2.3 TRIBOLOGICAL HYBRID MATERIAL

In tribological applications, hybrid materials, which are one of the fastest-growing forms of materials, are gradually being utilised [119]. In the last two decades, research has shifted from monolithic materials to composite materials to meet the global demand for environmentally friendly, high-performance, low-weight, and corrosion- and wear-resistant materials [120]. The main focus here is on how different materials and compositions of hybrid materials affect friction and wear behaviour [121].

Hybrid composites are multifunctional materials made up of two or more distinct constituents separated by a discontinuous interface on a macroscale. Multiple discrete phases are integrated into a matrix material to form a hybrid composite. Hybrid reinforcement refers to the discrete phase, which is usually harder and stronger than the matrix material. Matrix materials in metallic, polymeric, and ceramic forms are available [122]. The classification of tribological hybrid materials is shown in Figure 2.2 [123].

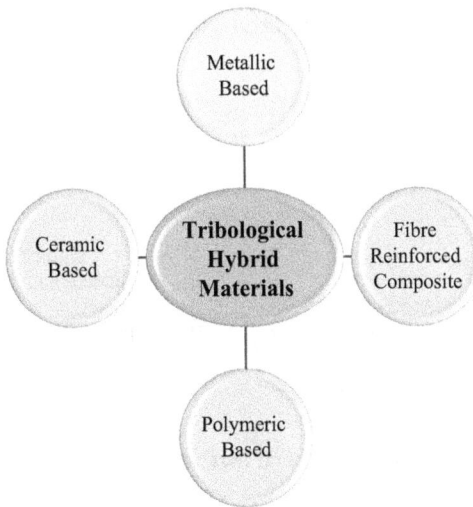

Figure 2.2 Classification of tribological hybrid materials.

2.3.1 Metallic hybrid material

Metal matrix composites (MMCs) have gotten a lot of interest recently because they outperform monolithic metals in terms of wear resistance, corrosion resistance, stiffness and strength. MMCs are made by mixing a hard reinforcement component with a metallic matrix in most circumstances. Copper, magnesium, aluminium, and their alloys are the various matrix materials employed [124–126]. CNT, Al_2O_3, SiC, BN, MoS_2, TiC, WC, and graphite are the reinforcement materials used with the metal matrix to form MMCs [127–133]. HMMCs (Hybrid MMCs) are used in a variety of applications which are as described in Figure 2.3 [134].

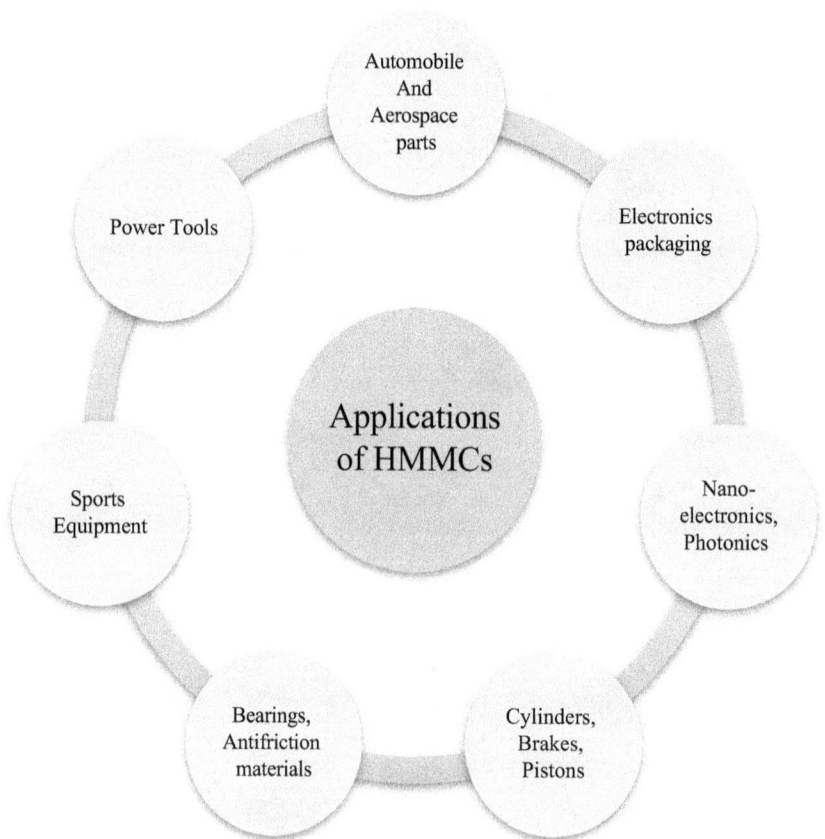

Figure 2.3 Applications of hybrid metal matrix composites.

2.3.1.1 Aluminium-based hybrid material

Aluminium and its alloys are lightweight, ductile, robust, and long-lasting materials that have good corrosion resistance, great electricity conductivity, and are suitable for reuse [135]. It has minimal wear resistance amidst these beneficial properties. However, specific applications demand stronger hardness, modulus of elasticity, and tensile strength than a conventional aluminium alloy with excellent tribological properties. To improve tribological properties while reducing weight, the aluminium alloy is reinforced with a hard as well as a brittle second phase, and the newly created material has improved its mechanical and tribological capabilities [136].

Organic and inorganic components make up the majority of hybrid composites. In comparison to standard alloys, aluminium hybrid MMCs (AHMMCs) are gaining popularity because they produce improved hardness, increased wear resistance, and longer fatigue life while remaining lightweight [137]. More research is being done these days on the development of aluminium-based hybrid composites to see if they may be used in various technical and automotive fields [138].

In comparison to pure aluminium metal, the strength and tribological capabilities of the aluminium matrix reinforced with particles like silicon carbide, alumina, and others have been improved. Particulate AMCs (aluminium matrix composites) have improved abrasive wear resistance as well as seizure resistance [139–143]. The electronic packaging sector has adopted Particulates-reinforced AMCs since their coefficient of thermal expansion may be changed to meet individual needs [144]. As a result, there has been a greater focus on particulate AMCs in recent years, which are utilised in the automotive sector due to their high load-carrying capacity, good sliding wear resistance, and lightweight combined with high thermal stability. Due to their improved mechanical qualities over unreinforced alloys, AMCs reinforced with particulates are frequently utilised in the fabrication of railway and car parts such as cylinders, brakes, discs/drums, piston insert rings, and pistons [145, 146].

Several factors influence the mechanical as well as tribological properties of composites, including operational test settings, environmental conditions, synthesis technique, kind of reinforcement, form and size, quantity of reinforcement, and matrix composition [147]. Even while single-reinforced aluminium matrix hybrid composites (AMHCs) have good wear and friction resistance, adding two or more reinforcements to a single aluminium matrix, referred to as AMHCs improves these qualities [148–153]. AMHCs filled using a combination of reinforcements such as oxide/oxide, carbide/oxides, oxide/graphite, carbide/boride, oxide/boride, and carbide/graphite have been documented in numerous studies [154–157]. The reinforcement of aluminium alloy is studied with SiC and various other fillers and the effect of various concentrations of these fillers on the properties is tabulated as seen in Table 2.1 [158].

Table 2.1 Performance attributes of tribological hybrid-reinforced Al 7075 alloy

Sr. No.	Formulation (wt%)	Vicker hardness	Ultimate tensile strength (MPa)	Ultimate compression strength (MPa)
1.	Al 7075	122	65.475	491.09
2.	Al 7075 5% SiC 5% Gr	176	247.21	617.74
3.	Al 7075 5% SiC 5% hBN	120	198.715	394.69
4.	Al 7075 5% SiC 5% MoS_2	135	137.355	445.21

2.3.1.2 Copper-based hybrid material

Due to key properties such as good thermal conductivity, temperature stability, corrosion resistance, electrical conductivity, and non-magneticity, copper and its alloys are utilised in a variety of industries [159, 160]. Copper is useful for components that need to disperse heat fast because of its increased thermal conductivity; nevertheless, its poor strength and high deformability make it unsuitable for many industrial uses [161].

Figure 2.4 depicts the characteristics and performance of copper matrix composites reinforced by the rGO-MoS$_2$ hybrid [162]. An extra reinforcement must be applied to the copper matrix to boost strength, wear, and frictional resistance. Even in normal environmental conditions, pure copper is easily oxidised, generating a nonconductive ceramic coating that reduces its electrical conductivity and strength. To boost copper's strength and lower its oxidation propensity, researchers have recently focused on reinforcing it with other ceramics such as Al_2O_3 [163–166], diamond [167], graphite [168, 169], graphene [170], SiC [171], and zirconium oxide (ZrO_2) [172–174].

The electrical properties, thermal conductivity, and ductility of copper are all well-known [175–177]. Copper's ability to be used in a broad array of applications is hampered by its poor frictional properties, severe wear, and low hardness. Different reinforcement materials have lately been included into the copper matrix to enhance its tribological and mechanical qualities. A number of nanomaterials, including WC, carbon nanotubes, SiC, graphene, Al_2O_3, and others, are used to make Cu matrix composites [178–181]. When MWCNTs and SiC were incorporated, hybrid composites outperformed pure copper in terms of wear resistance. First, the improved wear resistance exhibited in hybrid materials can be attributed to the higher hardness of MWCNTs and SiC as well as strengthened interfacial bonding and the development of a carbon-rich tribolayer. Second, the formation of a

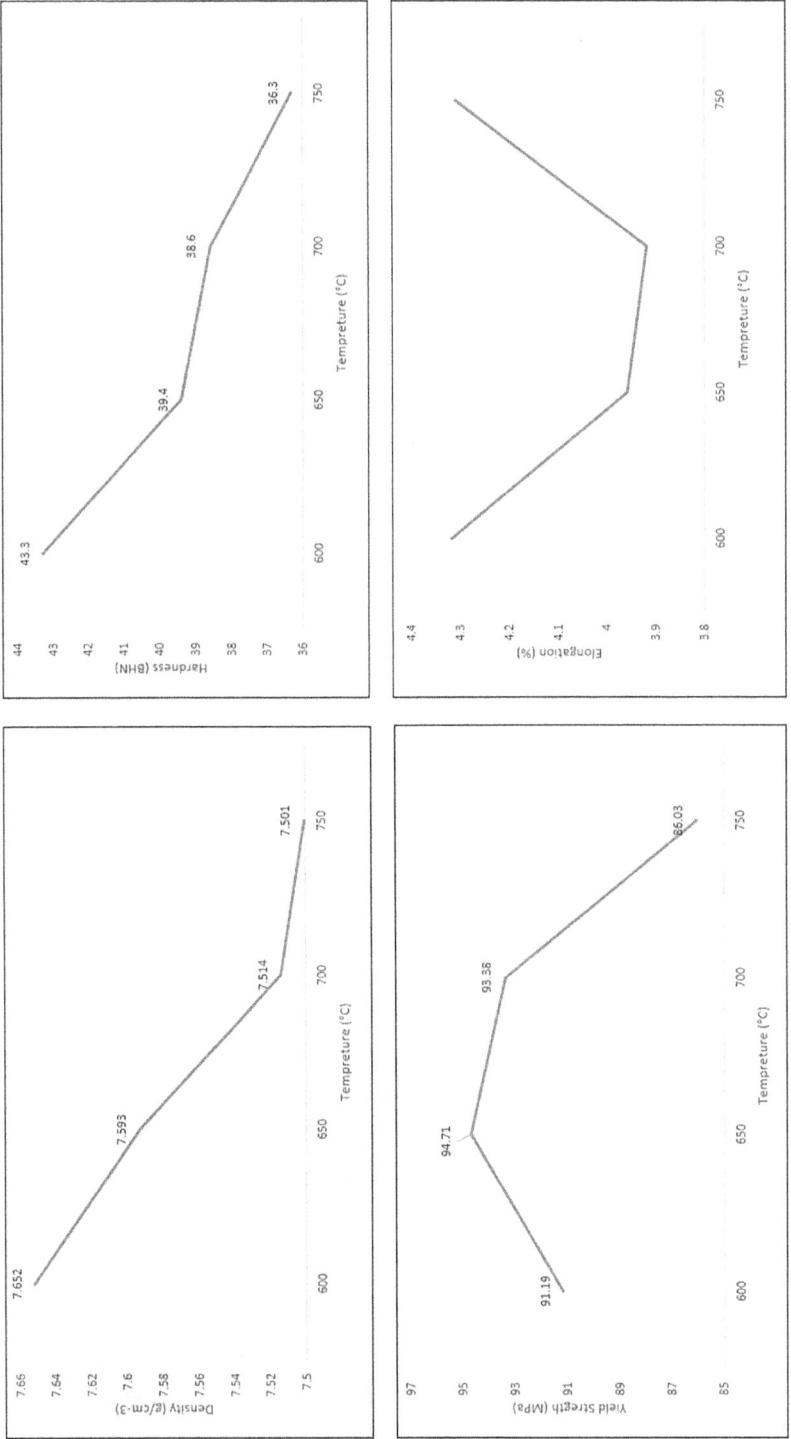

Figure 2.4 The effect of rGO-MoS$_2$ hybrid filler reinforcement on copper matrix composites.

Table 2.2 The performance of cast copper–TiO₂–boric acid hybrid composites

Sr. no.	Formulation	Vickers hardness number	Tensile strength (MPa)
1.	Pure Cu	80	~175–200
2.	Cu – 3 wt% TiO₂ – 1 wt% Boric acid	~100–110	~200–225
3.	Cu – 5 wt% TiO₂ – 1 wt% Boric acid	~110–120	~225–250
4.	Cu – 3 wt% TiO₂ (Cu-coated) – 1 wt% Boric acid	140	~225–250
5.	Cu – 5 wt% TiO₂ (Cu-coated) – 1 wt% Boric acid	~150	~250–275

lubricating tribolayer is caused by the breakdown of MWCNTs, which act like a solid lubricant. The carbon-rich tribolayer serves a critical function in avoiding direct contact in continuous contact areas [182].

Furthermore, as the applied stress increases, the soft and ductile copper matrix suffers significant plastic deformation, leading to the creation of a tribolayer between mating surfaces that contains iron and Fe oxides as well as CNT. The existence of these layers lubricates the interface and prevents actual contact between the surfaces. The tribolayer layer loses its stability because as load increases, pushing it out and putting a thin oxide coating into touch with the steel disc, which behaves as a lubricant and minimises the CoF [183].

MoS₂ has been proven to be a useful reinforcing phase for improving tribological characteristics in a copper matrix [184]. Graphene, the decade's most rapidly evolving material, has sparked intense interest in a wide variety of applications, including tribological ones [185, 186]. Graphene is increasingly being used as a reinforcing phase in copper composites to increase their tribological and mechanical properties [187, 188]. Unreinforced copper has a greater CoF and wear rate than hybrid composites [189]. MMCs based on copper have a lot of potential for usage in this type of application [190]. Table 2.2 presents the performance of hybrid copper composites [191].

2.3.1.3 Molybdenum-based hybrid material

Solid lubricating compounds in MMCs, as is well known, are amongst the most effective techniques to control friction. Various fillers, such as MoS₂, Al₂O₃, graphite, and others, were explored to enhance the tribological properties of the metal matrix [192–196]. Among these, MoS₂, a representative of the transition metal sulphide family, is a very solid lubricant for use in dry settings. The distinctive structures comprising covalently bonded S–Mo–S trilayers and passive basal planes present in particular crystallites are largely responsible for their remarkable tribological capabilities [197]. However, in order to achieve optimal friction coefficient, most composites developed in the past had a high percentage of MoS₂ (15–40%), which weakened the mechanical qualities of composites. One option to fix this problem is to add a small amount of MoS_2 to the mixture. As a result, during sintering, the majority of MoS_2 is dissolved, permitting molybdenum to diffuse into the alloy and become useful [198].

Nanosized MoS_2, in particular, is thought to be a potential lubricant addition. MoS_2 monolayers are sandwich structures made up of covalently bound S–Mo–S. The van der Waals interaction also bonds single layers of MoS_2, leading to a decrease in friction coefficient. MoS_2 has a variety of uses due to its unique feature, including solid lubricants, lubricant additives, and self-lubricating coatings [199].

One strategy for enhancing the tribological and rheological properties of MoS_2 nanolubricants is to combine MoS_2 nanoparticles with CNMs. CNMs, such as fullerene-like materials, have been intensively investigated as a solid or colloidal liquid lubricant due to their increased tensile strength and flexibility [200]. Only the MoS_2/CNMs hybrid nanostructure's synergetic action may have a significant impact on engine oil properties. Numerous researches have also shown that graphene and MoS_2 may be used as standalone fillers in metal-matrix-based composites to improve tribological as well as mechanical properties. The graphene-MoS_2 hybrid is also gaining popularity as a lubricant additive and composite matrix reinforcement filler for tribological applications [201].

MoS_2 is a 2D nanoparticle having not only the attributes listed above, but also good economic benefits, abundant resources, and environmental friendliness. As a result, nanoelectronics and photonics, catalysts, tribology, and other disciplines have used it [202, 203]. Because MoS_2 is an inorganic solid with inactive basal planes, low conductivity, and insufficient active sites, it cannot be used as a modifier in organic matrices. As a result, considerable work has been expended to improve its activity and stability [204].

2.3.1.4 Magnesium-based hybrid material

Because of its small weight, magnesium is a popular material for automotive applications. Magnesium and its alloys are the most popular lightweight structural parts required by the car industry to conserve energy [205]. Magnesium materials are employed in various industries, especially aerospace, cars, defence, and sports, due to their appealing properties. Despite its positive features, it has some flaws that limit its application, including poor wear, insufficient high temperature creep resistance, insufficient ductility, stiffness, and corrosion resistance. The low thermal stability, ductility, and wear resistance of magnesium limit its wide range of industrial uses [206]. These materials' exceptional qualities, including refractoriness, wear resistance, hardness, and compressive strength, make them ideal for applications such as reinforcement in composite matrixes [207].

It is a critical item to achieve because commercial uses such as electrical sliding contacts, bushes, bearings, and other parts used in anti-friction require materials with better anti-friction, anti-wear, and strength properties. It has been shown that adding reinforcements to magnesium to form composites and alloys reduces the boundaries of the metal [208]. Many researchers employed ceramic reinforcements like SiC, B_4C, Al_2O_3, and TiC to improve the wear and mechanical characteristics of materials. The wear resistance of magnesium-based hybrid materials is greatly improved when reinforcement materials such as Al_2O_3 and MoS_2 are included [209].

The magnesium hybrid material reinforced with SiC and graphite is a promising contender material for wear resistance operations where lighter-weight metals and composites are subjected to sliding motion [210]. The SiC particles are sheared during sliding contact wear, and the sheared layer attaches to the metal surfaces producing a thin film. Furthermore, the SiC hard film has minimal ductility under low load conditions and can bear stress without breakage. Limiting the plastic brittle fracture at the contact surface can reduce wear and surface abrasion [211, 212]. The oxide layer between the particles can be compacted more easily using SiC particle reinforcement. Fragmented SiC particles are mixed with the oxide layer to increase its film's hardness and wear resistance. Finally, the composite's enhanced hardness of oxide film results in increased capacity of bearing a load and stability [213].

Furthermore, adding graphite to a material improves the material's chemical stability, mechanical behaviour, and thermal properties. The damping behaviour of magnesium composites is influenced by TiC particles. TiC particle reinforcement greatly improves yield strength, tensile strength, and elastic modulus; however, ductility is lowered to some extent [214]. Furthermore, CNMs have been employed to reinforce Mg materials in recent decades. It increases a material's strength and is commonly used in structural load-bearing applications [215].

Magnesium Metal Matrix Hybrid Composites (MMMHCs) are used automotive and aerospace industries. Corrosion resistance is strong with this material. Furthermore, the material's performance is largely determined by the suitable reinforcement material mix. MMMHCs are useful in a variety of sectors [216].

2.3.2 Ceramic-based hybrid material

As a second phase matter in a metal matrix, ceramics provide better combinations of critical qualities such as specific strength, regionally controlled thermal expansions, efficient damping, good tribology performance, creep resistance, corrosion resistance, stiffness, etc. In ceramics, some issues remain unsolved, such as low ductility, low fracture toughness, and unexpected corrosive behaviour, which have been addressed through effective composition fixing [217–220]. Ceramics are lightweight and have great hardness and stiffness, as well as heat resistance and refractoriness, making them a potential wear-resistant material for aerospace and automotive technologies, as well as mechanical engineering in general and metallurgy. When ceramics are employed in composite materials, the brittleness of the material is reduced, and reliable tribotechnical materials that provide low wear as well as friction in many applications can be created. As a result, tribotechnical nanomaterials such as nanostructured composite ceramic and ceramic–metal materials for fabrication of tribotechnical constructions, nanostructured tribotechnical coatings, nanopowder solid lubricants, and friction modifiers for liquid and plastic lubricants have received a lot of attention recently [221, 222].

In comparison to unmodified ceramics, the incorporation of external components into matrices can result in higher dependability, flexural strength, and toughness. Simultaneously, tribological performance may be enhanced [223]. An example of ceramic-based composite is graphite-fibre-reinforced glass which reduces the CoF of the composite to values comparable to resins, while the wear resistance is comparable to that of glasses [224].

2.3.2.1 Carbide-based hybrid material

Composites comprising silicon and graphite carbide particles are well-known and well-examined materials in this area; such solutions are detailed in refs. [225, 226]. In this regard, several hard carbide particles such as boron carbide, silicon carbide, titanium carbide, and alumina have been used to improve the tribological and mechanical behaviour of MMCs [227–229].

Silicon carbide (SiC) is a synthetic material with excellent mechanical, electrical, tribological, and thermal properties, making it ideal for industrial use. Its abrasive properties are crucial in the production of abrasive wheels and high-wear products [230]. Due to SiC density, which is equivalent to that of aluminium, it is the most extensively employed of these ceramic particles. Furthermore, at low temperatures, SiC does not react with aluminium, limiting undesirable brittle reactions at surfaces [231]. Boron carbide is also one of the lightest materials and, after diamond and cubic boron nitride, the third hardest. This ceramic material is employed in low-weight composite applications and has greater wear resistance and abrasive property. Because of its enhanced hardness and low density, it can be used in both residential and commercial products [232]. The addition of boron carbide particulates to a matrix promotes thermal stability because the interaction of B_4C with oxygen produces products that minimise volumetric shrinkage. These residues also reduce the rate of deterioration of interior layers and increase thermal stability [233].

In the carbide family, boron carbide is also explored as a reinforcing agent and the effect of B_4C along with cow dung in the aluminium metal matrix is presented in Table 2.3 [234].

Alumina, which is notorious for its brittleness, is a good option for reinforcement. SiC whiskers placed in alumina have been found to act as a mechanical barrier to reduce the fracture erosion rate of the composite in

Table 2.3 Reinforcement of boron carbide and cow dung on Al 7075 metal matrix composite

Sr. No.	Formulation (wt%)	Hardness (BHN)	Tensile strength (MPa)	Impact (J)	Average flexural strength (MPa)	Wear rate (mm³/m)	Coefficient of friction
1.	Al 7075 – 100 B4C – 0 CDA – 0	110	~180–190	3.2	320	0.0050	~0.39–0.40
2.	Al 7075 – 90 B4C – 0 CDA – 10	~100–105	~230	3.0	~340–345	0.0040	~0.37–0.38
3.	Al 7075 – 90 B4C – 2.5 CDA – 7.5	~115–120	~250	2.9	~355–360	~0.0035–0.00375	~0.35–0.36
4.	Al 7075 – 90 B4C – 5 CDA – 5	~1301–135	~260–270	~2.8–2.9	~335	~0.00275–0.0030	~0.34–0.35
5.	Al 7075 – 90 B4C – 7.5 CDA – 2.5	~145	~290	~2.6	320	0.0020	0.31
6.	Al 7075 – 90 B4C – 10 CDA – 0	~150–155	~270–280	2.2	300	~0.0023	~0.32–0.33

comparison to pure alumina [235]. Low wear occurs at high temperatures due to an extra tribochemical process [236]. However, because silicon carbide whiskers are potent carcinogens, they are rapidly being phased out of composite ceramics manufacture. The abrasive wear resistance of ceramic matrix composites is influenced by the porosity of the matrix, the type of inclusions, and the interfacial bonding between the matrix and the second phase [237].

2.3.2.2 Oxide-based hybrid material

The addition of micro-/nano-Al_2O_3, SiC, CuO, TiO_2, ZnO, and ZrO_2 improves the wear resistance of a hybrid polymer. Different morphology ceramic fillers (TiO_2 and ZrO_2) were used for the reinforcement of bamboo–glass hybrid polymer composites [238].

Furthermore, GO's 2D structure, high strength, and ease of shear capability endow it with exceptional tribological properties, making it a desirable additive for anti-wear and friction reduction [239–241]. The addition of GOs grafted with Tungsten disulphide in the epoxy composite decrease the rate of microcrack formation, and the mechanism for the same is shown in Figure 2.5 [242].

TiO_2, a notable inorganic transition metal dioxide, has stimulated various investigations in recent decades due to its enticing functional properties and intriguing uses. Several hybrid modifications, such as carbon coating and surface functionalisation with porous materials have been conducted to improve the performance of TiO_2 nanostructures. Several attempts have been undertaken at the moment to synthesise and characterise TiO_2 hybrid nanostructures [243]. TiO_2/Ti_3C_2Tx hybrid can effectively enhance the tribological properties by forming a film on the surface, which avoids direct contact between the surfaces [244]. Because of their structure, size, and

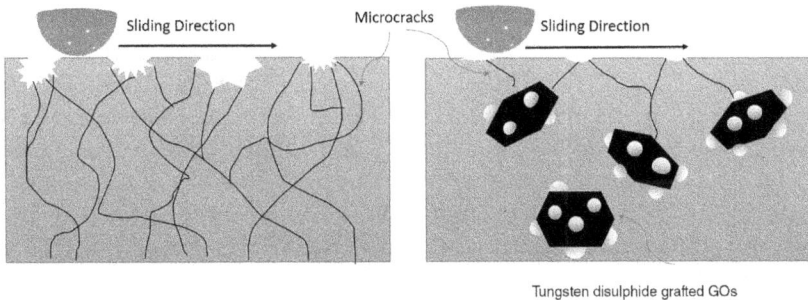

Sliding Direction   Microcracks   Sliding Direction

Tungsten disulphide grafted GOs

Figure 2.5 Reinforcement of epoxy composite with Tungsten-disulphide-grafted GO.

remarkable tribological and environmental qualities, nanoadditives have recently gained a lot of attention in different energy-related disciplines.

Nanoparticles like silicon dioxide (SiO_2), graphene oxide (GO), and a GO/SiO_2 hybrid were explored by a number of researchers for their tribological and machining properties. Nano-silicon dioxide has a huge potential for improving lubricating oil and grease tribological properties [245]. The findings show that MoS_2 incorporated hybrid nanoparticles have a big future as lubricant additives. In comparison to the previously stated nanoparticles, SiO_2 nanoparticles have gained a lot of study interest [246].

2.3.2.3 Nitride-based hybrid material

Recent research has found that the addition of conductive nanoparticles to polymer composites enhanced their thermal stability and improved high-temperature tribological performance significantly [247]. h-BN, also known as "white graphite" because of its crystalline honeycomb structure that is similar to graphite, has high oxidation resistance, self-lubricating properties, and thermal conductivity [248–250]. Many studies have found that adding a higher percentage of solid lubricant to composites diminishes their mechanical qualities while increasing their wear resistance. As a result, h-BN has emerged as a promising lubricant additive [251].

In comparison to standard coarse-grained ceramics, nanostructured materials have numerous advantages. Nanocomposites constructed of non-oxide ceramic compounds have low specific gravity, high hardness, low wear rate, and mechanical properties that do not oxidise at high temperatures [252–255]. One of the strongest structural ceramics, Si_3N_4, has emerged as a key tribological material, particularly in rolling applications. It possesses a low elastic modulus, high strength, and great fatigue resistance, as well as excellent oxidation resistance and very strong thermal shock resistance. Si_3N_4 is only stable in oxidising situations under very low partial pressures of oxygen; in air, it develops a silicon oxide surface layer rapidly. This layer prevents further oxidation; but, if it is destroyed, such as through wear, oxidation occurs rapidly [256]. Unique features of Si_3N_4-based nanocomposites have been used to develop novel wear-resistant components that can endure extreme working environments like acids and bases, high temperatures, etc. [257, 258]. The tribological properties of Si_3N_4-based ceramics are influenced by their microstructure. The addition of nanowhiskers and nanorods to Si_3N_4-based microstructures is a unique technique to enhance the properties of their composites [259, 260]. Another method for improving the tribological characteristics of Si_3N_4-based ceramics is to include titanium nitride particles into the Si_3N_4 matrix. When TiN is added to Si_3N_4, the CoF is lowered [261].

To address the insufficient tribological sliding behaviour of Si_3N_4 materials, secondary phases with low friction and strong wear resistance were introduced [262]. Secondary phases should possess higher mechanical as well as wear resistance. The addition of h-BN or titanium nitride to Si_3N_4-based materials minimises friction and wear under highly stressed dry sliding situations [263, 264].

Drills are made of MMC due to their excellent heat conductivity. Examples include silica nitride particles incorporated in cobalt matrix, tank armours made of steel reinforced with boron nitride, and power electronic modules made of Al-graphite. These composites have a lot of uses in space systems due to their numerous working temperatures and resistance to absorbing moisture [265].

2.3.3 Polymeric-based hybrid material

Polymeric materials have recently piqued interest in engineering applications, not only because of their simplicity of manufacture and inexpensive production costs but also because of their better tribological performance in composite forms. Because of their superior qualities, polymeric composites have gotten a lot of attention [266, 267]. Polymer matrix composites (PMCs) feature high energy efficiency and strength-to-weight ratio, resistance to corrosion, low processing temperature, ease of manufacture, and recyclability. The matrix element in PMCs homogeneous material into which reinforcements are embedded to change the function of the matrix material. Thermoplastics and thermosets are common matrix materials in PMCs [268].

Fibres/particulate fillers are inserted as Reinforcing elements in the matrix to improve the performance of PMCs, resulting in hybrid PMCs (HPMCs). The reinforcements are surrounded and closely bound by the polymer matrix element [269, 270]. The tribological behaviour of composites is influenced by the kind of reinforcement fibres and the fillers utilised [271, 272]. HPMCs are widely used in the aircraft, sports, military, thermal power, automotive, and aerospace industries [273].

Polymer nanocomposites are increasingly being employed as engineering materials in applications where tribological characteristics are critical. Glass, carbon, and textile fibres are used as fillers in a variety of polymers to improve their tribological properties. Nanoparticles such as carbon nanotubes, nanoclay, titanium oxide (TiO_2), and SiC are added in amounts ranging from 2% to 5% by weight, drastically improving mechanical characteristics [274, 275].

2.3.3.1 Epoxy-based hybrid material

Epoxy resin is a rising engineering plastic with corrosion resistance and low curing shrinkage and that has found widespread application in the scientific and technological fields [276]. Pure EP has poor wear and lubrication resistance because of its intrinsic brittleness and low-temperature resistance, severely limiting its usage in motion [277]. The addition of strengthening and lubricating fillers to a polymer matrix increases its tribological capabilities, as is well known. Many researches have found that various fillers like graphene [278], CNT-ZnSO$_4$ (zinc sulphate) [279], and MoS$_2$ [280] may significantly improve the EP's friction and wear properties.

In recent years, materials scientists, technicians, and industrialists have been interested in using nanoparticles in epoxy matrices to increase the materials' tribological behaviour. The addition of uniformly sized SiO$_2$-TiO$_2$ nanoparticles to the epoxy matrix inhibited the growth of cracks. In addition, if the applied stress is less than a specific threshold, the reinforcing elements will reduce the rate of wear. These factors may contribute to a slower rate of epoxy matrix wearing [281]. According to the results of the experiment, the wear mechanism of the composites changed from adhesive to abrasive wear as the SiO$_2$-TiO$_2$ concentration increased, and the wear behaviour of composites was improved dramatically [282].

The addition of metallic additives to epoxy increases its wear resistance [283]. Epoxy's wear resistance is improved when it is strengthened with PTFE [284]. The wear rate of epoxy is reduced when silicon carbide particles are added to glass-fibre-reinforced epoxy [285]. Adding mica particles and tricalcium phosphate (TCP) to E-glass-filled epoxy boosted its wear resistance even more [286]. A smidgeon of potassium titanate bonded with epoxy improves the specific wear rate [287]. Epoxy wear resistance was enhanced using a combination of titanium dioxide nanoparticles, short carbon fibre, graphite flakes, etc. [288].

Epoxies may not provide a significant level of sliding wear resistance because of their limited impact resistance. Reinforcements in the resin may improve load-bearing capacity, reduce CoFs, improve wear resistance, improve thermal characteristics, and improve mechanical strength [289]. The reinforcement of epoxy composite by SiO$_2$ and SCF (short carbon fibre) is shown in Table 2.4 [290].

Table 2.4 Reinforcement of epoxy with microsphere, silicon dioxide, and short carbon fibre

Sr. no.	Hybrid filling	Maximal loading ability (MPa)
1	Epoxy Plain	~3
2	Oil-loaded microsphere – 8 phr SiO$_2$ – 0 phr SCF – 0 phr	~4
3	Oil-loaded microsphere – 8 phr SiO$_2$ – 5 phr SCF – 0 phr	~7
4	Oil-loaded microsphere – 8 phr SiO$_2$ – 0 phr SCF – 1 phr	~6
5	Oil-loaded microsphere – 8 phr SiO$_2$ – 5 phr SCF – 1 phr	~7

2.3.3.2 PEEK-based hybrid materials

Polymer composites exhibit strong tribological behaviour at high operating conditions depending on the matrix and fillers used [291]. The most commonly utilised polymer matrix for tailoring composites for tribological purposes is PEEK [292, 293]. PEEK is a semi-crystalline engineering thermoplastic with outstanding good tribological behaviour as well as mechanical and thermal properties, structural stability, minimal flammability and low gas emission, chemical resistance, etc. [294].

However, in its purest state, to make it acceptable for tribological applications, it needs to be reinforced with various reinforcing agents, like inorganic nanoparticles or other forms of fibres. One method for reducing the friction and wear behaviour of PEEK composites is to incorporate a range of reinforcing agents, such as short carbon fibres [295–297], lubricants in form of solids [298–300], and carbon-containing particles [301, 302]. PEEK filled with SCF/PTFE/graphite is widely recognised as an excellent friction material, with the SCF/PTFE/graphite fillers having synergistic effects on tribological behaviour [303].

PEEK hybrid composites reinforced with nano-SiO_2 particles and SCF had significantly lower friction coefficients. Wear of matrix, interlayer debonding, and fibre thinning and its removal are the most common phenomena responsible for composite deterioration [304]. Surface abrading and ploughing occur in the existence of hard agglomerates which creates plastic deformation in the part of the surface. The eroded areas of the hybrid composites become softer when the applied pressure is increased, and the agglomeration abrading impact is decreased. These two contributing components substantially increase the composite's wear resistance [305].

Adding modest amounts of nanofillers to traditional fibre-reinforced polymers like carbon nanotubes [306] opens up new possibilities, resulting in structured composites with enhanced wear behaviour [307, 308]. PEEK composites are frequently used to substitute metal components in bearings, gears, valves, nozzles, and prosthetic sectors (automotive, industrial industries, aerospace, and so forth) [309].

2.3.3.3 Polyimide-based hybrid material

Aromatic diamines and aromatic dianhydrides are used to make polyimides. Processing polyimides necessitates high temperatures. Polyimides (PI) are utilised in copiers, business machines, and space vehicle components such as piston rings, valves, rolling contact bearing retainers, non-lubricated seals, and bearings [310]. Fillers are commonly made of polyimides. Polyimides with regulated porosity are impregnated with liquid lubricants. Because of their higher heat-deflection temperature, they outperform lubricant-impregnated nylon [311]. TPI (thermosetting polyimide) has piqued the interest of researchers working on lubricating substances. due to its good mechanical qualities, extraordinary heat resistance, and self-lubrication [312, 313]. On the other side, TPI self-lubricating hybrids with increased wear resistance and decreased friction at high temperatures remain a major challenge. TPI composites having Ag-Mo hybrids and two distinct reinforced fibres are suggested to promote the load-carrying capabilities of the polymer matrix [314].

Despite polyimides' exceptional capabilities, multiple researches have been carried out to increase their tribological behaviour by including several nanofillers into the polyimide matrix, allowing PI to adapt to the harsh working environment. Due to its specific qualities, such as the boundary action, tiny size impact, environmentally friendly nature, and quantum effect, nano-carbon materials are regarded to be the most feasible reinforcement phases [315, 316]. Many studies have shown that PI nanocomposites incorporating carbon nanotubes are an efficient way for improving the tribological properties of polyimide films. CNTs readily bind with one another due to molecular inter-atomic interactions between structures; hence, their industrial usage is hampered by their dispersity and compatibility in the matrix [317–321].

The combination of CNT with graphene is explored as reinforcement in PI film and its mechanism of action is presented in Figure 2.6 [322]. The sequential representation of the figure shows the PI film on a metal surface, the film is reinforced with GO and CNT. At the microscopic level, the presence of CNT and graphene in combination is shown at the right end of the figure.

Superior thermal stability, bonding, mechanical, and tribological capabilities are among the physical and chemical properties of aromatic polyimide

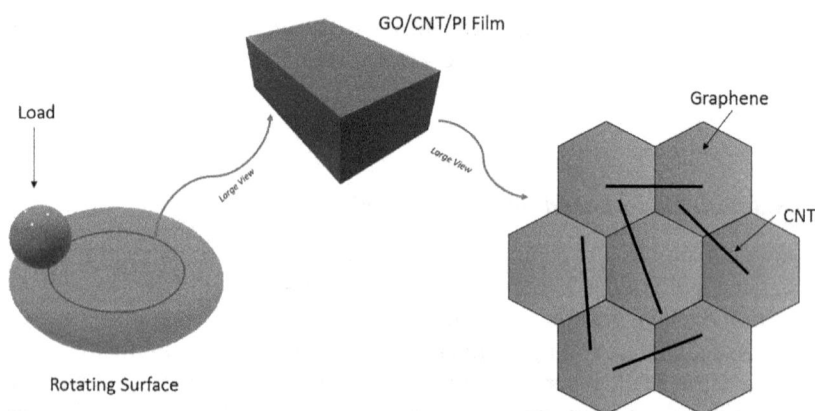

Figure 2.6 The reinforcement of PI film with graphene–CNT filler.

[323]. Its excellent properties make it an excellent adhesive resin for fabric–resin composites. Fabric–resin hybrids are bonded to metal surfaces to create high-performance composites [324, 325]. Fabric–resin composites are made composed of fabric as a matrix and resin as a binder. Due to their exceptional mechanical and tribological qualities, they have become appealing materials in a broad array of applications including aerospace, automotive, and railway transport systems [326, 327].

2.3.3.4 PMMA-based hybrids

PMMA is a popular amorphous thermoplastic that offers UV protection, chemical resistance, stretchability, high strength, and stability. Furthermore, PMMA is resistant to both acidic and basic environments [328, 329]. At high temperatures, however, PMMA's thermal stability and mechanical–dynamical qualities are limited. The addition of nanoparticles to the polymer matrix, such as clays, silica, or carbon nanotubes, is one technique to increase polymer performance [330–333]. PMMA nanocomposites often offer lower gas permeability, higher physical performance, and increased heat resistance while maintaining optical purity [334].

Hybrid materials can sometimes produce surprising new features that are not seen in individual components. The ability to modify the properties of organic–inorganic hybrid materials by altering the contents of organic and inorganic ingredients is one of their key advantages [335]. PMMA/inorganic fillers are one of the most investigated hybrid materials. PMMA is a widely used commercial plastic with applications in a variety of engineering fields. As an organic matrix, it has piqued interest due to its remarkable thermal and mechanical stability, which has been applied to a variety of sectors including bone replacement and optical devices [336]. It has also been widely utilised as bone cement due to its superior self-hardening and mechanical qualities when compared to other polymers [337]. Unfortunately, PMMA has low adhesive and abrasion resistance, which is one of the reasons for its limited applicability in other sectors such as dentistry. The addition of inorganic compounds to polymers is one technique to increase their performance [338].

PMMA is used in a variety of industries and could potentially be utilised in prosthetics, drug dispensers, food processing plants, throat bulbs, and spectacles [339, 340]. Unfortunately, this polymer has a low abrasion resistance when it comes to glass, which limits its possible applications. Efforts to increase PMMA's scratch and abrasion resistance [341] have resulted in plenty of other downsides, including a loss of impact strength, prompting experts to concentrate on the creation of PMMA nanocomposites. The addition of dispersed carbon nanotubes or graphene nanofillers as neat, modified, or functionalised nanomaterials to polymer matrices as neat, modified, or functionalised nanomaterials is attracting increasing attention in the hopes of improving the stability and mechanical, tribological, or degradation resistance of polymer matrices [342, 343].

2.3.4 Fibre-reinforced hybrid materials

Recent developments have shown that fibre-reinforced composites are among the most proficient advanced materials for replacing old metals as well as alloys in a variety of tribological applications [344]. The matrix, synthetic or natural fibres as reinforcements, and low concentrations of inorganic or organic fillers are used in these FRP composites. Because of their high strength and wear resistance, they have exceptional properties [345, 346].

Reinforcing fibres in advanced composite materials adjust the material's mechanical characteristics to meet the dimensional reliance of the loading environment, while also providing high strength and stiffness. Fibres are arranged in a particular way which can state their surface and shape quality while also protecting them from the weather and transferring stress. Matrix has a more homogenous and continuous structure than fibres, although it has less strength and rigidity [347]. Even though synthetic fibre composites are better for the environment, most academics prefer natural fibre composites owing to cost and pollution concerns. As a result, research into the tribological behaviour of composite materials is required. As illustrated in Figure 2.7 [348], various fibres can be classified in different ways.

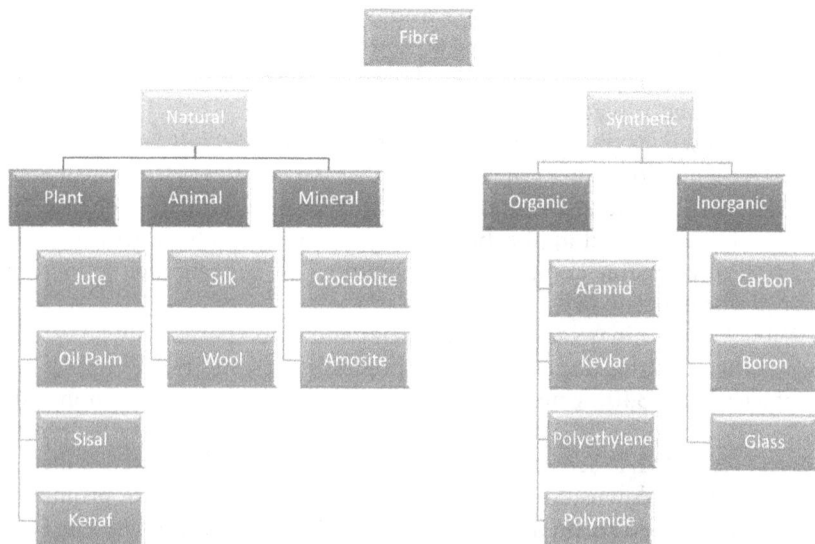

Figure 2.7 Classification of fibres.

2.3.4.1 Natural-fibre-reinforced hybrid material

Environmentally friendly and renewable materials are in need of development [348]. Natural fibre composites are a type of technology that fits into this category. In these composites, organically sourced fibre is employed as reinforcement. Plant, animal, or mineral materials could be the source of such fibres [349]. The possibility for plant-based fibre-reinforced polymer composites to replace synthetic reinforcement has attracted curiosity [350]. Plant fibres have emerged as an attractive reinforcement [351], because of their high availability, non-irritation to body parts, non-toxicity, and non-corrosive quality.

According to some studies, the tribological behaviour of natural-fibre-reinforced hybrids is not inherent and is heavily influenced by a variety of processing parameters like adhesion characteristics of fibres, additives, interaction situations, etc. [352, 353].

There have been few attempts to investigate the tribological behaviour of hybrid materials derived from natural fibres such as kenaf [354], oil palm [355], sisal [356], cotton [357], jute [358], betelnut [359], and bamboo [360]. Single fibres can overcome their limits while retaining their benefits by combining them into a matrix material. Along with better qualities, this approach also results in significant cost savings by eliminating the need for extra materials. Fibre hybrid biocomposites have lighter qualities than hybrid synthetic fibre composites [361, 362].

2.3.4.2 Synthetic-fibre-reinforced hybrid material

Synthetic fibres are fibres manufactured from chemicals, such as glass, carbon, aramid, boron, ceramic fibres, and so on, as well as thermosets and thermoplastics, which have been employed extensively to make FRP materials [363]. Fibre length, fibre percentage, and interphase adhesion between matrix and fibres are the major parameters that contribute to the FRP composite's exceptional performance. The insertion of fibres into the matrix is critical for producing the composites' desirable strength, stiffness, and tribological qualities [364].

Because of their extremely specialised material properties, synthetic fibres are widely used in structural applications, allowing constructions to be lighter than metallic ones. Glass fibre [365] is a nice example of a low-cost synthetic fibre. By combining glass fibre with a more expensive material such as carbon/graphite, the hybrid composite will be more economically viable. Carbon/graphite fibres are chosen because of their enhanced mechanical properties, lightweight, and high modulus. Enhanced hybrid systems for a number of events and applications have been developed thanks to hybridisation [366].

Glass, Kevlar, and carbon fibres are the three types of synthetic fibres. They are strong, stiff, and fire-resistant. Such fibres are expensive and pollute the environment when they are produced [367]. PEEK, SiC, graphite, and silicon graphite are some of the fillers that, when gently combined with fibre-reinforced composites, help to improve mechanical qualities. Over the last two decades, the usage of additives in glass fibre polymeric composites has increased considerably, decreasing the weight and cost of products. However, they have a lot of benefits, including being lighter and cheap, high strength, and enhanced flexibility [368]. The optimisation of matrix composition and filler parameters requires a thorough understanding of the tribological characteristics of fillers and fibre-reinforced hybrid materials [369].

2.3.5 Miscellaneous hybrid materials

In a variety of tribological applications, polyethylene with an ultra-high molecular weight is used. Reinforcements are applied to UHMWPE to overcome the polymer's significant wear problems [370]. The glass and poly-p-phenylene terephthalamide (PPDT) fibres are explored as a reinforcing agent in PTFE [371]. The PTFE and graphene are studied as a reinforcing agent in Nylon 66 [372]. Ruminant animal excrement fibres are innovative biopolymer that is widely accessible and seems to have a lot of potential for meeting raw material demand in a variety of industrial applications [373]. Researchers have proposed incorporating cow dung waste as a reinforcing material to form composites with improved physical and mechanical properties in recent years [374]. The physio-mechanical characteristics of polylactic acid biocomposites reinforced with cow dung were studied. The addition of cow manure increased flexural strength and dynamic mechanical behaviour while lowering tensile and impact characteristics, confirming the biocomposites for load-bearing capabilities [375].

2.4 MODELLING CONCEPT OF HYBRID MATERIAL

The fabrication and evaluation of desired features of hybrid composites as well as the optimisation of machining settings are the main concerns of researchers [376]. The design of experiment (DOE) is one of the most essential and effective statistical methodologies for analysing the impact of various factors at the same time [377]. It entails a series of actions that must be carried out in a specific order in order for the experiment to provide a more accurate picture of process performance. DOE techniques like Taguchi, ANNs (artificial neural networks), and surface response have increased in popularity since they were useful in exposing the impact on several parameters in a hierarchical rank order [378].

2.4.1 Taguchi method

The Taguchi technique is a strong modelling system that may be used to increase cost, performance, and quality improvement in a simple, effective, and systematic manner [379]. It is a multistep technique that optimises parameters with the fewest number of repetitions possible [380]. The methodology employs an empirical measurement tool known as the signal-to-noise ratio, which is a logarithmic function of the intended outcome and works as the optimisation objective function. The experiment's findings have been subjected to Analysis of Variance (ANOVA), a statistical approach for identifying the effect of researched wear factors, such as sliding speed, distance, and applied load. Using ANOVA [381], it's feasible to identify which independent variable outperforms the others and the cumulative percentage of that variable.

2.4.2 Artificial neural networks

ANNs, motivated by the functioning of the human brain, were utilised extensively in recent years for associating material properties with processing parameters and for modelling, predicting, and optimising complicated physical and nonphysical systems. In this study, to build a model that can estimate wear rate more accurately than a regression model, ANN was utilised to link the weight fraction of hybrid reinforcement, force, sliding distance, and velocity with wear rate [382]. To model ANN, a feed-forward back-propagation multilayer network is used [383]. They estimated the wear rate and CoF using ANNs [384].

2.4.3 Response surface methodology

Response surface methodology (RSM) is a unique blend of statistical, graphical, and mathematical methods used to model, analyse, and simulate situations with one or even more output variables influenced by multiple independent factors. By correlating empirically recorded input and output components, it also generates statistical models to optimise the process [385]. Many researchers have created a regression equation to anticipate a response using the response surface approach for optimising process parameters.

Ideal conditions, interactions, and variable square terms can all be predicted using the RSM. RSM is a tool for estimating the relationships between changing variables. The Box–Behnken design (BBD) was created by George Box and Donald Behnken as one of the experimental designs for RSM [386]. The BBD of RSM analysis is a multi-objective optimisation methodology that combines ANOVA, a statistical tool for assessing the effect of each factor on the combined responses, to identify the contribution of each parameter to the overall goal [387]. In conjunction with RSM, BBD was chosen to do statistical optimisation and build the regression model. This strategy has lately gained popularity among researchers due to its capacity to minimise unfavourable outcomes by rejecting trials conducted in extreme conditions [388].

2.4.4 Linear regression model

A multivariate linear regression analysis was performed using MINITAB 16 [383]. The created model establishes a linear relationship between the variables and the unknown amount. The regression model connects the ANOVA-derived significant factors, namely the sliding speed and distance and its interaction with applied load. A typical statistical strategy for creating a link between input factors and output responses is regression analysis. It is also easy to examine how each input factor affects the result [389].

2.5 APPLICATIONS

Applications of the tribological hybrid materials can be summarised as given in Table 2.5.

Table 2.5 Applications of tribological hybrid materials

Tribological hybrid material	Subtype of hybrid material	Applications	References
Metallic	Aluminium-based hybrids	Electronic packaging, fabricating railway and auto parts, cylinders, brakes, pistons	[390, 145, 146]
	Copper-based hybrids	Catalyst for carbon nanostructure, antimicrobial coating, corrosion resistant coupling	[391, 392, 393]
	Molybdenum-based hybrids	Nanoelectronics, photonics, catalyst	[394, 202, 203]
	Magnesium-based hybrids	Aerospace parts, automobile parts, biomedical, structural application	[216, 395, 396]
Ceramic	Carbide-based hybrids	Catalyst, used in capacitors, super capacitors	[397, 398]
	Oxide-based hybrids	Super capacitors, solar cells, various fuel cells, biosensors	[399, 400, 401]
	Nitride-based hybrids	Drills, tank armours, power electronics modules, optical biosensor, nano-photonic wave guide	[262, 402, 403]

(Continued)

Table 2.5 (Continued) Applications of tribological hybrid materials

Tribological hybrid material	Subtype of hybrid material	Applications	References
Polymeric	Epoxy-based hybrids	Load bearing application in automotive, aerospace and oil and gas, industrial application	[404, 405, 406]
	PMMA-based hybrids	Biomimetic application, scaffolds, aeroplane glazing	[407, 408, 409]
	PEEK-based hybrids	Biomedical application, dentistry, implants, gears, nozzles, seal rings	[309, 410, 411]
	Polyimide-based hybrids	Automotive, bearings, aerospace material, railway transport, Optoelectronic application	[326, 327, 412]
Fibre-reinforced	Natural-fibre-reinforced hybrids	Window frames, Automotive Applications, dashboards, door panels, seat cushions	[413, 414, 415]
	Synthetic-fibre-reinforced hybrids	Aircraft components, automotive parts, construction equipment	[416, 417, 418]

2.6 CONCLUSION

Many sectors have taken steps to reduce wear and friction in recent years. Actually, it is an area of surface engineering that provides outstanding mechanical, chemical, and tribological qualities as well as improved performance. Surface engineering studies the structure and design of diverse materials, including composites, as well as difficulties involving the augmentation or modification of material surface qualities. As the demand for modern engineering grows, new advanced materials are being developed that include biomaterials, smart materials, and hybrid materials in the manufacturing process. The utilisation of hybrid materials is one of the most sophisticated approaches to addressing tribological challenges. Hybrid materials research has been one of the most fascinating fields of science and technology study in the last three decades. The hybrid materials can be classified as metal-based, ceramic-based, polymeric-based, and fibre-reinforced-based materials. The synergistic effect of two or more materials is a key feature of tribological materials. Their performance is explored in various applications such as pistons, supercapacitors, solar arrays, automotive parts, and aeroplane components.

REFERENCES

1. Stachowiak, Gwidon W., and Andrew W. Batchelor, *Engineering tribology*, 4th edition, Butterworth-heinemann, 2013.
2. Holmberg, Kenneth, Peter Andersson, and Ali Erdemir. "Global energy consumption due to friction in passenger cars." *Tribology International*, Vol 47, Page 221–234, 2012.
3. Bhushan, Bharat, *Introduction to tribology*, 2nd edition, John Wiley & Sons, 2002.
4. Meng, Y., Xu, J., and Jin, Z., "A review of recent advances in tribology." *Friction*, Volume 8, Page 221–300, 2020.
5. Jost, P. "Economic impact of tribology." In *Proceeding Mechanical Failures Prevention Group*, Page 117–139, 1976.
6. Guo, Hong Fei, Hao Jiang, Yun Gong Cai, and Zeng Qi Zhao. "Research and analysis of graphene coatings based on knowledge mapping." *Materials Science Forum*, Vol 1036, Page 93–103, 2021.
7. Swapnil Chandgude, and Sachin Salunkhe, "Biofiber-reinforced polymeric hybrid composites: An overview on mechanical and tribological performance." *Polymer Composites*, Vol 41, Issue 10, 2020.
8. Faustini, Marco, Lionel Nicole, Eduardo Ruiz-Hitzky, and Clément Sanchez. "History of organic–inorganic hybrid materials: Prehistory, art, science, and advanced applications." *Advanced Functional Materials*, Vol 28, 2018.
9. Friedrich, Klaus, and Ulf Breuer, *Multifunctionality of polymer composites: Challenges and new solutions*. William Andrew, 2015.
10. Czichos, Horst, *Tribology: A systems approach to the science and technology of friction, lubrication, and wear*, Elsevier, Vol 1, 2009.
11. Bhushan, Bharat, *Principles and applications of tribology*, John Wiley & Sons, 1999.
12. Hussain, Omar, Shahid Saleem, and Babar Ahmad. "Implant materials for knee and hip joint replacement: A review from the tribological perspective." *IOP Conference Series: Materials Science and Engineering*, Vol 561, Page 012007, 2019.
13. Bonek, M., "The investigation of microstructures and properties of high-speed steel HS6-5-2-5 after laser alloying." *Archives of Metallurgy and Materials* Vol 59, Page 1647–1651, 2014.
14. Lisiecki, A., "Study of optical properties of surface layers produced by laser surface melting and laser surface nitriding of titanium alloy." *Materials*, Vol 12, Page 3112, 2019.
15. Lisiecki, A., and Piwnik, J., "Tribological characteristic of titanium alloy surface layers produced by diode laser gas nitriding." *Archives of Metallurgy and Materials*, Vol 61, Page 543–552, 2016.
16. Janicki, D., "Fabrication of high chromium white iron surface layers on ductile cast iron substrate by laser surface alloying." *Strojniski Vestnik*, Vol 63, Page 705–714, 2017.
17. Yao, Lin, and Junhui He. "Recent progress in antireflection and self-cleaning technology–From surface engineering to functional surfaces." *Progress in Materials Science*, Vol 61, Page 94–143, 2014.
18. Bell, T., K. Mao, and Y. Sun. "Surface engineering design: Modelling surface engineering systems for improved tribological performance." *Surface and Coatings Technology*, Vol 108, Page 360–368, 1998.

19. N. Dingremont, E. Bergmann, P. Collignon, and H. Michel, "Optimization of duplex coatings built from nitriding and ion plating with continuous and discontinuous operation for construction and hot working steels." *Surface and Coatings Technology*, Vol 72, Issue 3, Page 163–168, 1995.
20. M. Van Stappen, B. Malliet, L. Stals, L. De Schepper, J.R. Roos, and J.P. Celis, "Characterization of TiN coatings deposited on plasma nitrided tool steel surfaces." *Materials Science and Engineering A*, Vol 140, Page 554–562, 1991.
21. E.I. Meletis, A. Erdemir, and G.R. Fenske, "Tribological characteristics of DLC films and duplex plasma nitriding/DLC coating treatments." *Surface and Coatings Technology*, Vol 73, Issues 1–2, Page 39–45, 1995.
22. U. Huchel, S. Bramers, J. Crummenauer, S. Dressler, and S. Kinkel, "Single cycle, combination layers with plasma assistance." *Surface and Coatings Technology*, Vol 76–77, Part 1, Page 211–217, 1995.
23. Bell, T., K. Mao, and Y. Sun. "Surface engineering design: Modelling surface engineering systems for improved tribological performance." *Surface and Coatings Technology* 108, Page 360–368, 1998.
24. Bell, T., Y. Sun, K. Mao, and P. Buchhagen. "Modeling plasma nitriding." *Advanced Materials & Processes*, Vol 150, Issue 2, Page 40, 1996.
25. W. Ames, and A.T. Alpas, "Wear mechanism in hybrid composites of graphite-20 pct SiC in A356 aluminium alloy (Al-7pct Si 0,3 pct Mg), *Metallurgical and Materials Transactions A*, Vol 26, A1, Page 85–98, 1995.
26. M.L. Ted Guo, and C.-Y.A. Tsao, "Tribological behavior of self-lubricating aluminum/SiC/graphite hybrid composites synthesized by the semi-solid powder-desiccations method." *Composites Science and Technology*, Vol 60, Issue 1, Page 65–74, 2000.
27. J.I. Song, S.I. Bae, K.C. Ham, and K.S. Han, "Abrasive wear behavior of hybrid metal matrix composites." *Key Engineering Materials*, Vol 183, Page 1267–1272, 2000.
28. Yoshimi Watanabe, and Tatsuru Nakamura, "Microstructures and wear resistances of hybrid Al–(Al3Ti+Al3Ni) FGMs fabricated by a centrifugal method." *Intermetallics*, Vol 9, Issue 1, Page 33–43, 2001.
29. Myalski, J., J. Wieczorek, and A. Dolata-Grosz, "Tribological properties of heterophase composites with an aluminium matrix." *Journal of Achievements in Materials and Manufacturing Engineering*, Vol 15, Page 53–57, 2006.
30. Dolata-Grosz, A., J. Śleziona, and B. Formanek, "Structure and properties of aluminium cast composites strengthened by dispersion phases." *Journal of Materials Processing Technology*, Vol 175, Issues 1–3, Page 192–197, 2006.
31. Dolata-Grosz, A., and J. Wieczorek "Tribological properties of composite working under dry technically friction condition." *Journal of Achievements in Materials and Manufacturing Engineering*, Vol 18, Issue 1–2, 2006.
32. Myalski, J., J. Wieczorek, and A. Dolata-Grosz. "Tribological properties of heterophase composites with an aluminium matrix." *Journal of Achievements in Materials and Manufacturing Engineering*, Vol 15, Issue 1–2, Page 53–57, 2006.
33. Takadoum, Jamal, *Materials and surface engineering in tribology*, John Wiley & Sons, 2013.
34. Glaeser, William. *Materials for tribology*, Vol 20. Elsevier, 1992.

35. Patel, Nitesh R., and Piyush P. Gohil, "A review on biomaterials: Scope, applications & human anatomy significance." *International Journal of Emerging Technology and Advanced Engineering*, Vol 2, Issue 4, Page 91–101, 2012.

36. Hench, Larry L, "Bioceramics: From concept to clinic." *Journal of the American Ceramic Society*, Vol 74, Issue 7, Page 1487–1510, 1991.

37. Dangsheng, Xiong. "Friction and wear properties of UHMWPE composites reinforced with carbon fiber." *Materials Letters*, Vol 59, Page 175–179, 2005.

38. Shi, Yijun, Liwen Mu, Xin Feng, and Xiaohua Lu. "Friction and wear behavior of CF/PTFE composites lubricated by choline chloride ionic liquids." *Tribology Letters*, Vol 49, Page 413–420, 2013.

39. Wang, Zhijiang, Lina Wu, Yulin Qi, Wei Cai, and Zhaohua Jiang. "Self-lubricating Al2O3/PTFE composite coating formation on surface of aluminium alloy." *Surface and Coatings Technology*, Vol 204, Page 3315–3318, 2010.

40. Segan, Sören, Meike Jakobi, Paree Khokhani, Sascha Klimosch, Florian Billing, Markus Schneider, Dagmar Martin et al. "Systematic investigation of polyurethane biomaterial surface roughness on human immune responses in vitro." *BioMed Research International*, Vol 2020, Page 2314–6133, 2020.

41. Bellon, J. M., L. A. Contreras, J. Bujan, and A. Carrera-San Martin. "The use of biomaterials in the repair of abdominal wall defects: A comparative study between polypropylene meshes (Marlex) and a new polytetrafluoroethylene prosthesis (Dual Mesh)." *Journal of Biomaterials Applications*, Vol 12, Page 121–135, 1997.

42. Fini, Milena, Gianluca Giavaresi, Nicolo Nicoli Aldini, Paola Torricelli, Rodolfo Botter, Dario Beruto, and Roberto Giardino. "A bone substitute composed of polymethylmethacrylate and α-tricalcium phosphate: Results in terms of osteoblast function and bone tissue formation." *Biomaterials*, Vol 23, Page 4523–4531, 2002.

43. Ma, Zuwei, Masaya Kotaki, Thomas Yong, Wei He, and Seeram Ramakrishna. "Surface engineering of electrospun polyethylene terephthalate (PET) nanofibers towards development of a new material for blood vessel engineering." *Biomaterials*, Vol 26, Page 2527–2536, 2005.

44. Hsiue, Ging-Ho, Shyh-Dar Lee, Patricia Chuen-Thuen Chang, and Chen-Yu Kao. "Surface characterization and biological properties study of silicone rubber membrane grafted with phospholipid as biomaterial via plasma induced graft copolymerization." *Journal of Biomedical Materials Research: An Official Journal of The Society for Biomaterials, The Japanese Society for Biomaterials, and the Australian Society for Biomaterials*, Vol 42, Page 134–147, 1991.

45. Toth, Jeffrey M., Mei Wang, Bradley T. Estes, Jeffrey L. Scifert, Howard B. Seim III, and A. Simon Turner. "Polyetheretherketone as a biomaterial for spinal applications." *Biomaterials*, Vol 27, Page 324–334, 2006.

46. Huttunen-Saarivirta, Elina. "Microstructure, fabrication and properties of quasicrystalline Al–Cu–Fe alloys: A review." *Journal of Alloys and Compounds*, Vol 363, Page 154–178, 2004.

47. Kumar, Keshav, "Sliding wear studies in epoxy containing alumina powders." *High Temperature Materials and Processes*, Vol 17, Issue 4, Page 271–274, 1998.

48. Rieu J, and Goeuriot P, "Ceramic composites for biomedical applications." *Clinical Materials*, Vol 12, Page 211, 1993.

49. Pelletier, Jacques, and André Anders. "Plasma-based ion implantation and deposition: A review of physics, technology, and applications." *IEEE Transactions on Plasma Science*, Vol 33, Page 1944–1959, 2005.
50. Han, Shanying, Jie Sun, Shuangba He, Mingliang Tang, and Renjie Chai. "The application of graphene-based biomaterials in biomedicine." *American Journal of Translational Research*, Vol 11, Page 3246, 2019.
51. Uzoma, Paul C., Huan Hu, Mahdi Khadem, and Oleksiy V. Penkov. "Tribology of 2D nanomaterials: A review." *Coatings*, Vol 10, Page 897, 2020.
52. Liqiang Mai, Yajie Dong, Lin Xu, and Chunhua Han, "Single nanowire electrochemical devices." *Nano Letters*, Vol 10, Issue 10, Page 4273–4278, 2010.
53. Holmberg, Kenneth, and Ali Erdemir, "Influence of tribology on global energy consumption, costs and emissions." *Friction*, Vol 5, Issue 3, Page 263–284, 2017.
54. Gulzar, M., H. H. Masjuki, M. A. Kalam, M. Varman, N. W. M. Zulkifli, R. A. Mufti, and Rehan Zahid, "Tribological performance of nanoparticles as lubricating oil additives." *Journal of Nanoparticle Research*, Vol 18, Issue 8, Page 1–25, 2016.
55. Padgurskas, Juozas, Raimundas Rukuiza, Igoris Prosyčevas, and Raimondas Kreivaitis. "Tribological properties of lubricant additives of Fe, Cu and Co nanoparticles." *Tribology International*, Vol 60, Page 224–232, 2016.
56. Bogunovic, Lukas, Sebastian Zuenkeler, Katja Toensing, and Dario Anselmetti, "An oil-based lubrication system based on nanoparticular TiO 2 with superior friction and wear properties." *Tribology Letters*, Vol 59, Issue 2, Page 1–12, 2015.
57. Alves, Salete Martins, Bráulio Silva Barros, Marinalva Ferreira Trajano, Kandice Suane Barros Ribeiro, and E. J. T. I. Moura, "Tribological behavior of vegetable oil-based lubricants with nanoparticles of oxides in boundary lubrication conditions" *Tribology International*, Vol 65, Page 28–36, 2013.
58. Gupta, Rajeev Nayan, and A. P. Harsha, "Tribological evaluation of calcium-copper-titanate/cerium oxide-based nanolubricants in sliding contact" *Lubrication Science*, Vol 30, Issue 4, Page 175–187, 2018.
59. Kalin, Mitjan, Janez Kogovšek, Janez Kovač, and Maja Remškar "The formation of tribofilms of MoS 2 nanotubes on steel and DLC-coated surfaces." *Tribology Letters*, Vol 55, Issue 3, Page 381–391, 2014.
60. Chen, Lijuan, and Dingyi Zhu, "Preparation and tribological properties of unmodified and oleic acid-modified CuS nanorods as lubricating oil additives" *Ceramics International*, Vol 43, Issue 5, Page 4246–4251, 2017.
61. Rabaso, Pierre, Fabrice Ville, Fabrice Dassenoy, Moussa Diaby, Pavel Afanasiev, Jérôme Cavoret, Béatrice Vacher, and Thierry Le Mogne "Boundary lubrication: Influence of the size and structure of inorganic fullerene-like MoS2 nanoparticles on friction and wear reduction" *Wear*, Vol 320, Page 161–178, 2014.
62. Nan, Feng, Yi Xu, Binshi Xu, Fei Gao, Yixiong Wu, and Zhuguo Li, "Tribological behaviors and wear mechanisms of ultrafine magnesium aluminum silicate powders as lubricant additive" *Tribology International*, Vol 81, Page 199–208, 2015.
63. He, Anshun, Shuiquan Huang, Jung-Ho Yun, Zhengyi Jiang, Jason R. Stokes, Sihai Jiao, Lianzhou Wang, and Han Huang, "Tribological characteristics of aqueous graphene oxide, graphitic carbon nitride, and their mixed suspensions" *Tribology Letters*, Vol 66, Issue 1, Page 1–12, 2018.

64. Kumar, Parveen, and Mohammad Farooq Wani, "Tribological characterisation of graphene oxide as lubricant additive on hypereutectic Al-25Si/steel tribopair" *Tribology Transactions*, Vol 61, Issue 2, Page 335–346, 2018.

65. Ali, Mohamed Kamal Ahmed, Hou Xianjun, Liqiang Mai, Cai Qingping, Richard Fiifi Turkson, and Chen Bicheng, "Improving the tribological characteristics of piston ring assembly in automotive engines using Al2O3 and TiO2 nanomaterials as nano-lubricant additives." *Tribology International*, Vol 103, Page 540–554, 2016.

66. Mosleh, M., and M. Ghaderi "Deagglomeration of transfer film in metal contacts using nanolubricants", *Tribology Transactions*, Vol 55, Issue 1, Page 52–58, 2012.

67. Xia, Wenzhen, Jingwei Zhao, Hui Wu, Sihai Jiao, Xianming Zhao, Xiaoming Zhang, Jianzhong Xu, and Zhengyi Jiang, "Analysis of oil-in-water based nanolubricants with varying mass fractions of oil and TiO2 nanoparticles." *Wear*, Vol 396, Page 162–171, 2018.

68. Kumara, Chanaka, Huimin Luo, Donovan N. Leonard, Harry M. Meyer, and Jun Qu. "Organic-modified silver nanoparticles as lubricant additives." *ACS Applied Materials & Interfaces*, Vol 9, Issue 42, Page 37227–37237, 2017.

69. Meng, Yuan, Fenghua Su, and Yangzhi Chen, "Supercritical fluid synthesis and tribological applications of silver nanoparticle-decorated graphene in engine oil nanofluid." *Scientific Reports*, Vol 6, Issue 1, Page 1–12, 2016.

70. Spikes, Hugh. "Friction modifier additives." *Tribology Letters*, Vol 60, Page 1–26, 2005.

71. Zhmud, Boris, and Bogdan Pasalskiy. "Nanomaterials in lubricants: An industrial perspective on current research." *Lubricants*, Vol 1, Page 95–101, 2013.

72. Kotia, Ankit, Ravindra Kumar, Abhisek Haldar, Piyush Deval, and Subrata Kumar Ghosh, "Characterization of Al2O3 SAE 15W40 engine oil nanolubricant and performance evaluation in 4-stroke diesel engine." *Journal of the Brazilian Society of Mechanical Sciences and Engineering*, Vol 40, Issue 1, Page 1–9, 2018.

73. Thottackkad, Manu V., P. K. Rajendrakumar, and K. Prabhakaran Nair, "Experimental studies on the tribological behaviour of engine oil (SAE15W40) with the addition of CuO nanoparticles." *Industrial Lubrication and Tribology*, 2014.

74. Thottackkad, Manu Varghese, Rajendrakumar Krishnan Perikinalil, and Prabhakaran Nair Kumarapillai, "Experimental evaluation on the tribological properties of coconut oil by the addition of CuO nanoparticles." *International Journal of Precision Engineering and Manufacturing*, Vol 13, Issue 1, Page 111–116, 2012.

75. Esfe, Mohammad Hemmat, Fatemeh Zabihi, Hossein Rostamian, and Saeed Esfandeh, "Experimental investigation and model development of the non-Newtonian behavior of CuO-MWCNT-10w40 hybrid nano-lubricant for lubrication purposes." *Journal of Molecular Liquids*, Vol 249, Page 677–687, 2018.

76. Esfe, Mohammad Hemmat, Hossein Rostamian, and Mohammad Reza Sarlak, "A novel study on rheological behavior of ZnO-MWCNT/10w40 nanofluid for automotive engines." *Journal of Molecular Liquids*, Vol 254, Page 406–413, 2018.

77. Rashin, M., Nabeel, J., and Hemalatha S, "Synthesis and viscosity studies of novel ecofriendly ZnO–coconut oil nanofluid." *Experimental Thermal and Fluid Science*, Vol 51, Page 312–318, 2013.
78. Xia, Wenzhen, Jingwei Zhao, Hui Wu, Sihai Jiao, Xianming Zhao, Xiaoming Zhang, Jianzhong Xu, and Zhengyi Jiang, "Analysis of oil-in-water based nanolubricants with varying mass fractions of oil and TiO2 nanoparticles." *Wear*, Vol 396, Page 162–171, 2018.
79. Aldana, Paula Ussa, Fabrice Dassenoy, Beatrice Vacher, Thierry Le Mogne, and Benoît Thiebaut. "WS2 nanoparticles anti-wear and friction reducing properties on rough surfaces in the presence of ZDDP additive." *Tribology International*, Vol 102, Page 213–221, 2016.
80. Thottackkad, Manu V., P. K. Rajendrakumar, and N. K. Prabhakaran. "Tribological analysis of surfactant modified nanolubricants containing CeO2 nanoparticles." *Tribology - Materials, Surfaces and & Interfaces*, Vol 8, Page 125–130, 2014.
81. Goodarzi, Marjan, Davood Toghraie, Mahdi Reiszadeh, and Masoud Afrand, "Experimental evaluation of dynamic viscosity of ZnO–MWCNTs/engine oil hybrid nanolubricant based on changes in temperature and concentration." *Journal of Thermal Analysis and Calorimetry*, Vol 136, Issue 2, Page 513–525, 2019.
82. Vardhaman, B.S.A., M. Amarnath, J. Ramkumar, and K. Mondal. "Enhanced tribological performances of zinc oxide/MWCNTs hybrid nanomaterials as the effective lubricant additive in engine oil." *Materials Chemistry and Physics*, Vol 253, Page 123447, 2020.
83. Jamaluddin, Nor Athira, Norfazillah Talib, and Amiril Sahab Abdul Sani. "Tribological analyses of modified jatropha oil with hbn and graphene nanoparticles as an alternative lubricant for machining process." *Journal of Advanced Research in Fluid Mechanics and Thermal Sciences*, Vol 76, Page 1–10, 2020.
84. Zin, V., S. Barison, F. Agresti, L. Colla, C. Pagura, and M. Fabrizio. "Improved tribological and thermal properties of lubricants by graphene based nano-additives." *RSC Advances*, Vol 6, Page 59477–59486, 2016.
85. Awang, N. W., D. Ramasamy, K. Kadirgama, G. Najafi, and Nor Azwadi Che Sidik. "Study on friction and wear of Cellulose Nanocrystal (CNC) nanoparticle as lubricating additive in engine oil." *International Journal of Heat and Mass Transfer*, Vol 131, Page 1196–1204, 2019.
86. Shang, Wangji, Tao Cai, Yunxiao Zhang, Dan Liu, and Shenggao Liu, "Facile one pot pyrolysis synthesis of carbon quantum dots and graphene oxide nano-materials: All carbon hybrids as eco-environmental lubricants for low friction and remarkable wear-resistance." *Tribology International*, Vol 118, Page 373–380, 2018.
87. Zin, V., F. Agresti, S. Barison, L. Colla, E. Mercadelli, M. Fabrizio, and C. Pagura. "Tribological properties of engine oil with carbon nano-horns as nano-additives." *Tribology Letters*, Vol 55, Page 45–53, 2014.
88. Xiao, Huaping, and Shuhai Liu. "2D nanomaterials as lubricant additive: A review." *Materials & Design*, Vol 135, Page 319–332, 2017.
89. Upadhyay, Ram Krishna, and Arvind Kumar, "Boundary lubrication properties and contact mechanism of carbon/MoS2 based nanolubricants under steel/steel contact." *Colloid and Interface Science Communications*, Vol 31, Page 100186, 2019.

90. Kim, Kwang-Seop, Hee-Jung Lee, Changgu Lee, Seoung-Ki Lee, Houk Jang, Jong-Hyun Ahn, Jae-Hyun Kim, and Hak-Joo Lee, "Chemical vapor deposition-grown graphene: The thinnest solid lubricant." *ACS Nano*, Vol 5, Issue 6, Page 5107–5114, 2011.

91. Berman, Diana, Ali Erdemir, and Anirudha V. Sumant, "Graphene: A new emerging lubricant." *Materials Today*, Vol 17, Issue 1, Page 31–42, 2014.

92. Mungse, Harshal P., and Om P. Khatri, "Chemically functionalized reduced graphene oxide as a novel material for reduction of friction and wear." *The Journal of Physical Chemistry C*, Vol 118, Issue 26, Page 14394–14402, 2014.

93. Ye, Xiangyuan, Limin Ma, Zhigang Yang, Jinqing Wang, Honggang Wang, and Shengrong Yang, "Covalent functionalization of fluorinated graphene and subsequent application as water-based lubricant additive." *ACS Applied Materials & Interfaces*, Vol 8, Issue 11, Page 7483–7488, 2016.

94. Rosentsveig, Rita, A. Gorodnev, N. Feuerstein, Hilla Friedman, A. Zak, N. Fleischer, J. Tannous, F. Dassenoy, and Reshef Tenne, "Fullerene-like MoS2 nanoparticles and their tribological behavior." *Tribology Letters*, Vol 36, Issue 2, Page 175–182, 2019.

95. Xu, Yufu, Yubin Peng, Karl D. Dearn, Xiaojing Zheng, Lulu Yao, and Xianguo Hu, "Synergistic lubricating behaviors of graphene and MoS2 dispersed in esterified bio-oil for steel/steel contact." *Wear*, Vol 342, Page 297–309, 2015.

96. Choudhary, Shivani, Harshal P. Mungse, and Om P. Khatri, "Dispersion of alkylated graphene in organic solvents and its potential for lubrication applications." *Journal of Materials Chemistry*, Vol 22, Issue 39, Page 21032–21039, 2012.

97. Cho, Dae-Hyun, Jin-Seon Kim, Sang-Hyuk Kwon, Changgu Lee, and Young-Ze Lee, "Evaluation of hexagonal boron nitride nano-sheets as a lubricant additive in water" *Wear*, Vol 302, Issue 1–2, Page 981–986, 2013.

98. Senatore, Adolfo, Vincenzo D'Agostino, Vincenzo Petrone, Paolo Ciambelli, and Maria Sarno, "Graphene oxide nanosheets as effective friction modifier for oil lubricant: Materials, methods, and tribological results." *International Scholarly Research Notices*, 2013.

99. Watanabe, Kenji, Takashi Taniguchi, and Hisao Kanda, "Direct-bandgap properties and evidence for ultraviolet lasing of hexagonal boron nitride single crystal." *Nature Materials*, Vol 3, Issue 6, Page 404–409, 2004.

100. Paul, Gayatri, Harish Hirani, Tapas Kuila, and N. C. Murmu, "Nanolubricants dispersed with graphene and its derivatives: An assessment and review of the tribological performance." *Nanoscale*, Vol 11, Issue 8, Page 3458–3483, 2019.

101. Chouhan, Ajay, Harshal P. Mungse, and Om P. Khatri. "Surface chemistry of graphene and graphene oxide: A versatile route for their dispersion and tribological applications." *Advances in Colloid and Interface Science*, Vol 283, Page 102215, 2020.

102. Lei, Fan, Meng Yang, Feng Jiang, He Zhang, Zhi Zhang, and Dazhi Sun. "Microwave-assisted liquid phase exfoliation of graphite fluoride into fluorographene." *Chemical Engineering Journal*, Vol 360, Page 673–679, 2019.

103. Meng, Yuan, Fenghua Su, and Yangzhi Chen, "Synthesis of nano-Cu/graphene oxide composites by supercritical CO2-assisted deposition as a novel material for reducing friction and wear." *Chemical Engineering Journal*, Vol 281, Page 11–19, 2015.

104. Hilton, Michael R., Reinhold Bauer, Stephen V. Didziulis, Michael T. Dugger, John M. Keem, and James Scholhamer, "Structural and tribological studies of MoS2 solid lubricant films having tailored metal-multilayer nanostructures." *Surface and Coatings Technology*, Vol 53, Issue 1, Page 13–23, 1991.

105. Chen, Baiming, Qinling Bi, Jun Yang, Yanqiu Xia, and Jingcheng Hao. "Tribological properties of solid lubricants (graphite, h-BN) for Cu-based P/M friction composites." *Tribology International*, Vol 41, Issue 12, Page 1145–1152, 2008.

106. Berman, Diana, Ali Erdemir, and Anirudha V. Sumant, "Few layer graphene to reduce wear and friction on sliding steel surfaces." *Carbon*, Vol 54, Page 454–459, 2013.

107. Song, Haojie, Zhiqiang Wang, Jin Yang, Xiaohua Jia, and Zhaozhu Zhang, "Facile synthesis of copper/polydopamine functionalized graphene oxide nanocomposites with enhanced tribological performance"., *Chemical Engineering Journal*, Vol 324, Page 51–62, 2017.

108. Murty, B. S., P. Shankar, Baldev Raj, B. B. Rath, and James Murday. "Applications of nanomaterials." In *Textbook of nanoscience and nanotechnology*, Springer, Berlin, Heidelberg, Page 107–148, 2013.

109. Bengisu, Murat, and Marinella Ferrara, "Materials that move: Smart materials." *Intelligent Design*, 2018.

110. Lee, Deuk Won, Jae Young Choi, Myeong Woo Cho, Chul Hee Lee, Won Oh Cho, and Hyuk Chae Yun, "Tribological characteristics in modified magneto-rheological fluid" *Applied Mechanics and Materials*, Vol 110, Page 225–231, 2012.

111. Bossis, Georges, Sandris Lacis, Alain Meunier, and O. Volkova. "Magnetorheological fluids." *Journal of Magnetism and Magnetic Materials*, Vol 252, Page 224–228, 2002.

112. Ali, Mohamed Kamal Ahmed, and Hou Xianjun, "M50 matrix sintered with nanoscale solid lubricants shows enhanced self-lubricating properties under dry sliding at different temperatures." *Tribology Letters*, Vol 67, Issue 3, Page 1–16, 2019.

113. Gandhi, Mukesh V., and B. D. Thompson. *Smart materials and structures.* Springer Science & Business Media, 1992.

114. Sliney, Harold E., "Solid lubricant materials for high temperatures: A review." *Tribology International*, Vol 5, Page 303–315, 1982.

115. Huai, Wenjuan, Chenhui Zhang, and Shizhu Wen. "Graphite-based solid lubricant for high-temperature lubrication." *Friction*, Vol 9, Page 1660–1672, 2021.

116. Murray, S. Frank, and Salvadore J. Calabrese, "Effect of solid lubricants on low speed sliding behavior of silicon nitride at temperatures to 800° C." *Lubrication Engineering*, Vol 49, Issue 12, Page 955–964, 1993.

117. Fasihi, Panahsadat, Olivia Kendall, Ralph Abrahams, Peter Mutton, Quan Lai, Cong Qiu, and Wenyi Yan. "Effect of graphite and MoS2 based solid lubricants for application at wheel-rail interface on the wear mechanism and surface morphology of hypereutectoid rails." *Tribology International*, Vol 157, Page 106886, 2021.

118. Braithwaite, Eric Reeves. *Solid lubricants and surfaces.* Elsevier, 2013.

119. Friedrich, Klaus, Li Chang, and Frank Haupert. "Current and future applications of polymer composites in the field of tribology." In *Composite materials*, Page 129–167. Springer, London, 2011.

120. Wadham. *Module 1-basic material properties and how they can be altered,* Canteach, 1993.
121. Briscoe, B. J., and S. K. Sinha. "Tribology of polymeric solids and their composites." In *Wear—Materials, mechanisms and practice,* Page 223–267, 2015.
122. Gururaja, M. N., and A. N. Hari Rao. "A review on recent applications and future prospectus of hybrid composites." *International Journal of Soft Computing and Engineering,* Vol 1, Page 352–355, 2012.
123. Friedrich, Klaus, ed. *Advances in composite tribology.* Elsevier, 2012.
124. Li, Z., C. Maldanado, T. H. North, and B. Altshuller. "Mechanical and metallurgical properties of MMC friction welds." *Welding Journal-Including Welding Research Supplement,* Vol 00076, Issue 9, Page 367, 1997.
125. Senapati, Ajit Kumar, Viplav Saumya Manas, Akash Singh, Shaktipada Dash, and Pratik Kumar Sahoo, "A comparative investigation on physical and mechanical properties of mmc reinforced with waste materials." *International Journal of Research in Engineering and Technology,* Vol 5,Issue 3, Page 172–178, 2016.
126. Pramanik, Alokesh. "Effects of reinforcement on wear resistance of aluminum matrix composites." *Transactions of Nonferrous Metals Society of China,* Vol 26, Issue 2, 348–358, 2016.
127. Premnath, A. Arun, T. Alwarsamy, T. Rajmohan, and R. Prabhu. "The influence of alumina on mechanical and tribological characteristics of graphite particle reinforced hybrid Al-MMC." *Journal of Mechanical Science and Technology,* Vol 28, Issue 11, 4737–4744, 2014.
128. Ozben, Tamer, Erol Kilickap, and Orhan Cakır. "Investigation of mechanical and machinability properties of SiC particle reinforced Al-MMC." *Journal of Materials Processing Technology,* Vol 198, Issue 1–3, 220–225, 2008.
129. Shyu, R. F., and C. T. Ho. "In situ reacted titanium carbide-reinforced aluminum alloys composite." *Journal of Materials Processing Technology,* Vol 171, Issue 3, 411–416, 2006
130. Lou, D., J. Hellman, D. Luhulima, J. Liimatainen, and V. K. Lindroos. "Interactions between tungsten carbide (WC) particulates and metal matrix in WC-reinforced composites." *Materials Science and Engineering A,* Vol 340, Issue 1–2, 155–162, 2003.
131. Bradbury, Christopher R., Jaana-Kateriina Gomon, Lauri Kollo, Hansang Kwon, and Marc Leparoux. "Hardness of multi wall carbon nanotubes reinforced aluminium matrix composites." *Journal of Alloys and Compounds,* Vol 585, Issue 362–367, 2014.
132. Premnath, A. Arun, T. Alwarsamy, T. Rajmohan, and R. Prabhu. "The influence of alumina on mechanical and tribological characteristics of graphite particle reinforced hybrid Al-MMC." *Journal of Mechanical Science and Technology,* Vol 28,Issue 11, Page 4737–4744, 2014.
133. Rajesh Kumar, L., A. Saravanakumar, V. Bhuvaneswari, M. P. Jithin Karunan, N. Karthick Raja, and P. Karthi. "Tribological behaviour of AA2219/MOS2 metal matrix composites under lubrication." *AIP Conference Proceedings,* Vol 2207, Issue 1, Page 020005, 2020
134. Taha, Mohamed A, "Practicalization of cast metal matrix composites (MMCCs)." *Materials & Design,* Vol 22, Issue 6, Page 431–441, 2001.

135. Singh, Jaswinder, and Amit Chauhan, "Characterization of hybrid aluminum matrix composites for advanced applications–A review." *Journal of Materials Research and Technology*, Vol 5, Issue 2, Page 159–169, 2016
136. Suthar, Jigar, and K. M. Patel. "Processing issues, machining, and applications of aluminum metal matrix composites." *Materials and Manufacturing Processes*, Vol 33, Page 499–527, 2018
137. Dharmalingam, S., R. Subramanian, K. Somasundara Vinoth, and B. Anandavel. "Optimization of tribological properties in aluminum hybrid metal matrix composites using gray-Taguchi method." *Journal of Materials Engineering and Performance*, Vol 20, Issue 8, Page 1457–1466, 2011
138. Sharma, Arun Kumar, Rakesh Bhandari, Amit Aherwar, Rūta Rimašauskienė, and Camelia Pinca-Bretotean. "A study of advancement in application opportunities of aluminum metal matrix composites." *Materials Today: Proceedings*, Vol 26, Page 2419–2424, 2020.
139. James, S. Johny, Kb Venkatesan, Pc Kuppan, and R. Ramanujam. "Hybrid aluminium metal matrix composite reinforced with SiC and TiB2." *Procedia Engineering*, Vol 97, Page 1018–1026, 2014.
140. Hossain, Shakil, Mamunur Rahman, Devesh Chawla, Alip Kumar, Prem Prakash Seth, Pallav Gupta, Devendra Kumar, Rajeev Agrawal, and Anbesh Jamwal. "Fabrication, microstructural and mechanical behavior of Al-Al2O3-SiC hybrid metal matrix composites." *Materials Today: Proceedings*, Vol 21, Page 1458–1461, 2020.
141. Das, Shubhajit, M. Chandrasekaran, Sutanu Samanta, Palanikumar Kayaroganam, and Paulo Davim. "Fabrication and tribological study of AA6061 hybrid metal matrix composites reinforced with SiC/B4C nanoparticles." *Industrial Lubrication and Tribology*, 2019.
142. Chand, Sudipta, and Polymersetty Chandrasekhar, "Influence of B4C/BN on solid particle erosion of Al6061 metal matrix hybrid composites fabricated through powder metallurgy technique." *Ceramics International*, Vol 46, Issue 11, Page 17621–17630, 2020.
143. Baradeswaran, A., and A. Elaya Perumal, "Study on mechanical and wear properties of Al 7075/Al2O3/graphite hybrid composites" *Composites Part B: Engineering*, Vol 56, Page 464–471, 2014.
144. Kumar, N.G. Siddesh, G.S. Shiva Shankar, S. Basavarajappa, and R. Suresh. "wear behaviour of hybrid metal matrix composites at high temperature." *Journal of Manufacturing Technology Research*, Vol 9, Issue 3/4, Page 151–167, 2017.
145. Ayar, M.S., P.M. George, and R.R. Patel. "Advanced research progresses in aluminium metal matrix composites: An overview." *AIP Conference Proceedings*, Vol 2317, Page 020026, 2021.
146. Jawaid, Mohammad, Rajini Nagarajan, Jacob Sukumaran, and Patrick De Baets, eds. *Synthesis and tribological applications of hybrid materials*, John Wiley & Sons, 2018.
147. Babu, B. Suresh, A. S. Maniratnam, T. Balaji, G. Chandramohan, T. Ambedkar, B. Arunvarman, A. Hari Prasath, and R. Hari Venkatesh. "Investigations of the effect of the tool rotational speed on friction stir welded joint on aluminium metal matrix hybrid composite." *IOP Conference Series: Materials Science and Engineering*, Vol 988, Page 012045, 2020.

148. Elango, G., and B.K. Raghunath, "Tribological behavior of hybrid (LM25Al+ SiC+ TiO2) metal matrix composites." *Procedia Engineering*, Vol 64, Page 671–680, 2013.

149. Mitrović, Slobodan, Miroslav Babić, Blaža Stojanović, Nenad Miloradović, M. Pantić, and D. Džunić, "Tribological potential of hybrid composites based on zinc and aluminium alloys reinforced with SiC and graphite particles" *Tribology in Industry*, Vol 34, Issue 4, Page 177–185, 2012

150. Nayim, S.T.I., Hasan, M.Z., Seth, P.P., Gupta, P., Thakur, S., Kumar, D., and Jamwal, A., "Effect of CNT and TiC hybrid reinforcement on the micro-mechano-tribo behaviour of aluminium matrix composites, *Materials Today: Proceedings*, Vol 21, Page 1421–1424, 2020.

151. Pavithran, B., J. Swathanandan, N. Praveen, S. R. Prasanna Kumar, and D. Senthil Kumaran, "Study of mechanical and tribological properties of Al6061 reinforced with silicon carbide and graphite particles" *International Journal of Technology Enhancements and Emerging Engineering Research*, Vol 3, Page 60–64, 2015

152. James, S. J., Venkatesan, K., Kuppan, P., and Ramanujam, R. "Hybrid aluminium metal matrix composite reinforced with SiC and TiB2." *Procedia Engineering*, Vol 97, Page 1018–1026, 2014.

153. Elango, G., B. K. Raghunath, K. Palanikumar, and K. Thamizhmaran, "Sliding wear of LM25 aluminium alloy with 7.5% SiC+ 2.5% TiO2 and 2.5% SiC+ 7.5% TiO2 hybrid composites" *Journal of Composite Materials*, Vol 48, Issue 18, Page 2227–2236, 2014.

154. Kumar, K. Sunil Ratna Ch Ratnam, Ch Ramakrishna, and Ch Lakshmi Poornima. "Fabrication and corrosion behaviour of aluminium hybrid and non-hybrid MMCs additions reinforced by powder with metallurgy B4C and Gr." In *Recent Advances in Material Sciences: Select Proceedings of ICLIET 2018*, Page 217, 2018.

155. Shuvho, M. B. A., M. A. Chowdhury, M. Kchaou, B. K. Roy, A. Rahman, and M. A. Islam. "Surface characterization and mechanical behavior of aluminum based metal matrix composite reinforced with nano Al2O3, SiC, TiO2 particles." *Chemical Data Collections*, Vol 28, Page 100442, 2020.

156. Muthazhagan, Chinnasamy, A. Gnanavelbabu, G. B. Bhaskar, and K. Rajkumar. "Influence of graphite reinforcement on mechanical properties of aluminum-boron carbide composites." In *Advanced Materials Research*, Vol 845, Page 398–402, 2014.

157. Jadhav, Pankaj R., B. R. Sridhar, Madeva Nagaral, and Jayasheel I. Harti. "Mechanical behavior and fractography of graphite and boron carbide particulates reinforced A356 alloy hybrid metal matrix composites." *Advanced Composites and Hybrid Materials*, Vol 3, Issue 1, Page 114–119, 2020.

158. Devaganesh, S., PK Dinesh Kumar, N. Venkatesh, and R. Balaji. "Study on the mechanical and tribological performances of hybrid SiC-Al7075 metal matrix composites." *Journal of Materials Research and Technology*, Vol 9, Page 3759–3766, 2020.

159. Uddin, Sheikh M., Tanvir Mahmud, Christoph Wolf, Carsten Glanz, Ivica Kolaric, Christoph Volkmer, Helmut Höller, Ulrich Wienecke, Siegmar Roth, and Hans-Jörg Fecht, "Effect of size and shape of metal particles to improve hardness and electrical properties of carbon nanotube reinforced copper and copper alloy composites" *Composites Science and Technology*, Vol 70, Issue 16, Page 2253–2257, 2010.

160. Fathy, A., F. Shehata, M. Abdelhameed, and M. Elmahdy, "Compressive and wear resistance of nanometric alumina reinforced copper matrix composites" *Materials & Design*, Vol 36, Page 100–107, 2012.

161. Pourrajab, Rashid, Aminreza Noghrehabadi, Ebrahim Hajidavalloo, and Mohammad Behbahani. "Investigation of thermal conductivity of a new hybrid nanofluids based on mesoporous silica modified with copper nanoparticles: Synthesis, characterization and experimental study." *Journal of Molecular Liquids*, Vol 300, Page 112337, 2020.

162. Nautiyal, Hemant, Sangita Kumari, Om P. Khatri, and Rajnesh Tyagi. "Copper matrix composites reinforced by rGO-MoS2 hybrid: Strengthening effect to enhancement of tribological properties." *Composites Part B: Engineering*, Vol 173, Page 106931, 2019

163. Amirthagadeswaran, K. S. "Corrosion and wear behaviour of nano Al2O3 reinforced copper metal matrix composites synthesized by high energy ball milling." *Particulate Science and Technology*, 2019.

164. Jamaati, Roohollah, Mohammad Reza Toroghinejad, "Application of ARB process for manufacturing high-strength, finely dispersed and highly uniform Cu/Al₂O₃ composite." *Materials Science and Engineering*, Vol 527, Issue 27–28, Page 7430–7435, 2010.

165. Shehata, F., A. Fathy, M. Abdelhameed, and S. F. Moustafa, "Fabrication of copper–alumina nanocomposites by mechano-chemical routes" *Journal of Alloys and Compounds*, Vol 476, Issue 1–2, Page 300–305, 2009.

166. Singh, Jaswinder. "Fabrication characteristics and tribological behavior of Al/SiC/Gr hybrid aluminum matrix composites: A review." *Friction*, Vol 4, Page 191–207, 2016.

167. Andrey M. Abyzov, Fedor M. Shakhov, Andrey I. Averkin, and Vladimir Nikolaev, "Mechanical properties of a diamond–copper composite with high thermal conductivity." *Materials & Design*, Vol 87, Page 527–539, 2015.

168. Ben Liu, Dongqing Zhang, Xiangfen Li, Zhao He, Xiaohui Guo, Zhanjun Liu, and Quangui Guo, "Effect of graphite flakes particle sizes on the microstructure and properties of graphite flakes/copper composites." *Journal of Alloys and Compounds*, Vol 766, Page 382–390, 2018.

169. Venkatesh, R., and Vaddi Seshagiri Rao. "Thermal, corrosion and wear analysis of copper based metal matrix composites reinforced with alumina and graphite." *Defence Technology*, Vol 14, Page 346–355, 2018.

170. Guosen Shao, Ping Liu, Ke Zhang, Wei Li, Xiaohong Chen, and Fengcang Ma, "Mechanical properties of graphene nanoplates reinforced copper matrix composites prepared by electrostatic self-assembly and spark plasma sintering." *Materials Science and Engineering A*, Vol 739, Page 329–334, 2019.

171. S.C. Tjong, and K.C. Lau, "Tribological behaviour of SiC particle-reinforced copper matrix composites." *Materials Letters*, Vol 43, Issues 5–6, Page 274–280, 2000.

172. Fathy, Adel, Omayma Elkady, and Ahmed Abu-Oqail, "Microstructure, mechanical and wear properties of Cu–ZrO2 nanocomposites." *Materials Science and Technology*, Vol 33, Issue 17, Page 2138–2146, 2017.

173. Adel Fathy, Omayma Elkady, and Ahmed Abu-Oqail, "Synthesis and characterization of Cu–ZrO2 nanocomposite produced by thermochemical process." *Journal of Alloys and Compounds*, Vol 719, Page 411–419, 2017.

174. Abu-Oqail, A., A. Wagih, A. Fathy, O. Elkady, and A. M. Kabeel, "Effect of high energy ball milling on strengthening of Cu-ZrO2 nanocomposites." *Ceramics International*, Vol 45, Issue 5, Page 5866–5875, 2019.

175. Hassan, S. F., and M. Gupta. "Development of high strength magnesium copper based hybrid composites with enhanced tensile properties." *Materials Science and Technology*, Vol 19, Issue 2, Page 253–259, 2003.

176. Uddin, Sheikh M., Tanvir Mahmud, Christoph Wolf, Carsten Glanz, Ivica Kolaric, Christoph Volkmer, Helmut Höller, Ulrich Wienecke, Siegmar Roth, and Hans-Jörg Fecht. "Effect of size and shape of metal particles to improve hardness and electrical properties of carbon nanotube reinforced copper and copper alloy composites." *Composites Science and Technology*, Vol 70, Issue 16, Page 2253–2257, 2010.

177. Gao, Xin, Hongyan Yue, Erjun Guo, Hong Zhang, Xuanyu Lin, Longhui Yao, and Bao Wang. "Mechanical properties and thermal conductivity of graphene reinforced copper matrix composites." *Powder Technology*, Vol 301, Page 601–607, 2016.

178. Deshpande, P. K., and R. Y. Lin., "Wear resistance of WC particle reinforced copper matrix composites and the effect of porosity." *Materials Science and Engineering A*, Vol 418, Issue 1–2, Page 137–145, 2006.

179. Zhan, Yongzhong, and Guoding Zhang, "Friction and wear behavior of copper matrix composites reinforced with SiC and graphite particles" *Tribology Letters*, Vol 17, Issue 1, Page 91–98, 2004.

180. Tjong, S. C., and K. C. Lau. "Tribological behaviour of SiC particle-reinforced copper matrix composites." *Materials Letters*, Vol 43, Issue 5–6, Page 274–280, 2000.

181. Chen, Fanyan, Jiamin Ying, Yifei Wang, Shiyu Du, Zhaoping Liu, and Qing Huang. "Effects of graphene content on the microstructure and properties of copper matrix composites." *Carbon*, Vol 96, Page 836–842, 2016.

182. Mallikarjuna, H. M., C. S. Ramesh, P. G. Koppad, R. Keshavamurthy, and D. Sethuram. "Nanoindentation and wear behaviour of copper based hybrid composites reinforced with SiC and MWCNTs synthesized by spark plasma sintering." *Vacuum*, Vol 145, Page 320–333, 2017.

183. Lin, C. B., Zue-Chin Chang, Y. H. Tung, and Yuan-Yuan Ko. "Manufacturing and tribological properties of copper matrix/carbon nanotubes composites." *Wear*, Vol 270, no. 5–6 (2011): 382–394.

184. Furlan, Kaline Pagnan, José Daniel Biasoli de Mello, and Aloisio Nelmo Klein. "Self-lubricating composites containing MoS2: A review." *Tribology International*, Vol 120, Page 280–298, 2018.

185. Kim, W. J., T. J. Lee, and S. H. Han., "Multi-layer graphene/copper composites: Preparation using high-ratio differential speed rolling, microstructure and mechanical properties" *Carbon*, Vol 69, Page 55–65, 2014.

186. Barani, Zahra, Amirmahdi Mohammadzadeh, Adane Geremew, Chun-Yu Huang, Devin Coleman, Lorenzo Mangolini, Fariborz Kargar, and Alexander A. Balandin, "Thermal properties of the binary-filler hybrid composites with graphene and copper nanoparticles." *Advanced Functional Materials*, Vol 30, Issue 8, Page 1904008, 2020.

187. Berman, Diana, Ali Erdemir, and Anirudha V. Sumant. "Graphene: A new emerging lubricant." *Materials Today*, Vol 17, Issue 1, Page 31–42, 2014.

188. Chu, Ke, Jing Wang, Ya-ping Liu, and Zhong-rong Geng. "Graphene defect engineering for optimizing the interface and mechanical properties of graphene/ copper composites." *Carbon*, Vol 140, Page 112–123, 2018.

189. Coupard, D., M. C. Castro, J. Coleto, Ana Garcia, Javier Goni, and J. K. Palacios. "Wear behavior of copper matrix composites." *Key Engineering Materials*, Vol 127, Page 1009–1016, 1997.

190. Kaczmar, J. W., K. Pietrzak, and W. Włosiński. "The production and application of metal matrix composite materials." *Journal of Materials Processing Technology*, Vol 106, Issue 1–3, Page 58–67, 2007.

191. Ramesh, C. S., R. Noor Ahmed, M. A. Mujeebu, and M. Z. Abdullah. "Fabrication and study on tribological characteristics of cast copper–TiO2–boric acid hybrid composites." *Materials & Design*, Vol 30, Page 1632–1637, 2009.

192. Kanthavel, K., K. R. Sumesh, and P. Saravanakumar. "Study of tribological properties on Al/Al2O3/MoS2 hybrid composite processed by powder metallurgy." *Alexandria Engineering Journal 55*, Page 13–17, 2016.

193. Soleymani, S., A. Abdollah-Zadeh, and S. A. Alidokht. "Microstructural and tribological properties of Al5083 based surface hybrid composite produced by friction stir processing." *Wear*, Vol 278, Page 41–47.

194. Rajesh, S., C. Velmurugan, Ebenezer Jacob Dhas DS, and Samuel Ratna Kumar PS. "Studies on the tribological behaviour of exsitu-synthesized AlMg1SiCu/ titanium carbide/molybdenum disulfide hybrid composites." *Materials Research Express*, Vol 6, Page 1265, 2019.

195. Gupta, Manoj Kumar. "Analysis of tribological behavior of Al/Gr/MoS2 surface composite fabricated by friction stir process." *Carbon Letters*, Vol 30, Page 399–408, 2020.

196. Liang, Tao, W. Gregory Sawyer, Scott S. Perry, Susan B. Sinnott, and Simon R. Phillpot. "First-principles determination of static potential energy surfaces for atomic friction in MoS2 and MoO3." *Physical Review B*, Vol 77, Page 104105, 2018.

197. Anokhin, V. M., O. M. Ivasishin, and A. N. Petrunko. "Structure and properties of sintered titanium alloyed with aluminium, molybdenum and oxygen." *Materials Science and Engineering A*, Vol 243, Page 269–272, 1998.

198. Ramesh, C. S., R. Noor Ahmed, M. A. Mujeebu, and M. Z. Abdullah. "Fabrication and study on tribological characteristics of cast copper–TiO2– boric acid hybrid composites." *Materials & Design*, Vol 30, Issue 5, Page 1632–1637, 2009.

199. Guo, Junde, Runling Peng, Hang Du, Yunbo Shen, Yue Li, Jianhui Li, and Guangneng Dong. "The application of nano-MoS2 quantum dots as liquid lubricant additive for tribological behavior improvement." *Nanomaterials*, Vol 10, Page 200, 2020.

200. Zou, T. Z., J. P. Tu, H. D. Huang, D. M. Lai, L. L. Zhang, and D. N. He. "Preparation and Tribological Properties of Inorganic Fullerene-like MoS2." *Advanced Engineering Materials*, Vol 8, Page 289–293, 2016.

201. Mutyala, Kalyan C., Yimin A. Wu, Ali Erdemir, and Anirudha V. Sumant. "Graphene-MoS2 ensembles to reduce friction and wear in DLC-Steel contacts." *Carbon*, Vol 146, Page 524–527, 2019.

202. Ansari, N., and E. Mohebbi. "Increasing optical absorption in one-dimensional photonic crystals including MoS2 monolayer for photovoltaics applications." *Optical Materials*, Vol 62, Page 152–158, 2016.

203. Yang, Xianguang, and Baojun Li. "Monolayer MoS2 for nanoscale photonics." *Nanophotonics*, Vol 9, Page 1557–1577, 2020.

204. Jayan, Jitha S., Saritha Appukuttan, and Kuruvilla Joseph. "MoS2: Advanced nanofiller for reinforcing polymer matrix." *Physica E: Low-dimensional Systems and Nanostructures* Page 114716, 2021.

205. Ye, Hai Zhi, and Xing Yang Liu. "Review of recent studies in magnesium matrix composites." *Journal of Materials Science*, Vol 39, Issue 20, Page 6153–6171, 2004.

206. Luo, Alan A. "Magnesium casting technology for structural applications." *Journal of Magnesium and Alloys*, Vol 1, Issue 1, Page 2–22, 2013.

207. Saranu, Ravikumar, Ratnam Chanamala, and Srinivasa Rao Putti. "Corrosion and tribological behavior of magnesium metal matrix hybrid composites-A review." In *AIP Conference Proceedings*, Vol 2259, Page 020018, 2020

208. Eacherath, Suneesh, and Sivapragash Murugesan. "Synthesis and characterization of magnesium-based hybrid composites–A review." *International Journal of Materials Research*, Vol 109, Page 661–672, 2018.

209. Victor, M. T., Selvakumar, G., Surendarnath, S., and Ravindran, P. "Mechanical properties of magnesium hybrid composite reinforced with Al_2O_3 and MoS_2 particles through PM route." *Materials Today: Proceedings*, Vol 37, Page 2396–2400, 2021.

210. Girish, B. M., B. M. Satish, Sadanand Sarapure, D. R. Somashekar, and Basawaraj. "Wear behavior of magnesium alloy AZ91 hybrid composite materials." *Tribology Transactions*, Vol 58, Page 481–489, 2015.

211. Wang, H. Y., Q. C. Jiang, Y. Q. Zhao, Feng Zhao, Bao-Xia Ma, and Yan Wang. "Fabrication of TiB2 and TiB2–TiC particulates reinforced magnesium matrix composites." *Materials Science and Engineering A*, Vol 372, Page 109–114, 2004.

212. Zhang, Mei-Juan, Yong-Bing Liu, Xiao-Hong Yang, A. N. Jian, and Ke-Shuai Luo. "Effect of graphite particle size on wear property of graphite and Al2O3 reinforced AZ91D-0.8% Ce composites." *Transactions of Nonferrous Metals Society of China*, Vol 18, Page s273–s277, 2008.

213. Kondoh, Katsuyoshi, Ritsuko Tsuzuki, and Eiji Yuasa. "Tribological properties of magnesium matrix composite alloys dispersed with Mg2Si particles." *Advances in Technology of Materials and Materials Processing Journal*, Vol 7, Page 33–36, 2005.

214. Anbuchezhiyan, G., B. Mohan, S. Kathiresan, and R. Pugazenthi. "Influence of microstructure and mechanical properties of TiC reinforced magnesium nano composites." *Materials Today: Proceedings*, Vol 27, Page 1530–1534, 2020.

215. Goh, C. S., J. Wei, L. C. Lee, and M. Gupta. "Development of novel carbon nanotube reinforced magnesium nanocomposites using the powder metallurgy technique." *Nanotechnology*, Vol 17, Page 7, 2005.

216. Dey, Abhijit, and Krishna Murari Pandey. "Magnesium metal matrix composites–A review." *Reviews on Advanced Materials Science*, Vol 42, 2015.

217. Ryan, William. *Properties of ceramic raw materials*. Elsevier, 2013.

218. Kumar, Binod, and Lawrence G. Scanlon. "Polymer–ceramic composite electrolytes: Conductivity and thermal history effects." *Solid State Ionics*, Vol 124, Page 239–254, 1999.

219. Liu, Yingguang, Jianqiu Zhou, and Tongde Shen. "Effect of nano-metal particles on the fracture toughness of metal–ceramic composite." *Materials & Design*, Vol 45, Page 67–71, 2013.

220. Gogotsi, George A. "Fracture toughness of ceramics and ceramic composites." *Ceramics International*, Vol 29, Page 777–78, 2003.
221. Zhang, Chaohui. "Understanding the wear and tribological properties of ceramic matrix composites." *Advances in Ceramic Matrix Composites*, Page 312–339, 2014.
222. Sahoo, Prasanta, and J. Paulo Davim. "Tribology of ceramics and ceramic matrix composites." *Tribology for Scientists and Engineers*, Page 211–231, 2013.
223. Hyde, A. R. "Ceramic matrix composites." *Materials & Design*, Vol 11, Page 30–36, 1990.
224. Nieto, Andy, Ankita Bisht, Debrupa Lahiri, Cheng Zhang, and Arvind Agarwal. "Graphene reinforced metal and ceramic matrix composites: A review." *International Materials Reviews*, Vol 62, Page 241–302, 2017.
225. Zhou, W., and Z.M. Xu, "Casting of SiC reinforced metal matrix composites." *Journal of Materials Processing Technology*, Vol 63, Page 358–363, 1997.
226. Suresha, S., and B. K. Sridhara. "Effect of addition of graphite particulates on the wear behaviour in aluminium–silicon carbide–graphite composites." *Materials & Design*, Vol 31, Page 1804–1812, 2010.
227. Shin, Sangmin, Donghyun Lee, Yeong-Hwan Lee, Seongmin Ko, Hyeonjae Park, Sang-Bok Lee, Seungchan Cho, Yangdo Kim, Sang-Kwan Lee, and Ilguk Jo. "High temperature mechanical properties and wear performance of B4C/Al7075 metal matrix composites." *Meta*, Vol 9, Page 1108, 2019.
228. Mohanavel, V., P. Periyasamy, M. Balamurugan, and T. Sathish. "A review on mechanical and tribological behaviour of aluminium based metal matrix composites." *International Journal of Mechanical and Production Engineering Research and Development*, Page 473–478, 2018.
229. Agrawal, Eshan, and Vinod Tungikar. "Study on tribological properties of Al-TiC composites by Taguchi method." *Materials Today: Proceedings*, Vol 26, Page 2242–2247, 2020.
230. Choyke, W. J., and G. Pensl. "Physical properties of SiC." *MRS Bulletin*, Vol 22, Page 25–29, 1997.
231. Agrawal, Eshan, and Vinod Tungikar. "Study on tribological properties of Al-TiC composites by Taguchi method." *Materials Today: Proceedings*, Vol 26, Page 2242–2247, 2020.
232. Moshtaghioun, Bibi Malmal, Diego Gomez-Garcia, Arturo Dominguez-Rodriguez, and Richard I. Todd. "Abrasive wear rate of boron carbide ceramics: Influence of microstructural and mechanical aspects on their tribological response." *Journal of the European Ceramic Society*, Vol 36, Page 3925–3928, 2016.
233. Rallini, Marco, Luigi Torre, Josè M. Kenny, and Maurizio Natali. "Effect of boron carbide nanoparticles on the thermal stability of carbon/phenolic composites." *Polymer Composites*, Vol 38, Page 1819–1827, 2017.
234. Manikandan, R. and, T. V. Arjunan. "Studies on micro structural characteristics, mechanical and tribological behaviours of boron carbide and cow dung ash reinforced aluminium (Al 7075) hybrid metal matrix composite." *Composites Part B: Engineering*, Vol 183, Page 107668, 2020.
235. C.S. Yust, J.M. Leitnaker, and C.E. Devore, "Wear of an Alumina-Silicon Carbide Whisker Composite." *Wear*, Vol 122, Page 151–164, 1988.

236. C.S. Yust, and L.F. Allard, "Wear characteristics of an alumina-silicon carbide whisker composite at temperatures to 800°C in air." *Tribology Transactions*, Vol 32, Page 331–338, 1989.
237. Holz, Dietmar, Rolf Janssen, Klaus Friedrich, and Nils Claussen, "Abrasive wear of ceramic-matrix composites." *Journal of the European Ceramic Society*, Vol 5, Issue 4, Page 229–232, 1989.
238. Latha, P. Sneha, and M. Venkateswara Rao. "Investigation into effect of ceramic fillers on mechanical and tribological properties of bamboo-glass hybrid fiber reinforced polymer composites." *SILICON* 10, Page 1543–1550, 2018.
239. Butler, Sheneve Z., Shawna M. Hollen, Linyou Cao, Yi Cui, Jay A. Gupta, Humberto R. Gutiérrez, Tony F. Heinz et al. "Progress, challenges, and opportunities in two-dimensional materials beyond graphene." *ACS Nano*, Vol 7, Page 2898–2926, 2013.
240. Seymour, Matthew, and Nikolas Provatas. "Structural phase field crystal approach for modeling graphene and other two-dimensional structures." *Physical Review B*, Vol 93, Page 035447, 2016.
241. Zhao, Jun, Yingru Li, Yongfu Wang, Junyuan Mao, Yongyong He, and Jianbin Luo. "Mild thermal reduction of graphene oxide as a lubrication additive for friction and wear reduction." *RSC Advances*, Vol 7, Page 1766–1770, 2017.
242. Li, Yong, Yongjun Zhou, Yanling Wang, Meng Liu, Junya Yuan, and Xuehu Men. "Facile synthesis of WS2@ GO nanohybrids for significant improvement in mechanical and tribological performance of EP composites." *Tribology International*, Vol 163, Page 107148, 2021.
243. Jaiswal, Vinay, Sima Umrao, Rashmi B. Rastogi, Rajesh Kumar, and Anchal Srivastava. "Synthesis, characterization, and tribological evaluation of TiO2-reinforced boron and nitrogen co-doped reduced graphene oxide based hybrid nanomaterials as efficient antiwear lubricant additives." *ACS Applied Materials & Interfaces*, Vol 8, Page 11698–11710, 2016.
244. Xue, Maoquan, Zhiping Wang, Feng Yuan, Xianghua Zhang, Wei Wei, Hua Tang, and Changsheng Li, "Preparation of TiO_2/Ti_3C_2Tx hybrid nanocomposites and their tribological properties as base oil lubricant additives." *RSC Advances*, Vol 7, Issue 8, Page 4312–4319, 2017.
245. He, Qiang, Anling Li, Yachen Guo, Songfeng Liu, and L. H. Kong. "Effect of nanometer silicon dioxide on the frictional behavior of lubricating grease." *Nanomaterials and Nanotechnology*, Vol 7, 2017.
246. Srivyas, P. D., and M. S. Charoo. "A review on tribological characterization of lubricants with nano additives for automotive applications." *Tribology in Industry*, Vol 40, 2018.
247. Pan, Chen, Kaichang Kou, Qian Jia, Yu Zhang, Guanglei Wu, and Tiezheng Ji. "Improved thermal conductivity and dielectric properties of hBN/PTFE composites via surface treatment by silane coupling agent." *Composites Part B: Engineering*, Vol 111, Page 83–90, 2017.
248. Yu, Jingjing, Wenjie Zhao, Yinghao Wu, Deliang Wang, and Ruotao Feng, "Tribological properties of epoxy composite coatings reinforced with functionalized C-BN and H-BN nanofillers." *Applied Surface Science*, Vol 434, Page 1311–1320, 2018.
249. Guo, Yongqiang, Zhaoyuan Lyu, Xutong Yang, Yuanjin Lu, Kunpeng Ruan, Yalan Wu, Jie Kong, and Junwei Gu, "Enhanced thermal conductivities and decreased thermal resistances of functionalized boron nitride/polyimide composites." *Composites Part B: Engineering*, Vol 164, Page 732–739, 2019.

250. Li, Yang, Genjiu Xu, Yongqiang Guo, Tengbo Ma, Xiao Zhong, Qiuyu Zhang, and Junwei Gu, "Fabrication, proposed model and simulation predictions on thermally conductive hybrid cyanate ester composites with boron nitride fillers." *Composites Part A: Applied Science and Manufacturing*, Vol 107, Page 570–578, 2018.

251. Shahnazar, Sheida, Samira Bagheri, and Sharifah Bee Abd Hamid. "Enhancing lubricant properties by nanoparticle additives." *International Journal of Hydrogen Energy*, Vol 41, Page 3153–3170, 2016.

252. Somiya, Shigeyuki, *Handbook of advanced ceramics: Materials, applications, processing, and properties*, Academic Press, 2013.

253. Kato, Koji, and Koshi Adachi, "Wear of advanced ceramics." *Wear*, Vol 253, Issue 11–12, Page 1097–1104, 2002.

254. Tatami, Junichi, Hiromi Nakano, Toru Wakihara, and Katsutoshi Komeya, "Development of advanced ceramics by powder composite process." *Kona Powder and Particle Journal*, Vol 28, Page 227–240, 2010.

255. Herrmann, M., J. Schilm, G. Michael, and J. Adler. "Corrosion behaviour of different technical ceramics in acids, basic solutions and under hydrothermal condition." In *CFI (Ceramic Forum International/Berichte der DKG)*, Vol 80, 2003.

256. Stachowiak, Gwidon W., ed. *Wear: Materials, mechanisms and practice*, 2006.

257. Zgalat-Lozynskyy, O., V. Varchenko, N. Tischenko, A. Ragulya, M. Andrzejczuk, and A. Polotai. "Tribological behaviour of Si3N4-based nanocomposites." *Tribology International*, Vol 91, Page 85–93, 2015.

258. Li, Ming Li, Qiong Yu, Ying Xu, Qing Guo Lu, and Chun Jiang Zhou. "Properties of Si3N4 based nanocomposites prepared by pressureless sintering method." *Advanced Materials Research*, Vol 532, Page 53–56, 2012.

259. Schulz, Ingrid, Mathias Herrmann, Ingolf Endler, Ilmars Zalite, Bruno Speisser, and Johannes Kreusser, "Nano Si3N4 composites with improved tribological properties." *Lubrication Science*, Vol 21, Issue 2, Page 69–81, 2009.

260. Zgalat-Lozynskyy, O., M. Andrzejczuk, V. Varchenko, M. Herrmann, A. Ragulya, and A. Polotai, "Superplastic deformation of Si3N4 based nanocomposites reinforced by nanowhiskers." *Materials Science and Engineering A*, Vol 606, Page 144–149, 2014.

261. Huang, Jow-Lay, Ming-Tung Lee, Horng-Hwa Lu, and Ding-Fwu Lii. "Microstructure, fracture behavior and mechanical properties of TiN/Si3N4 composites." *Materials Chemistry and Physics*, Vol 45, Page 203–209, 1996.

262. Xing, Youqiang, Jianxin Deng, Ze Wu, and Hongwei Cheng. "Effect of regular surface textures generated by laser on tribological behavior of Si3N4/TiC ceramic." *Applied Surface Science*, Vol 265, Page 823–832, 2013.

263. Carrapichano, J. M., J. R. Gomes, and R. F. Silva. "Tribological behaviour of Si3N4–BN ceramic materials for dry sliding applications." *Wear*, Vol 253, Page 1070–1076, 2002.

264. Hadad, M., G. Blugan, J. Kübler, Eric Rosset, L. Rohr, and J. Michler. "Tribological behaviour of Si3N4 and Si3N4–% TiN based composites and multi-layer laminates." *Wear*, Vol 260, Page 634–641, 2006.

265. Eichler, Jens, and Christoph Lesniak. "Boron nitride (BN) and BN composites for high-temperature applications." *Journal of the European Ceramic Society*, Vol 28, Page 1105–1109, 2008.

266. Briscoe, B, "Wear of polymers: An essay on fundamental aspects." *Tribology International*, Vol 14, Issue 4, Page 231–243, 1981.

267. Tanaka, Kyuichiro, and Takashi Miyata, "Studies on the friction and transfer of semicrystalline polymers." *Wear*, Vol 41, Issue 2, Page 383–398, 1977.
268. Shalin, Radiĭ Evgen'evich, ed. *Polymer matrix composites*. Vol 4. Springer Science & Business Media, 2012.
269. Arpitha, G. R., M. R. Sanjay, and B. Yogesha. "Review on comparative evaluation of fiber reinforced polymer matrix composites." *Carbon*, Vol 4000, Page 30, 2014.
270. Boggarapu, Vasavi, Raghavendra Gujjala, and Shakuntla Ojha. "A critical review on erosion wear characteristics of polymer matrix composites." *Materials Research Express*, Vol 7, Page 022002, 2020.
271. Koniuszewska, Anna G., and Jacek W. Kaczmar. "Application of polymer based composite materials in transportation." *Progress in Rubber, Plastics and Recycling Technology*, Vol 32, Page 1–24, 2016.
272. Hussain, Farzana, Mehdi Hojjati, Masami Okamoto, and Russell E. Gorga. "Polymer-matrix nanocomposites, processing, manufacturing, and application: An overview." *Journal of Composite Materials*, Vol 40, Page 1511–1575, 2006.
273. Chen, Xiangbao, Baoyan Zhang, and Liying Xing. "Application and development of advanced polymer matrix composites." *Materials China*, Vol 6, Page 2–12.
274. Larsen, Thomas Ølholm. *Tribological studies of polymer-matrix-composites*. Department of Chemical Engineering, Technical University of Denmark, 2007.
275. Dmitriev, Andrey I., Werner Osterle, Bernd Wetzel, and Ga Zhang. "The impact of nanoparticles on matrix properties in PMCS."
276. Jin, Fan-Long, Xiang Li, and Soo-Jin Park. "Synthesis and application of epoxy resins: A review." *Journal of Industrial and Engineering Chemistry*, Vol 29, Page 1–11, 2015.
277. Jin, Fan-Long, Xiang Li, and Soo-Jin Park. "Synthesis and application of epoxy resins: A review." *Journal of Industrial and Engineering Chemistry*, Vol 29, Page 1–11, 2015.
278. Zhou, Helezi, Hongjian Wang, Xusheng Du, Yingyan Zhang, Huamin Zhou, Hong Yuan, Hong-Yuan Liu, and Yiu-Wing Mai, "Facile fabrication of large 3D graphene filler modified epoxy composites with improved thermal conduction and tribological performance." *Carbon*, Vol 139, Page 1168–1177, 2018.
279. Li, Xiang, Beibei Chen, Yuhan Jia, Xiaofang Li, Jin Yang, Changsheng Li, and Fengyuan Yan, "Enhanced tribological properties of epoxy-based lubricating coatings using carbon nanotubes-ZnS hybrid." *Surface and Coatings Technology*, Vol 344, Page 154–162, 2018.
280. Chen, Beibei, Xiang Li, Yuhan Jia, Lin Xu, Hongyu Liang, Xiaofang Li, Jin Yang, Changsheng Li, and Fengyuan Yan. "Fabrication of ternary hybrid of carbon nanotubes/graphene oxide/MoS2 and its enhancement on the tribological properties of epoxy composite coatings." *Composites Part A: Applied Science and Manufacturing*, Vol 115, Page 157–165, 2015.
281. Lu, Shao-rong, Jing Hongyu, Hai-liang Zhang, and Xia-yu Wang. "Wear and mechanical properties of epoxy/SiO2-TiO2 composites." *Journal of Materials Science*, Vol 40, Page 2815–2821, 2005.
282. Suh, Nam P. "The delamination theory of wear." *Wear*, Vol 25, Page 111–124, 1973.
283. Vasconcelos, Pedro V., F. Jorge Lino, Antonio M. Baptista, and Rui JL Neto, "Tribological behaviour of epoxy based composites for rapid tooling." *Wear*, Vol 260, Issue 1–2, Page 30–39, 2006.

284. Larsen, Thomas Ø., Tom L. Andersen, Bent Thorning, Andy Horsewell, and Martin E. Vigild, "Changes in the tribological behavior of an epoxy resin by incorporating CuO nanoparticles and PTFE microparticles." *Wear*, Vol 265, Issue 1–2, Page 203–213, 2008.

285. Suresha, B., G. Chandramohan, P. R. Sadananda Rao, P. Sampathkumaran, and S. Seetharamu, "Influence of SiC filler on mechanical and tribological behavior of glass fabric reinforced epoxy composite systems." *Journal of Reinforced Plastics and Composites*, Vol 26, Issue 6, Page 565–578, 2007.

286. Srivastava, V. K., and S. Wahne, "Wear and friction behaviour of soft particles filled random direction short GFRP composites." *Materials Science and Engineering A*, Vol 458, Issue 1–2, Page 25–33, 2007.

287. Sudheer, M., K. M. Subbaya, Dayananda Jawali, and Thirumaleshwara Bhat. "Mechanical properties of potassium titanate whisker reinforced epoxy resin composites." *Journal of Minerals and Materials Characterization and Engineering*, Vol 11, Page 193, 2012.

288. Alhazmi, W. H., Y. Jazaa, S. Mousa, A. A. Abd-Elhady, and H. E. M. Sallam. "Tribological and mechanical properties of epoxy reinforced by hybrid nanoparticles." *Latin American Journal of Solids and Structures*, Vol 18, 2021.

289. Matykiewicz, Danuta. "Hybrid epoxy composites with both powder and fiber filler: A review of mechanical and thermomechanical properties." *Materials*, Vol 13, Page 1802, 2020.

290. Guo, Qing Bing, Kin Tak Lau, Min Zhi Rong, and Ming Qiu Zhang. "Optimization of tribological and mechanical properties of epoxy through hybrid filling." *Wear*, Vol 269, Page 13–20, 2010.

291. Zhang, G., and A. K. Schlarb, "Correlation of the tribological behaviors with the mechanical properties of poly-ether-ether-ketones (PEEKs) with different molecular weights and their fiber filled composites." *Wear*, Vol 266, Issue 1–2, Page 337–344, 2009.

292. Lu, Z. P., and K. Friedrich, "On sliding friction and wear of PEEK and its composites." *Wear*, Vol 181, Page 624–631, 1995.

293. Jones, D. P., D. C. Leach, and D. R. Moore, "Mechanical properties of poly (ether-ether-ketone) for engineering applications." *Polymer*, Vol 26, Issue 9, Page 1385–1393, 1985.

294. P.A. Staniland, "Poly(ether ketone)s." In G. Allen, J.C. Bevington (Eds.), *Comprehensive polymer science*, Pergamon Press, Vol 5, Page 484–497, 1989.

295. Molazemhosseini, A., H. Tourani, A. Khavandi, and B. Eftekhari Yekta, "Tribological performance of PEEK based hybrid composites reinforced with short carbon fibers and nano-silica." *Wear*, Vol 303, Issue 1–2, Page 397–404, 2013.

296. Gebhard, Andreas, Thomas Bayerl, Alois K. Schlarb, and Klaus Friedrich, "Increased wear of aqueous lubricated short carbon fiber reinforced poly-etheretherketone (PEEK/SCF) composites due to galvanic fiber corrosion." *Wear*, Vol 268, Issue 7–8, Page 871–876, 2010.

297. Godara, A., D. Raabe, and S. Green, "The influence of sterilization processes on the micromechanical properties of carbon fiber-reinforced PEEK composites for bone implant applications." *Acta Biomaterialia*, Vol 3, Issue 2, Page 209–220, 2007.

298. Burris, David L., and W. Gregory Sawyer, "A low friction and ultra low wear rate PEEK/PTFE composite." *Wear*, Vol 261, Issue 3–4, Page 410–418, 2006.

299. Voort, J. V., and S. Bahadur, "The growth and bonding of transfer film and the role of CuS and PTFE in the tribological behavior of PEEK." *Wear*, Vol 181, Page 212–221, 1995.

300. Xie, G. Y., G. S. Zhuang, G. X. Sui, and R. Yang, "Tribological behavior of PEEK/PTFE composites reinforced with potassium titanate whiskers." *Wear*, Vol 268, Issue 3–4, Page 424–430, 2010.

301. Liu, Liu, Fei Yan, Fangyuan Gai, Linghan Xiao, Lei Shang, Ming Li, and Yuhui Ao, "Enhanced tribological performance of PEEK/SCF/PTFE hybrid composites by graphene." *RSC Advances*, Vol 7, Issue 53, Page 33450–33458, 2017.

302. Kalin, M., M. Zalaznik, and S. Novak, "Wear and friction behaviour of poly-ether-ether-ketone (PEEK) filled with graphene, WS2 and CNT nanoparticles." *Wear*, Vol 332, Page 855–862, 2015.

303. Zhang, G., Z. Rasheva, and A. K. Schlarb, "Friction and wear variations of short carbon fiber (SCF)/PTFE/graphite (10 vol.%) filled PEEK: Effects of fiber orientation and nominal contact pressure." *Wear*, Vol 268, Issue 7–8, Page 893–899, 2010.

304. Friedrich, Klaus, and Alois K. Schlarb, *Tribology of polymeric nanocomposites: Friction and wear of bulk materials and coatings.* Elsevier, 2011.

305. Molazemhosseini, A., H. Tourani, A. Khavandi, and B. Eftekhari Yekta. "Tribological performance of PEEK based hybrid composites reinforced with short carbon fibers and nano-silica." *Wear*, Vol 303, Page 397–404, 2013.

306. Díez-Pascual, Ana M., Mohammed Naffakh, Marián A. Gómez, Carlos Marco, Gary Ellis, M. Teresa Martínez, Alejandro Ansón, José M. González-Domínguez, Yadienka Martínez-Rubi, and Benoit Simard, "Development and characterization of PEEK/carbon nanotube composites." *Carbon*, Vol 47, Issue 13, Page 3079–3090, 2009.

307. Qian, Hui, Emile S. Greenhalgh, Milo SP Shaffer, and Alexander Bismarck, "Carbon nanotube-based hierarchical composites: A review." *Journal of Materials Chemistry*, Vol 20, Issue 23, Page 4751–4762, 2010.

308. Díez-Pascual, Ana M., José M. González-Domínguez, M. Teresa Martínez, and Marián A. Gómez-Fatou, "Poly (ether ether ketone)-based hierarchical composites for tribological applications." *Chemical Engineering Journal*, Vol 218, Page 285–294, 2013.

309. Mahajan, G. V., and V. S. Aher. "Composite material: A review over current development and automotive application." *International Journal of Scientific and Research Publications*, Vol 2, Page 1–5, 2012.

310. Dixit, B. C., R. B. Dixit, and D. J. Desai. "Synthesis, characterization and material application of novel polyimide." *International Journal of Polymeric Materials*, Vol 58, Page 229–242, 2009.

311. Glaeser, William, *Materials for tribology*, Elsevier, Vol 20, 1992.

312. Harvey, Benjamin G., Gregory R. Yandek, Jason T. Lamb, William S. Eck, Michael D. Garrison, and Matthew C. Davis, "Synthesis and characterization of a high temperature thermosetting polyimide oligomer derived from a non-toxic, sustainable bisaniline." *RSC Advances*, Vol 7, Issue 37, Page 23149–23156, 2017.

313. Song, Yan-jiang, Xiao-dong Wang, Gang Zhang, Pei Huang, and Jun Shi. "Effects of fillers on the properties of TPI composites." *Journal of Nanjing University of Technology (Natural Science Edition)* Vol 1, 2006.

314. Duan, Chunjian, Ren He, Song Li, Mingchao Shao, Rui Yang, Liming Tao, Chao Wang, Ping Yuan, Tingmei Wang, and Qihua Wang, "Exploring the friction and wear behaviors of Ag-Mo hybrid modified thermosetting polyimide composites at high temperature." *Friction*, Vol 8, Issue 5, Page 893–904, 2020.

315. Chen, Beibei, Xiaofang Li, Xiang Li, Yuhan Jia, Jin Yang, and Changsheng Li. "Hierarchical carbon fiber-SiO2 hybrid/polyimide composites with enhanced thermal, mechanical, and tribological properties." *Polymer Composites*, Vol 39, Page E1626–E1634, 2018.

316. Li, J., and X. H. Cheng. "Effect of rare earth solution on mechanical and tribological properties of carbon fiber reinforced thermoplastic polyimide composite." *Tribology Letters*, Vol 25, Page 207–214, 2007.

317. Chen, C. S., X. H. Chen, L. S. Xu, Z. Yang, and W. H. Li, "Modification of multi-walled carbon nanotubes with fatty acid and their tribological properties as lubricant additive." *Carbon*, Vol 43, Issue 8, Page 1660–1666, 2005.

318. Cai, Hui, Fengyuan Yan, and Qunji Xue, "Investigation of tribological properties of polyimide/carbon nanotube nanocomposites." *Materials Science and Engineering A*, Vol 364, Issue 1–2, Page 94–100, 2004.

319. Min, Chunying, Dengdeng Liu, Chen Shen, Qiaqia Zhang, Haojie Song, Songjun Li, Xiaojuan Shen, Maiyong Zhu, and Kan Zhang, "Unique synergistic effects of graphene oxide and carbon nanotube hybrids on the tribological properties of polyimide nanocomposites." *Tribology International*, Vol 117, Page 217–224, 2018.

320. Satyanarayana, N., KS Skandesh Rajan, Sujeet K. Sinha, and Lu Shen, "Carbon nanotube reinforced polyimide thin-film for high wear durability." *Tribology Letters*, Vol 27, Issue 2, Page 181–188, 2007.

321. Zhang, Xinrui, Xianqiang Pei, and Qihua Wang, "Study on the friction and wear behavior of surface-modified carbon nanotube filled carbon fabric composites." *Polymers for Advanced Technologies*, Vol 22, Issue 12, Page 2157–2165, 2011.

322. Min, Chunying, Dengdeng Liu, Chen Shen, Qiaqia Zhang, Haojie Song, Songjun Li, Xiaojuan Shen, Maiyong Zhu, and Kan Zhang. "Unique synergistic effects of graphene oxide and carbon nanotube hybrids on the tribological properties of polyimide nanocomposites." *Tribology International*, Vol 117, Page 217–224, 2018.

323. Sava, Ion, and Stefan Chisca. "Surface properties of aromatic polymer film during thermal treatment." *Materials Chemistry and Physics*, Vol 134, Page 116–121, 2012.

324. Progar, D. J., V. L. Bell, and T. L. Stclair. "Polyimide adhesives." 1977.

325. Johnson, Robert O., and Eric O. Teutsch. "Thermoplastic aromatic polyimide composites." *Polymer Composites*, Vol 4, Page 162–166, 1983.

326. Hicyilmaz, Ayse Sezer, and Ayse Celik Bedeloglu. "Applications of polyimide coatings: A review." *SN Applied Sciences*, Vol 3, Page 1–22, 2021.

327. Yang, Shi-yong, Lin Fan, Mian Ji, Ai-jun Hu, Hai-xia Yang, Jin-gang Liu, and Min-hui He. "Advances in high temperature polyimide materials." *Polymer Bulletin*, Vol 10, 2011.

328. Preston, J., H. F. Mark, N. Bikales, C. Overberger, and G. Menges, *Encyclopedia of polymer science and technology*, Vol 381, 1988.

329. Ali, Umar, Khairil Juhanni Bt Abd Karim, and Nor Aziah Buang. "A review of the properties and applications of poly (methyl methacrylate)(PMMA)." *Polymer Reviews*, Vol 55, Page 678–705, 2015.

330. Etienne, Stéphanie, Claude Becker, David Ruch, Bruno Grignard, Grégory Cartigny, Christophe Detrembleur, Cédric Calberg, and Robert Jérôme, "Effects of incorporation of modified silica nanoparticles on the mechanical and thermal properties of PMMA." *Journal of Thermal Analysis and Calorimetry*, Vol 87, Issue 1, Page 101–104, 2007.

331. Kashiwagi, Takashi, Alexander B. Morgan, Joseph M. Antonucci, Mark R. VanLandingham, Richard H. Harris Jr, Walid H. Awad, and John R. Shields, "Thermal and flammability properties of a silica–poly (methylmethacrylate) nanocomposite." *Journal of Applied Polymer Science*, Vol 89, Issue 8, Page 2072–2078, 2003.

332. Kashiwagi, Takashi, Fangming Du, Jack F. Douglas, Karen I. Winey, Richard H. Harris, and John R. Shields, "Nanoparticle networks reduce the flammability of polymer nanocomposites." *Nature Materials*, Vol 4, Issue 12, Page 928–933, 2005

333. Winey, Karen I., Takashi Kashiwagi, and Minfang Mu. "Improving electrical conductivity and thermal properties of polymers by the addition of carbon nanotubes as fillers." *MRS Bulletin*, Vol 32, Page 348–353, 2007.

334. Chan, Jia Xin, Joon Fatt Wong, Michal Petrů, Azman Hassan, Umar Nirmal, Norhayani Othman, and Rushdan Ahmad Ilyas. "Effect of nanofillers on tribological properties of polymer nanocomposites: A review on recent development." *Polymers*, Vol 13, Page 2867, 2021.

335. Judeinstein, Patrick, and Clément Sanchez. "Hybrid organic–inorganic materials: A land of multidisciplinarity." *Journal of Materials Chemistry*, Vol 6, Page 511–525, 1996.

336. Pal, Kaushik. "Effect of different nanofillers on mechanical and dynamic behavior of PMMA based nanocomposites." *Composites Communications*, Vol 1, Page 25–28, 2016.

337. Lv, Yang, Ailing Li, Fang Zhou, Xiaoyu Pan, Fuxin Liang, Xiaozhong Qu, Dong Qiu, and Zhenzhong Yang. "A novel composite PMMA-based bone cement with reduced potential for thermal necrosis." *ACS Applied Materials & Interfaces*, Vol 7, Page 11280–11285, 2015.

338. Karthick, R., P. Sirisha, and M. Ravi Sankar. "Mechanical and tribological properties of PMMA-sea shell based biocomposite for dental application." *Procedia Materials Science*, Vol 6, Page 1989–2000, 2014.

339. Camargo, Pedro Henrique Cury, Kestur Gundappa Satyanarayana, and Fernando Wypych. "Nanocomposites: Synthesis, structure, properties and new application opportunities." *Materials Research*, Vol 12, Page 1–39, 2009.

340. Njuguna, James, and Krzysztof Pielichowski. "Polymer nanocomposites for aerospace applications: Properties." *Advanced Engineering Materials*, Vol 5, Page 769–778, 2003.

341. Malucelli, Giulio, and Francesco Marino. "Abrasion resistance of polymer nanocomposites–a review." *Abrasion Resistance of Materials*, Page 1–18, 2012.

342. Gorga, Russell E., and Robert E. Cohen. "Toughness enhancements in poly (methyl methacrylate) by addition of oriented multiwall carbon nanotubes." *Journal of Polymer Science Part B: Polymer Physics*, Vol 42, Page 2690–2702, 2004.

343. Saravanan, N., R. Rajasekar, S. Mahalakshmi, T. P. Sathishkumar, K. S. K. Sasikumar, and S. Sahoo. "Graphene and modified graphene-based polymer nanocomposites–a review." *Journal of Reinforced Plastics and Composites*, Vol 33, Page 1158–1170, 2014.

344. Rajak, Dipen Kumar, Durgesh D. Pagar, Pradeep L. Menezes, and Emanoil Linul. "Fiber-reinforced polymer composites: Manufacturing, properties, and applications." *Polymers*, Vol 11, Page 1667, 2019.

345. Begum, Shahida, Sabrina Fawzia, and M. S. J. Hashmi. "Polymer matrix composite with natural and synthetic fibres." *Advances in Materials and Processing Technologies*, Vol 6, Page 547–564, 2020.

346. Mohd Nurazzi, N., A. Khalina, S. M. Sapuan, A. H. A. M. Dayang Laila, M. Rahmah, and Z. Hanafee. "A review: Fibres, polymer matrices and composites." *Pertanika Journal of Science & Technology*, Vol 25, 2017.

347. Gupta, M. K., and R. K. Srivastava. "Mechanical properties of hybrid fibers-reinforced polymer composite: A review." *Polymer-Plastics Technology and Engineering*, Vol 55, Page 626–642, 2016.

348. Chandgude, Swapnil, and Sachin Salunkhe. "Biofiber-reinforced polymeric hybrid composites: An overview on mechanical and tribological performance." *Polymer Composites*, Vol 41, Page 3908–3939, 2020.

349. Cheung, Hoi-yan, Mei-po Ho, Kin-tak Lau, Francisco Cardona, and David Hui. "Natural fibre-reinforced composites for bioengineering and environmental engineering applications." *Composites Part B: Engineering*, Vol 40, Issue 7, Page 655–663, 2009.

350. Wambua, Paul, Jan Ivens, and Ignaas Verpoest. "Natural fibres: Can they replace glass in fibre reinforced plastics?." *Composites Science and Technology*, Vol 63, Issue 9, Page 1259–1264, 2003.

351. Thomas, Sabu, and Laly A. Pothan, eds. Natural fibre reinforced polymer composites: From macro to nanoscale. *Archives Contemporaines*, 2009.

352. Mohanty, Akash, and V. K. Srivastava. "Tribological behavior of particles and fiber-reinforced hybrid nanocomposites." *Tribology Transactions*, Vol 58, Page 1142–1150.

353. Vigneshkumar, S., and T. Rajasekaran. "Experimental analysis on tribological behavior of fiber reinforced composites." *IOP Conference Series: Materials Science and Engineering*, Vol 402, Page 012198, 2018.

354. Chin, C. W., and B. F. Yousif. "Potential of kenaf fibres as reinforcement for tribological applications." *Wear*, Vol 267, Issue 9–10, Page 1550–1557, 2009.

355. Yousif, B. F., and N. S. M. El-Tayeb. "Adhesive wear performance of T-OPRP and UT-OPRP composites." *Tribology Letters*, Vol 32, Issue 3, Page 199–208, 2008.

356. Chand, Navin, and U. K. Dwivedi. "Sliding wear and friction characteristics of sisal fibre reinforced polyester composites: Effect of silane coupling agent and applied load." *Polymer Composites*, Vol 29, Issue 3, Page 280–284, 2008.

357. Hashmi, S. A. R., U. K. Dwivedi, and Navin Chand. "Graphite modified cotton fibre reinforced polyester composites under sliding wear conditions." *Wear*, Vol 262, Issue 11–12, Page 1426–1432, 2007.

358. Chand, Navin, and U. K. Dwivedi. "Effect of coupling agent on abrasive wear behaviour of chopped jute fibre-reinforced polypropylene composites." *Wear*, Vol 261, Issue 10, Page 1057–1063, 2006.

359. Yousif, B. F., Saijod TW Lau, and S. McWilliam. "Polyester composite based on betelnut fibre for tribological applications." *Tribology International*, Vol 43, Issue 1–2, Page 503–511, 2010.

360. Nirmal, Umar, Jamil Hashim, and K. O. Low. "Adhesive wear and frictional performance of bamboo fibres reinforced epoxy composite." *Tribology International*, Vol 47, Page 122–133, 2012.

361. Guna, Vijaykumar, Manikandan Ilangovan, M. G. Ananthaprasad, and Narendra Reddy. "Hybrid biocomposites." *Polymer Composites*, Vol 39, Page E30–E54, 2018.

362. Mohanty, Amar K., Manjusri Misra, and Lawrence T. Drzal, eds. *Natural fibers, biopolymers, and biocomposites*. CRC Press, 2005.

363. Sathishkumar, T. P., J. Naveen, and S. Satheeshkumar. "Hybrid fiber reinforced polymer composites–a review." *Journal of Reinforced Plastics and Composites*, Vol 33, Page 454–471, 2014.

364. JyotiKalita, Jiban, and Kalyan Kumar Singh. "Tribological properties of different synthetic fiber reinforced polymer matrix composites–A review." *IOP Conference Series: Materials Science and Engineering*, Vol 455, Page 012134, 2018.

365. Altoubat, Salah, Ardavan Yazdanbakhsh, and Klaus-Alexander Rieder. "Shear behavior of macro-synthetic fiber-reinforced concrete beams without stirrups." *ACI Materials Journal*, Vol 106, Page 381, 2009.

366. Karthik, K., and A. Manimaran. "Wear behaviour of ceramic particle reinforced hybrid polymer matrix composites." *International Journal of Ambient Energy*, Vol 41, Page 1608–1612, 2020.

367. Praveenkumara, J., P. Madhu, T. G. Yashas Gowda, M. R. Sanjay, and Suchart Siengchin. "A comprehensive review on the effect of synthetic filler materials on fiber-reinforced hybrid polymer composites." *The Journal of The Textile Institute*, Page 1–9, 2021.

368. Sathishkumar, T. P., S. Satheeshkumar, and Jesuarockiam Naveen. "Glass fiber-reinforced polymer composites–a review." *Journal of Reinforced Plastics and Composites*, Vol 33, Page 1258–1275, 2014.

369. Pathan, Firojkhan, Hemant Gurav, and Sonam Gujrathi. "Optimization for tribological properties of glass fiber-reinforced PTFE composites with grey relational analysis." *Journal of Materials*, 2016.

370. Orndorff, Roy L. Jr "New UHMWPE/rubber bearing alloy." *Journal of Tribology*, Vol 122, Page 367–373, 2000.

371. Salunkhe, Sachin, and Pavan Chandankar. "Friction and wear analysis of PTFE composite materials." In *Innovative design, analysis and development practices in aerospace and automotive engineering*, Page 415–425, 2019.

372. Masood, Muhammad T., Evie L. Papadopoulou, José A. Heredia-Guerrero, Ilker S. Bayer, Athanassia Athanassiou, and Luca Ceseracciu. "Graphene and polytetrafluoroethylene synergistically improve the tribological properties and adhesion of nylon 66 coatings." *Carbon*, Vol 123, Page 26–33, 2017.

373. Fasake, Vinayak, and Kavya Dashora. "A sustainable potential source of ruminant animal waste material (dung fiber) for various industrial applications: A review." *Bioresource Technology Reports*, Page 100693, 2021.

374. Ma, Yunhai, Siyang Wu, Jian Zhuang, Jin Tong, and Hongyan Qi. "Tribological and physio-mechanical characterization of cow dung fibers reinforced friction composites: An effective utilization of cow dung waste." *Tribology International*, Vol 131, Page 200–211, 2019.

375. Yusefi, Mostafa, Mohammad Khalid, Faizah Md Yasin, Luqman Chuah Abdullah, Mohammad Reza Ketabchi, and Rashmi Walvekar. "Performance of cow dung reinforced biodegradable poly (Lactic Acid) biocomposites for structural applications." *Journal of Polymers and the Environment*, Vol 26, Page 474–486, 2018.

376. Radhika, N., and R. Subramaniam. "Machining parameter optimisation of an aluminium hybrid metal matrix composite by statistical modelling." *Industrial Lubrication and Tribology*, 2013.

377. Taguchi, G. and Konishi, S., *Taguchi methods, orthogonal arrays and linear graphs, tools for quality engineering*, Dearborn, MI: American Supplier Institute, Page 35–38, 1987.

378. Chauhan, S. R., Anoop Kumar, I. Singh, and Prashant Kumar. "Effect of fly ash content on friction and dry sliding wear behavior of glass fiber reinforced polymer composites-a taguchi approach." *Journal of Minerals and Materials Characterization and Engineering*, Vol 9, Page 365, 2010.

379. Basavarajappa, S., G. Chandramohan, and J. Paulo Davim. "Application of Taguchi techniques to study dry sliding wear behaviour of metal matrix composites." *Materials & Design*, Vol 28, Page 1393–1398, 2007.

380. Trehan, R., Singh, S. and Garg, M., "Optimization of mechanical properties of polyester hybrid composite laminate using Taguchi methodology – Part 1." *Proceedings of the Institution of Mechanical Engineers, Part L: Journal of Materials: Design and Applications*, 2013.

381. Radhika, N., R. Subramanian, and S. Venkat Prasat. "Tribological behaviour of aluminium/alumina/graphite hybrid metal matrix composite using Taguchi's techniques." *Journal of Minerals and Materials Characterization and Engineering*, Vol 10, Page 427, 2011.

382. Haque, Mohammed E., and K. V. Sudhakar. "Prediction of corrosion–fatigue behavior of DP steel through artificial neural network." *International Journal of Fatigue*, Vol 23, Page 1–4, 2001.

383. Stojanović, Blaža, Aleksandar Vencl, Ilija Bobić, Slavica Miladinović, and Jasmina Skerlić. "Experimental optimisation of the tribological behaviour of Al/SiC/Gr hybrid composites based on Taguchi's method and artificial neural network." *Journal of the Brazilian Society of Mechanical Sciences and Engineering*, Vol 40, Page 1–14, 2018.

384. Stojanovic, Blaza, Jasmina Blagojevic, Miroslav Babic, Sandra Velickovic, and Slavica Miladinovic. "Optimization of hybrid aluminum composites wear using Taguchi method and artificial neural network." *Industrial Lubrication and Tribology*, 2017.

385. Nanda, Bishnu Prasad, and Alok Satapathy. "An analysis of the sliding wear characteristics of epoxy-based hybrid composites using response surface method and neural computation." *Journal of Natural Fibers*, Vol 18, Page 2077–2091, 2021.

386. Nor, Ariff Farhan Mohd, Mohamad Zaki Hassan, Zainudin A. Rasid, Shamsul Sarip, and Mohd Yusof Md Daud. "Optimization on tensile properties of kenaf/multi-walled CNT hybrid composites with box-behnken design." *Applied Composite Materials*, Page 1–26, 2021.

387. Senthilkumar, N., T. Tamizharasan, and S. Gobikannan. "Application of response surface methodology and firefly algorithm for optimizing multiple responses in turning AISI 1045 steel." *Arabian Journal for Science and Engineering*, Vol 39, Page 8015–8030, 2014.

388. Roslan, Siti Amni, Mohamad Zaki Hassan, Zainudin A. Rasid, Nurul Aini Bani, Shamsul Sarip, Mohd Yusof Md Daud, and Firdaus Muhammad-Sukki. "Mode I fracture toughness of optimized alkali-treated Bambusa vulgaris bamboo by box-behnken design." *Advances in Material Sciences and Engineering*, Page 565–575, 2020.

389. Radhika, N., and R. Raghu. "Investigation on mechanical properties and analysis of dry sliding Wear behavior of Al LM13/AlN metal matrix composite based on Taguchi's technique." *Journal of Tribology*, Vol 139, Page 041602, 2017.

390. Singh, Jaswinder, and Amit Chauhan. "Characterization of hybrid aluminum matrix composites for advanced applications–A review." *Journal of Materials Research and Technology*, Vol 5, Page 159–169, 2016.

391. Meng, Zhao, Xue-feng Zhang, Jingchao Zhang, Bin Hu, and Yun Yang. "Application study of nano-copper based composite anti-friction coating for corrosion resistant couplings." *Journal of Petroleum Science and Engineering*, Vol 157, Page 1143–1147, 2017.

392. Montero, David A., Carolina Arellano, Mirka Pardo, Rosa Vera, Ricardo Gálvez, Marcela Cifuentes, María A. Berasain, Marisol Gómez, Claudio Ramírez, and Roberto M. Vidal. "Antimicrobial properties of a novel copper-based composite coating with potential for use in healthcare facilities." *Antimicrobial Resistance and Infection Control*, Vol 8, Page 1–10, 2019.

393. Larionova, Tatiana. "Copper-based composite materials reinforced with carbon nanostructures." *Materials Science*, Vol 21, Page 364–368, 2015.

394. Kumar, Nanjundan Ashok, Mushtaq Ahmad Dar, Rukhsana Gul, and Jong-Beom Baek. "Graphene and molybdenum disulfide hybrids: Synthesis and applications." *Materials Today*, Vol 18, Page 286–298, 2015.

395. Gupta, Manoj, and W. L. E. Wong. "Magnesium-based nanocomposites: Lightweight materials of the future." *Materials Characterization*, Vol 105, Page 30–46, 2015.

396. Nguyen, Hai, Wael Zatar, and Hiroshi Mutsuyoshi. "Hybrid polymer composites for structural applications." *Hybrid Polymer Composite Materials*, Page 35–51, 2017.

397. Rabi, Osama, Erum Pervaiz, Rubab Zahra, Maryum Ali, and M. Bilal Khan Niazi. "An inclusive review on the synthesis of molybdenum carbide and its hybrids as catalyst for electrochemical water splitting." *Molecular Catalysis*, Vol 494, Page 111116, 2020.

398. Xiao, Ying, Jang-Yeon Hwang, and Yang-Kook Sun. "Transition metal carbide-based materials: Synthesis and applications in electrochemical energy storage." *Journal of Materials Chemistry A*, Vol 4, 27, Page 10379–10393, 2016.

399. El-Shafai, Nagi M., Mahmoud Abdelfatah, Ibrahim M. El-Mehasseb, Mohamed S. Ramadan, Mohamed M. Ibrahim, Abdelhamed El-Shaer, Maged A. El-Kemary, and Mamdouh S. Masoud. "Enhancement of electrochemical properties and photocurrent of copper oxide by heterojunction process as a novel hybrid nanocomposite for photocatalytic anti-fouling and solar cell applications." *Separation and Purification Technology*, Vol 267 Page 118631, 2021.

400. Iqbal, Muhammad Zahir, Saman Siddique, Abbas Khan, Syed Shabhi Haider, and Mohammad Khalid. "Recent developments in graphene based novel structures for efficient and durable fuel cells." *Materials Research Bulletin*, Vol 122, Page 110674, 2020.

401. Dakshayini, B. S., Kakarla Raghava Reddy, Amit Mishra, Nagaraj P. Shetti, Shweta J. Malode, Soumen Basu, S. Naveen, and Anjanapura V. Raghu. "Role of conducting polymer and metal oxide-based hybrids for applications in ampereometric sensors and biosensors." *Microchemical Journal*, Vol 147, Page 7–24, 2019.

402. Samudrala, Sarath C., Subrata Das, Kyu J. Lee, Mohammad G. Abdallah, Brett R. Wenner, Jeffery W. Allen, Monica S. Allen, Robert Magnusson, and Michael Vasilyev. "Silicon-nitride microring resonators for nonlinear optical and bio-sensing applications." *Applied Optics*, Vol 60, Page G132–G138, 2021.

403. Lu, Yang, Kesu Cai, Yuee Li, Zhizhen Duan, Yang Xi, and Zhong Wang. "A high speed optical modulator based on graphene-on-graphene hybrid nano-photonic waveguide." *Optik*, Vol 179, Page 216–221, 2019.

404. Sasidharan, Sarath, and Anoop Anand. "Epoxy-based hybrid structural composites with nanofillers: A review." *Industrial & Engineering Chemistry Research*, Vol 59, Page 12617–12631, 2020.

405. Ravishankar, B., Sanjay K. Nayak, and M. Abdul Kader. "Hybrid composites for automotive applications–A review." *Journal of Reinforced Plastics and Composites*, Vol 38, Page 835–845, 2019.

406. Bobby, Satheesan, and Mohammed Abdul Samad. "Enhancement of tribo-logical performance of epoxy bulk composites and composite coatings using micro/nano fillers: A review." *Polymers for Advanced Technologies*, Vol 28, Page 633–644, 2017.

407. Kang, Ting, Xiaoli Hua, Peiqing Liang, Minyu Rao, Qin Wang, Changyun Quan, Chao Zhang, and Qing Jiang. "Synergistic reinforcement of polydopa-mine-coated hydroxyapatite and BMP2 biomimetic peptide on the bioactivity of PMMA-based cement." *Composites Science and Technology*, Vol 123, Page 232–240, 2016.

408. Subadra, Sharath P., Paulius Griskevicius, and Samy Yousef. "Low velocity impact and pseudo-ductile behaviour of carbon/glass/epoxy and carbon/glass/PMMA hybrid composite laminates for aircraft application at service tempera-ture." *Polymer Testing*, Vol 89, Page 106711, 2020.

409. Radha, G., S. Balakumar, Balaji Venkatesan, and Elangovan Vellaichamy. "A novel nano-hydroxyapatite—PMMA hybrid scaffolds adopted by conjugated thermal induced phase separation (TIPS) and wet-chemical approach: Analysis of its mechanical and biological properties." *Materials Science and Engineering*, Page 221–228, 2017.

410. Kumar, Manjeet, Rajesh Kumar, and Sandeep Kumar. "Synergistic effect of carbon nanotubes and nano-hydroxyapatite on mechanical properties of polyetheretherketone based hybrid nanocomposites." *Polymers and Polymer Composites*, 2020.

411. Cekic-Nagas, Isil, Ferhan Egilmez, Gulfem Ergun, Pekka Kalevi Vallittu, and Lippo Veli Juhana Lassila. "Load-bearing capacity of novel resin-based fixed dental prosthesis materials." *Dental Materials Journal*, Page 2016–367, 2017.

412. Tsai, Chia-Liang, Hung-Ju Yen, and Guey-Sheng Liou. "Highly transparent polyimide hybrids for optoelectronic applications." *Reactive and Functional Polymers*, Vol 108, Page 2–30, 2020.

413. Ashik, K. P., and Ramesh S. Sharma. "A review on mechanical properties of natural fiber reinforced hybrid polymer composites." *Journal of Minerals and Materials Characterization and Engineering*, Vol 3, Page 420, 2020.

414. Kerni, Love, Sarbjeet Singh, Amar Patnaik, and Narinder Kumar. "A review on natural fiber reinforced composites." *Materials Today: Proceedings*, Vol 28, Page 1616–1621, 2020.

415. Keya, Kamrun N., Nasrin A. Kona, Farjana A. Koly, Kazi Madina Maraz, Md Naimul Islam, and Ruhul A. Khan. "Natural fiber reinforced polymer composites: History, types, advantages and applications." *Materials Engineering Research*, Vol 1, Page 69–85, 2019.

416. Puttegowda, Madhu, Sanjay Mavinakere Rangappa, Mohammad Jawaid, Pradeep Shivanna, Yogesha Basavegowda, and Naheed Saba. "Potential of natural/synthetic hybrid composites for aerospace applications." In *Sustainable composites for aerospace applications*, Page 315–351, 2018.

417. Rahman, Rozyanty, and Syed Zhafer Firdaus Syed Putra. "Tensile properties of natural and synthetic fiber-reinforced polymer composites." In *Mechanical and physical testing of biocomposites, fibre-reinforced composites and hybrid composites*, Page 81–102, 2019.

418. Rasheed, Mohammad Abdur, and S. Suriya Prakash. "Mechanical behavior of sustainable hybrid-synthetic fiber reinforced cellular light weight concrete for structural applications of masonry." *Construction and Building Materials*, Vol 98, Page 631–640, 2015.

Chapter 3

Nanotechnology and Surface Engineering

Pankaj Shrivastava, Arka Ghosh, and Syed Nasimul Alam
National Institute of Technology Rourkela, Rourkela, Odisha, India

CONTENTS

3.1 INTRODUCTION

Surface engineering refers to a field that involves designing and modification technologies that improve the desired properties of metals, alloys, or non-metals for practical or decorative objectives. It deals with particle or bulk surfaces. Due to its several advantages, it has applications in various science and engineering fields like chemistry, physics, civil engineering, and electrical engineering. In any system, surfaces have great importance as surfaces enable the transmission of power and motion. Surfaces, therefore, become a crucial component for the effective functioning of the system. The properties of the bulk material are very different with respect to that

DOI: 10.1201/9781003319375-3

obtained from its surface. The surface of a material can be completely reengineered. The surface can be functionalized to achieve a specific molecular configuration. Surface engineering techniques are both varied and complex which provides changes to the outermost material interface. The history of surface engineering can be traced from the time of Thomas Edison in the early 1900s when gold plating was applied in form of thin films. During the 1980s, surface modifications and surface vacuum coatings were initially done by Berghan via plasma and ion techniques [1–3]. During the 1960s, the ion planting technique was discovered for an industry-based surface modification process that was a prominent step toward plasma-assisted coated deposition. After the early 1970s, the major surface engineering processes were based on plasma and thin film deposition. Modification of the surface properties by films or coatings is used in industrial applications. Surface engineering includes a diversity of technologies that alter the chemistry and properties of just a thin surface layer of the substrate. All solids are composed of bulk materials known as the bulk phase whose outer body has a thin covering called the surface phase. The surface phase protects and bonds the inner bulk phase and behaves like an interphase to the external surrounding due to which surface degradation takes place with time. In surface engineering, the surface phase of a material is usually modified to obtain desired properties without modifying the bulk. Several surface engineering technologies can be administered to the surface to improvise, design, and modify the surface properties and hence increase the life span of the component. Surface engineering provides cost-effective materials for robust design [4–6]. In the last two decades, surface engineering has mainly been having three consecutive developing stages. In the initial stage, various single coating surface modification techniques were used like physical vapor deposition (PVD) and chemical vapor deposition (CVD), thermal spraying, laser treatment, ion-implantation, etc. In the second stage, the use of a combination of two surface coating techniques came into existence in which two distinct traditional surface modification techniques combine and result in the cooperative effect of the two processes. The combinations could be of laser remelting with thermal spraying, or electroplating technique combined with other surface modification treatments, etc. The third stage saw the introduction of nanomaterials and nanotechnologies in surface engineering. More and more nanomaterials and nanotechnologies are today being used in surface engineering [7–9]. The nanolevel surface treatment-based engineering tools have always been intensively studied. Nanosurface engineering introduces the morphological, interfacial, physical, and structural properties of nanosized particles including 2D surfaces concerning various applications. The surface properties of nanomaterials impact a large number of physical properties like optical, toxicity, magnetic, and electronic properties, etc. [10, 11]. Nanotechnology represents a new frontier in research and development across a broad spectrum of areas. Though nanotechnology is concerned with the minuscule, the scope of nanotechnology is undeniably vast. Coating processes, such as atomic layer deposition (ALD), have pushed the boundaries of thin-film nanotechnology by enabling the development of conformal films with strong control over thickness and composition at the atomic scale. By selectively and sequentially introducing precursor gases into an ALD vacuum chamber, complex substrates are exposed to individual gas phases in a series of alternating cycles. Gas molecules precipitate on the target surface in a self-limiting manner, which means reactions cease as soon as all reaction sites are occupied. Theoretically, this yields a precisely uniform and extremely thin coating of precipitate which acts as the new reaction surface for subsequent gas-phase molecules. Nanotechnology and advanced surface engineering offer attractive advantages. New and promising products are expected to be introduced in the market with the application of nanotechnology in surface engineering. Products like flat-panel displays, bioimplants, pressure bearings, biosensors, magnetic memories, thin-film coatings, etc., all use thin-film coatings [12, 13]. Figure 3.1(a, b) shows the wide application area and various techniques of surface engineering.

Most manufactured products undergo surface engineering which involves surface modification or surface coating. Various engineering tools can be applied to different fields of application to obtain higher and more effective outcomes from the usual results. Surface engineering is nowadays extensively employed to tailor the surfaces of various substrates such that it could sustain harsh and adverse conditions and deliver prolonged service life in applications. Soft or low-strength alloys like Mg alloy prominently fail due to fatigue and wear. Also, about 90% of the Ti alloy hip prosthesis fails due to surface fracture occurring due to fatigue. So it is highly desired to engineer the surfaces to avoid failure initiating from the surface. The surface-engineered substrates or components are different from the non-engineered homogeneous components in exhibiting resistance against the material degrading sources like corrosion, wear, fatigue, and other tribo-related material deteriorations. Nanotechnology deals with particles having sizes

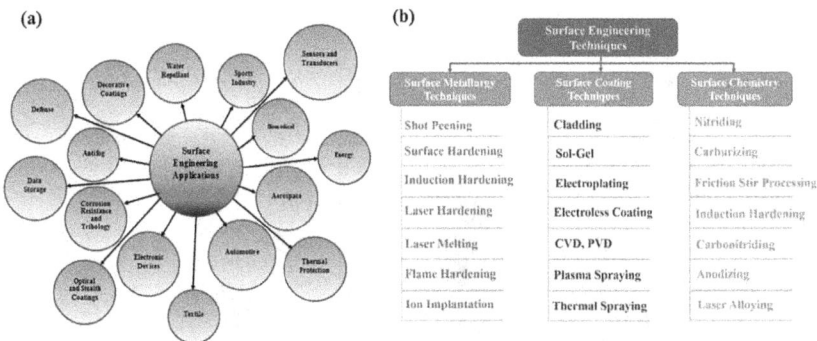

Figure 3.1 (a) Various applications of surface engineering and (b) various surface engineering techniques.

varying up to a few nanometers only. The small size of the nanoparticles results in a huge specific particle surface area as a result of which a whole spectrum of new properties of the material, emerges. Surface engineering with nanotechnology deals with the modification of surfaces using nano-technology. For instance, surfaces that have been modified by nanotechnol-ogy can create hydrophobic surfaces that repel water and also do not allow dirt particles to adhere to the surface. Nanometers-thick Ti surface layers improve the biocompatibility of implants. Extremely thin functional layers having a thickness of few nanometers can be used to selectively adjust opti-cal properties like color, reflection, or transparency and also improve strength and wear resistance. Figure 3.2 shows the various applications of nanotechnology.

The vast knowledge and understanding in the field of nanomaterials have been possible mainly due to the advanced analytical techniques that are available today. A given analytical technique may have a limited range of applications and, therefore, it is beneficial to employ more than one tech-nique and combine the results from these several techniques to get as much information as possible. The surface characterization techniques have the potential to analyze various types of surfaces. A number of these techniques are now being developed to understand surface properties even at the atomic level. The flaws in the surface generated are necessarily required to be pre-vented to avoid catastrophic failure or any severe damage to the component or component-assembled machinery. The various surface characterization techniques that are employed to define different surface properties and topography at micro- and nanoscales and to determine the surface struc-tures are optical microscopy, scanning, and tunneling electron microscopy (EM), X-ray diffraction (XRD), X-ray photoelectron spectroscopy (XPS), Raman spectroscopy, Fourier transform infrared spectroscopy (FTIR), 3D profilometry, ion scattering spectroscopy (ISS), atomic force microscopy (AFM), and scanning probe microscopy (SPM). In the present chapter, the basics of the surface engineering techniques like nanocoatings and thin film

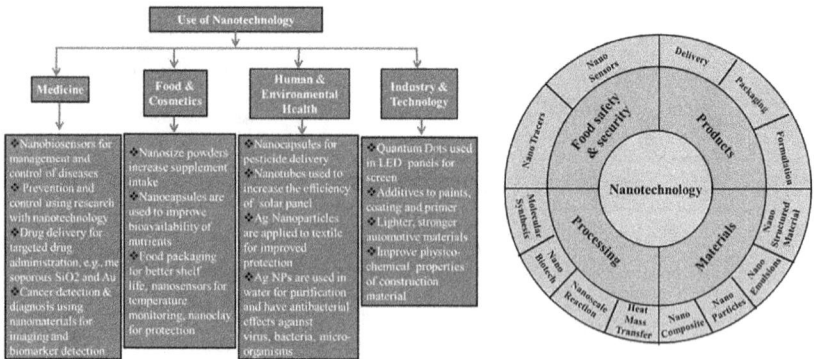

Figure 3.2 Applications of nanotechnology.

will be discussed based on their synthesis, industrial applications, and characterization. Efforts of researchers in surface engineering continue to push toward the atomic scale. Researchers are trying to give new properties to an array of commonly used materials like plastics, metals, ceramics, and glass. Nanotechnology and surface engineering also known as nanosurface engineering is a combination of nanomaterials and nanotechnology with surface engineering technologies. The nanotechnology involved in surface modification applies advanced techniques and various processes to characterize and analyze the different types of surfaces to acquire novel surface properties and functionalities.

3.2 PROCESSING TECHNIQUES FOR SURFACE ENGINEERING

3.2.1 Plasma surface treatments

Plasma-assisted processes have a wide range of application potential in the industry in areas like microelectronics, tool production, decoration purpose coatings, functional coatings on glass and plastic-made foils, decorative coatings, etc. Well-established surface coating techniques like electroplating and painting have a very high environmental and societal demand which has resulted in very high usage of plasma-assisted processes for surface engineering. Plasma surface engineering is being used to develop a wide range of products and refers to the techniques employed for developing tailor-made surfaces using plasma. It has majorly been used to modify and engineer various steel component surfaces that are used in automotive and manufacturing fields like gears, cutting tools, dies, crankshafts, etc. The highly reactive plasma has been effectively used for developing hard, thermal barrier, dielectric, semiconducting, optical, or tribological coatings for a wide range of applications [14–17]. Figure 3.3 is a schematic diagram illustrating

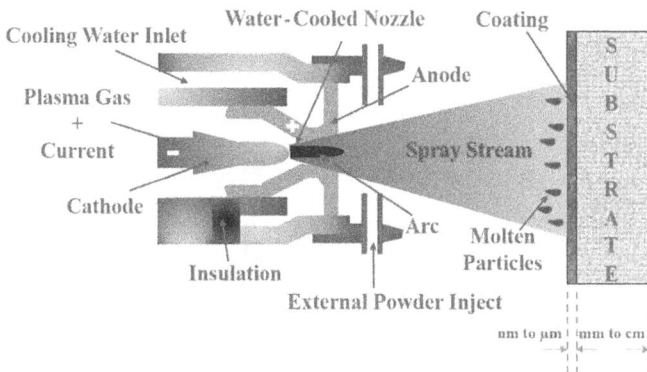

Figure 3.3 Schematic of plasma coating technique.

the plasma coating technique. The plasma spray coating technique includes a powder injection mechanism that injects the coating material directly into the plasma flame. The substrate is sprayed with molten as well as a semi-molten coating material to prepare a layer of coating that can protect the substrate from external wear and environmental corrosion. The coating material (in form of powder) is instantly heated and forced toward the material (substrate) surface that needs to be coated. After reaching the substrate, the molten coating cools down and disperses evenly through the force stream and forms a thin coating onto the surface. As the plasma spray coating technique uses high temperature, it is therefore suitable for ceramics and other materials that have a high melting point.

3.2.2 Chemical vapor deposition

The CVD technique is extensively used to make 2D thin films on solid substrates. This process includes the injection into the vacuum chamber of the reacted volatile precursors in gaseous or vapor form. After injecting the precursors, to enable to reaction, the chamber is heated to a certain temperature at which reaction initiates and gases disperse onto the preselected surface of the substrate. After a certain time, the material spreads and builds over the substrate surface and forms a coating. Figure 3.4 shows the schematic of the CVD process mentioning the different parts and sections of the instrument and process. CVD is widely employed in industries and has many products made by it used for various applications. CVD products are known for being durable surfaced products that are environmentally friendly and have high-temperature sustainability. The surfaces that need to be coated are required to be cleaned from contaminants before applying the coating. Due to the high temperature and reactivity of the precursor gases used during the process, CDV could limit the substrate material that is to be coated. The CVD technique has limited efficiency to coat particular areas of a targeted surface and the thickness of the coating is also limited due to the existence of the coating stresses [18, 19].

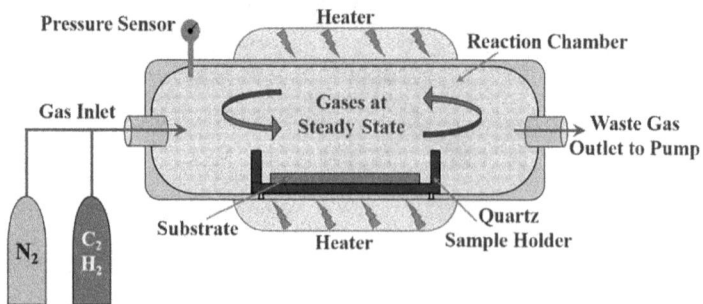

Figure 3.4 Schematic of CVD technique.

3.2.3 Physical vapor deposition

PVD process can provide a variety of coatings in which a solid depositing material is vacuum vaporized and atom by atom in form of a thin film deposited onto an electrically conductive substrate or the surface of a part by condensation from the vapor phase to the solid phase in the pure form or as an alloy composition coating. The PVD technique is employed to deposit thin-film coating for products having mechanical, electronic, or optical applications. The PVD process has been widely at the industrial scale and has also been combined with other processes to deposit coating having superior properties. The PVD process is performed in a high vacuum chamber (10^{-6} torr) incorporated with a source of cathode arc. PVD is one of the vacuum coating processes which has been widely used at the industrial scale to provide surfaces with thin lubricating films and coatings exhibiting superior properties. Advances in the PVD technique have made it possible to fabricate micro-and nanostructured coatings. For vapor generation, a few of the physical techniques like sputtering and evaporation are employed with the PVD process and hence coating is applied to the substrate surface. A comparative lower temperature of the substrate to that of the melting point of the coating material is used to effectively coat temperature-sensitive materials. Figure 3.5 shows the schematic of the PVD technique pointing to the different parts and mechanisms involved in the process. A few of the common types of PVD processes are sputtering, ion plating, evaporation coating, pulsed laser deposition, etc. [20, 21].

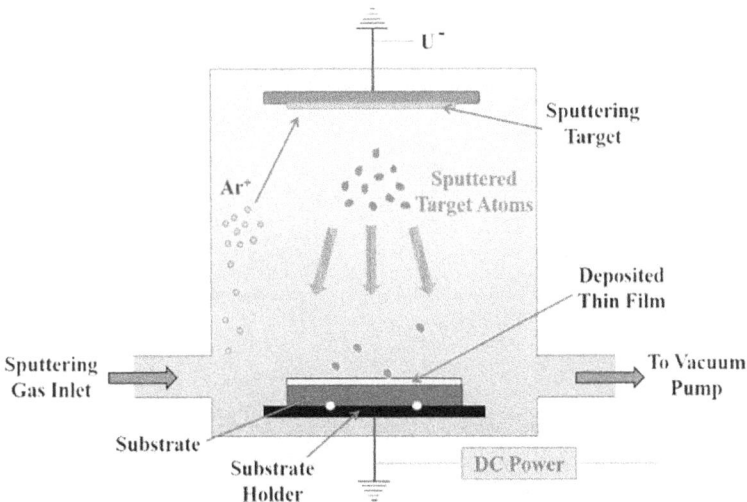

Figure 3.5 Schematic of PVD technique.

3.2.4 Laser surface modification

Laser surface modification is the modification of the surface properties of materials using processes that use lasers. Laser-based techniques are prominently employed for locally modifying the surfaces as they exhibit higher precision, lower heat generation, controlled heating, fast operation, and lower distortion. This area in the field of surface engineering is also referred to as laser-based surface engineering or modification. However, depending on the specific purposes of the process and sometimes on the physical mechanisms involved, surface modification techniques are named using specific terms like laser hardening, laser remelting, laser annealing, laser alloying, etc. In a laser-based surface modification, the surface properties of materials are modified using processes with laser beams. Some of the laser-based surface modification techniques like cladding, hardening, alloying, etc., along with their advanced properties of fast processing and minimal thermal gradient onto the substrate, are highly employed for altering the surface characteristics and to service the damaged sections and parts. A wide variety of processes in the context of laser material processing aims at the modification of various properties of surfaces of metallic, ceramic, glass, or polymer parts [22]. Typically, the process benefits from the direct interaction of the laser beam with substrates and workpieces and sometimes with additional material applied to the surfaces [23].

Figure 3.6 is a schematic representation of a typical laser-assisted surface modification of a substrate highlighting the mechanism by labeling the parts and sections in the setup.

Laser surface modification techniques can be precisely applied to specific parts of surfaces like narrow strips or small rectangles and areas which are hard to access like regions within drilled holes having a small diameter. Apart from this, the processing time for laser surface modification is also very short. Also, during this process, there is no direct contact contamination and wear

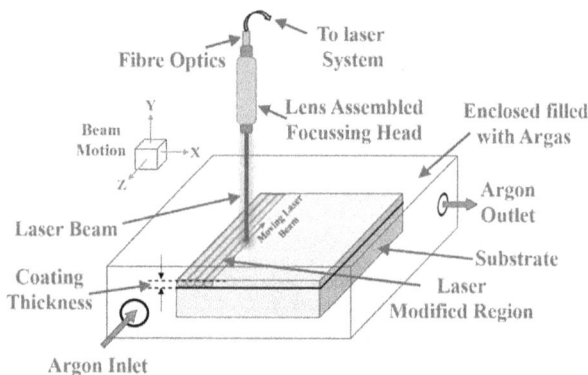

Figure 3.6 Laser surface modification.

of tools is also avoided. The laser surface modification processes can also be automated and can also be integrated with other processing steps.

3.2.5 Shot peening

Shot peening or shot blasting is a cold working process commonly applied to modify the material's mechanical properties by inducing a layer of compressive residual stress on the peripheral surface of the constituent. It is basically a stress-relieving technique that additionally provides strength to the components. The shot peening technique is also applied to impart fatigue strength to the components that have functioning in the field of high cyclic stresses. The tensile residual stresses that weaken a material internally are converted to compressive residual stress by the application of shots provided in the shot peening process, resulting in the enhancement in the sustainable life and load-bearing capacity along with this phenomenon, also preventing failure due to stress corrosion of the components. In a shot peening process, small spherical balls called shots are made to strike the surface with a force that is sufficient to produce a plastic deformation onto the surface. Shots are metallic, ceramic, or sometimes made of glass (depending on the application) spherical particles that act like a hammer or dimpling medium to the surfaces that create compressive stresses on the surfaces. When a bunch of small shots is impacted on a surface with a certain force and for a particular time, due to the multiple microindentations, the component undergoes a thin compressed layer on its surface. Due to the overlapping of the dimples on the surface, the compressive stress generated results in strengthening the sample surface, making the sample more resistive against fatigue loading, cracking, corrosion stress, galling, and cavitation erosion. An air blast system or a centrifugal blast wheel is generally used to perform the shot peening process [24, 25]. Figure 3.7 shows the schematic illustration of a typical

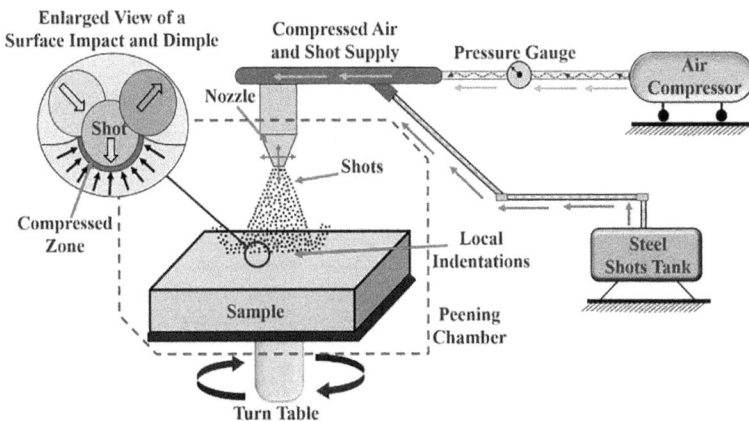

Figure 3.7 Shot peening technique.

shot peening process indicating the components involved in the process and an enlarged view shows the generation of the compressed zone that provides compressive stress to the surfaces of the sample.

3.2.6 Electroplating

Electroplating, also known as electrodeposition, is the process of plating material onto another by hydrolysis mainly to prevent the base metal from environmental corrosion and rusting and hence to increment the service life of the material or for decorative purposes. Electroplating is performed using an electrochemical cell. Electroplating is a surface-protecting process in which the surface of bulk material is protected by providing a layer of another material over it. It is a reverse electrochemical process in which the material that needs to be coated becomes cathode (also called the substrate) whereas the positive terminal of the circuit is attached to the material whose plating is to be done over the negative terminal attached material. Positive terminal attached material, i.e., anode gets dissolved in the electrolyte solution. There are some criteria for the selection of the anode material as a substrate depending on the type of properties like superior thermal and electrical conductivities as well as cost-effectiveness. Gold (Au) and Palladium (Pd) are the commonly used anode materials for electrodeposition. For the selection of the electrolyte, corrosion, strength, and wear resistance are some of the points that need to keep in mind. Generally, for metal-to-metal operation, copper sulfate ($CuSO_4$) is used as an electrolyte medium. The electrolyte carries the cations of the depositing substrate. These cations are reduced from the cathode in the zero valence state when the terminals are attached to the battery and the key is connected. Figure 3.8 depicts the schematic representation of the mechanism of the electroplating process showing the parts and bottom deposited anodic mud. Electroplating can be done in single pieces, small batches, or even pieces having large dimensions [26, 27].

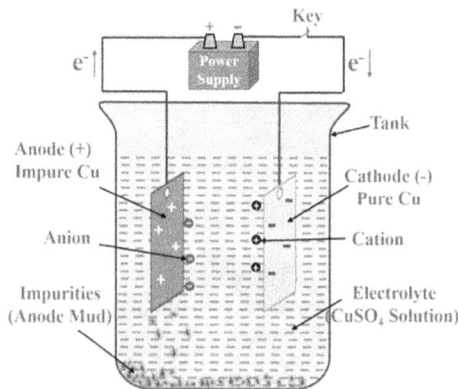

Figure 3.8 Electrodeposition process.

3.3 NANOTECHNOLOGY AND SURFACE ENGINEERING

Nanotechnology combines the fundamental as well as advanced understandings of the branches of science and technology of nanoscale entities. The term "nanotechnology" is used to represent the wide areas and subareas of science, technology, and engineering where the phenomenon that takes place at dimensions in the nanometer range are used in the characterization, synthesis, renovation, application, and design of systems, materials, and structural devices. Nanomaterials are typically less than 100 nm in size. Nanomaterials profoundly exhibit different properties due to multiple time increases in their surface area than to volume due to due to existence of a high size gradient. Developments in the field of engineering at the nanolevel have not been simple as scientists and engineers have to invent new techniques, approaches, and different solutions to build the structures using a 'bottom-up' approach instead of conventional 'top-down' manufacturing techniques. Nanotechnology governed surface engineering approach directs to the designing of interfacial, textural, and physical properties of 0D to 2D nanostructured surfaces for different applications. The surface interaction of nanomaterials influences different physical properties like magnetism, catalysis, optics, electronics, etc. Nanotechnology can be used for the functionalization and engineering of surfaces. Nanocoatings can modify the surface properties of the products and provide them with advanced features like anti-fog properties, antimicrobial and antiviral properties, and high wear resistance. These nanocoatings can be applied on the surfaces of a wide range of materials. Nanotechnology when used in surface coatings could lead to a host of surface engineering innovations. Nanocoating can be done on plastics, metals, ceramics, and glasses. Nanocoatings are often developed on surfaces to make them water-repellent. These nanocoatings provide hindrance from foreign particles and protect against dirt. The coating also became water repellent as a result of which water deposited over the coating takes away the dirt deposited over the coating due to which service life of the product increases to a great extent. Figure 3.9 explains the

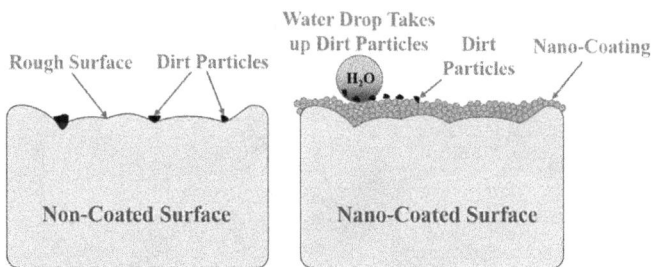

Figure 3.9 Nanocoated water repellent surface.

phenomenon clearly how a nanocoating surface helps in preventing the surface from foreign dirt and corrosion by repelling the water. Nanotechnology for surface engineering employs advanced analysis techniques to characterize the bulk material surfaces to obtain innovative surface features. These innovative art characterization techniques provide in-depth analysis results for the surface from micro- to nanoscale [28–31].

For instance, biocontamination is known to be prevalent widely in medical instruments, implants, and devices and is a leading issue. To stop biocontamination, it is essential to develop synthetic surfaces which repel the micro-organisms and viruses and provides a biocidal activity. The advanced research in the field of biological science and nanotechnology made it possible to program and design new smart surfaces that can decrease the proneness of infections. For example, a golf club head can be coated with a dense layer of nanostructured metal. This would result in a lighter and stronger clubface giving a better impact on the ball. Today, manufacturers can deposit a coating on metals, ceramics, or glasses having a thickness of a few microns to a few nanometers to give specific surface properties due to the improvement in process sophistication and quality. Hybrid materials which have a combination of properties derived from the synergistic behavior of the two or more combined materials could also be coated on materials. Some of the techniques used for developing a coating on various substrates are shown through schematic illustration in Figure 3.10(a–d). Below

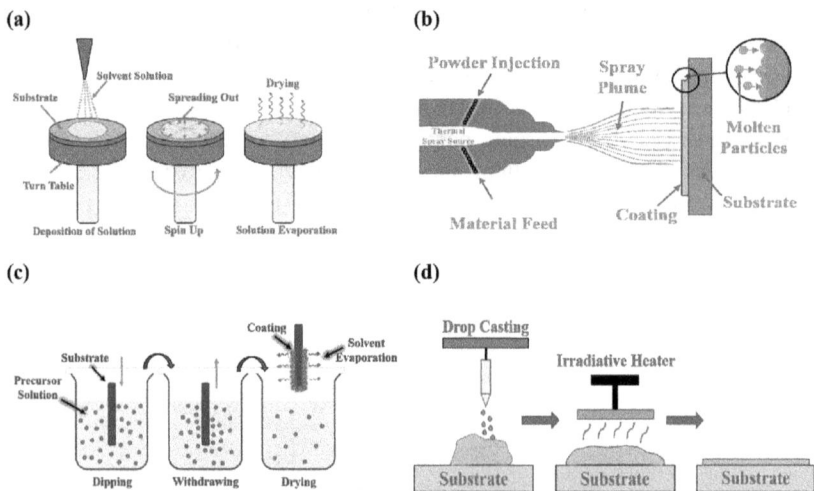

Figure 3.10 (a) Spin coating, (b) thermal spray coating, (c) dip coating, and (d) drop cast coating.

mentioned are advanced techniques that are generally employed to construct smart surfaces. In spin coating, initially, the coating material in form of solvent solution is poured or sprayed over the surface of the substrate and later, the substrate is given a spin due to which the coating spreads evenly throughout the surface of the substrate, and the thin film is coated. In the thermal spraying process, the coating material in form of powder is injected into the thermal spray source, followed by heating of powder due to which it melts, and through spray force, a coating is done on a vertically placed sample. In the dip coating technique, the substrate bulk material is dipped in a solution mixed with a precursor coating material. The substrate is kept in the solution for some time such that the coating material fully covers the surface that is to be coated. Later, the substrate is taken out and dried in air. In the drop cast coating technique, the coating is dropped or injected onto a substrate surface followed by heating, and sometimes heating and pressing, are also done for thick coatings. As a result, the material spreads throughout the substrate and develops a homogeneous coating on the substrate.

3.4 CHARACTERIZATION TECHNIQUES FOR SURFACE ENGINEERING

3.4.1 Nanoindentation

Nanoindentation is a technique developed during the middle of the 1970s to measure the bulk material properties like hardness, Young's modulus, yield stress, and toughness through surface indentations at the nanolevel. Nanoindentation is also called instrumented indentation testing [32, 33]. A nanoindenter can measure surface properties from the nanometer to the micrometer scale. It was initially used to determine the hardness of a small volume of materials. It is now a widely recognized technique for testing thin films and determining surface mechanical properties. Since the nanoindentation technique performs at the nanoscale, thus the technique is also widely employed to measure the properties of the thin films. The indentation depth is limited to a few nanometers and individual film properties are obtained reducing the effects from the substrate. The technique is very promising against thin-coated materials and heterogeneous materials. A nanoindent of $1/10^{th}$ of the film thickness or less is made while performing thin-film nanoindentation. Figure 3.11 shows the schematic diagram of a nanoindentation tester. The applied indent load (P) is increased gradually as the tip touches the film and starts to penetrate the samples until desired by the user. When the desired value is reached, further loading and penetration are stopped, and the loading is released. Now, the indented area (A_i)

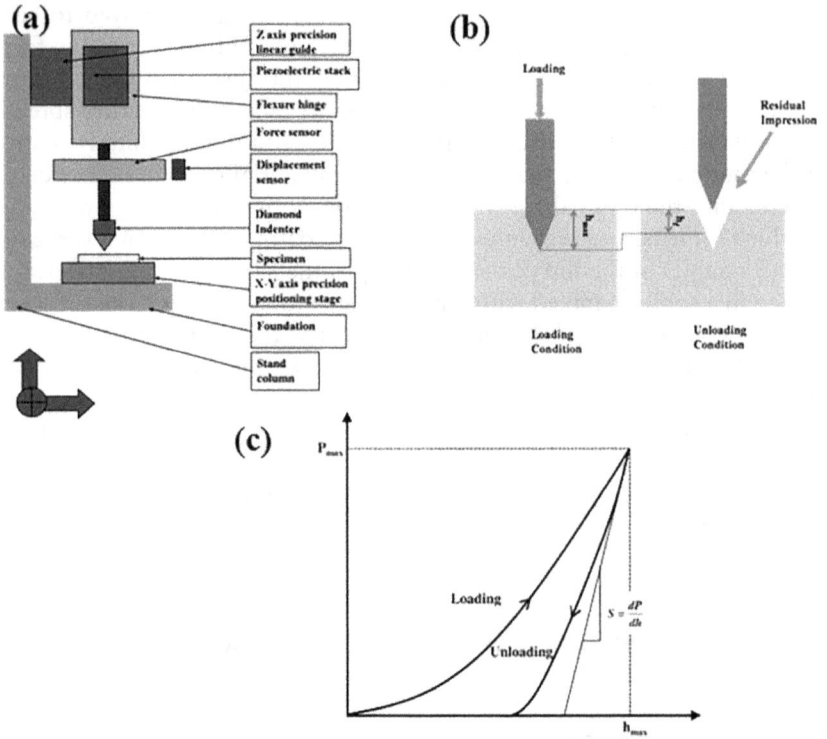

Figure 3.11 (a) Nanoindentation tester, (b) nanoindentation impression, and (c) a typical load–displacement curve for a nanoindentation process.

on the surface is measured and the hardness (H) value is calculated from the below expression:

$$H = \frac{P_{max}}{A_i} \tag{3.1}$$

where P_{max} = maximum indent load and A_i = Indent area

Figure 3.11(a) shows a typical nanoindentation tool labeled with the parts attached to it. Figure 3.11(b) schematically shows the loading and unloading of the indenter on a sample and the formation of an impression on the surface which is measured for hardness calculation and Figure 3.11(c) is a typical load–displacement curve for the nanoindentation process. During nanoindentation, the surface of a sample is displaced as pressure is applied by the tip of the indenter. Here, in Figure 3.12, h_{max} is the maximum penetration depth at maximum load (P_{max}) is the depth at peak load, h_f is the depth of the indent post unloading and h_e is the elastic displacement accompanied with recovery. Young's modulus (E) and hardness (H) are derived by calculating values from the unloading curve of the load–displacement plot [34].

Representation
of indentation

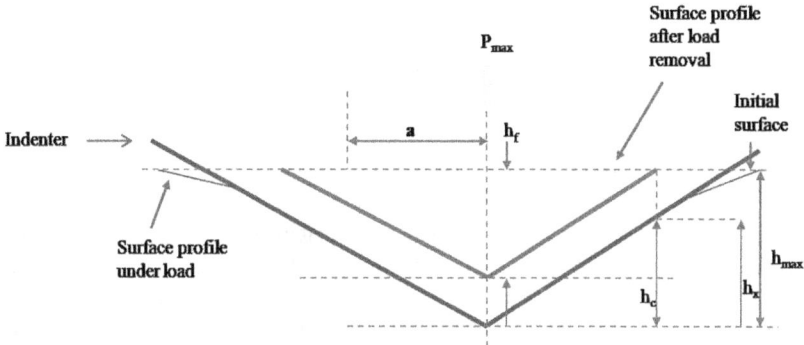

Figure 3.12 Representation of an indentation.

3.4.2 Wear test

The general purpose of surface engineering-based treatments is to enhance the component's service life against wear by either improving the wear resistance of the material or by minimizing the wear over the surface. With the evolution in the field of surface engineering innovative designs, it is desired to investigate the properties of newly developed materials used for the coating to establish a healthy combination of substrate-coating. Several standard techniques are used to test the samples against types of wear and hence to determine the wear performance of the material under defined conditions for a specific period. Nanocoatings are the thin layering on the samples that provide corrosion and wear resistance, high-temperature sustainability, specific strength and hardness, very low permeability, etc. There are several wear testing techniques like pin-on-disk or pin-on-drum wear test, ball-on-plate wear test, block-on-ring test, dry and rubber wheel test, adhesive test, repeated impact wear test, etc. The frequently employed test, i.e., the pin-on-disk wear test is discussed in the present chapter. Figure 3.13 shows the illustrative schematic figure of a pin-on-disk wear tester used to characterize the wear property between two materials. It consists of a pin that is fixed to an elastic arm. The pin is pressed against the sample called a disk under the required amount of weights. The stationary pin is pressed against a rotating disk, i.e., against the sample under the applied load. The sample can be rotated at a desired constant speed. The elastic arm stays fixed and maintains a sample and maintains the same contact point on the wear track between the pin and the disk. According to the type of the disk, the pin

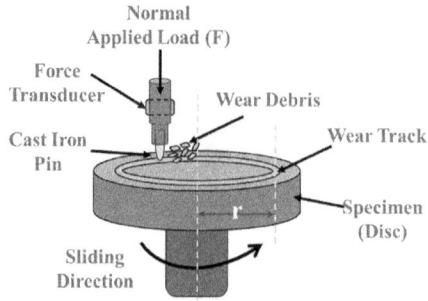

Figure 3.13 A pin-on-disc wear tester.

shape can be changed from spherical, cylindrical, to diamond-shaped. The shape of the pin helps in aligning it against the sample. Throughout the wear test, the software installed with the system continuously calculates the instant friction forces, wear depth, temperature, etc., and other desired data [35–37]. Load is applied either mechanically or pneumatically to the pin and the stationary pin starts to press the rotating disk with equivalent pressure. The transducer keeps track of the instant applied load. For the calculation of the wear on coatings and thin films, a high-end data acquisition system having a high frequency and high sampling rate is attached to the wear setup to detect the instant on-the-spot submicro- or nanofailure along with pin and film contact events in the various wear test events. The pins used in the pin-on-disc wear tester can be hardened bearing steel or nanocrystalline diamond (NCD) coated WC-Co flat-ended pins or sometimes cast iron pins are used and the disc is made of steel [38]. Micron-sized pins made of steel/cast iron are used with the spherical tip against the thin coatings to perform a wear test.

3.4.3 Electron microscopy

EM is the most frequently employed and prominent analytical tool to analyze samples for chemical and structural characterization at the nanoscale. They are very powerful instruments for analyzing nanomaterials. Basically, there are two types of EM are used namely scanning electron microscopy (SEM) and transmission electron microscopy (TEM) for the characterization of the nanomaterials and surfaces of the samples. EM is a highly recognized tool for nanomaterial characterization. In today's material science and characterization field, EM is an important and versatile analytical tool with a range of applications and methodologies to analyze micro- to nanosized samples like thin films to understand their features and properties. With the help of a powerful TEM tool, nanophase identification of a material can be obtained. It can perform a variety of functions effectively not only chemical and crystallographic analysis but it is known to be a prime tool

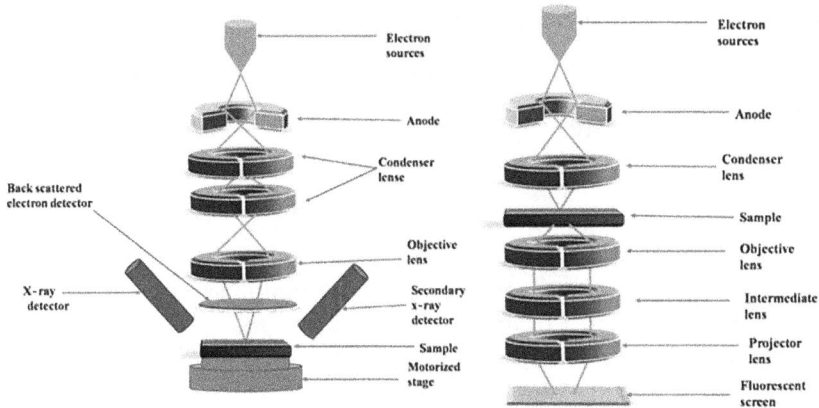

Figure 3.14 Schematic diagram of SEM and TEM.

for nanolevel measurements. On the other hand, SEMs are effectively been used for decades to characterize nanosized materials and have provision to integrate newer advanced techniques to perform other operations such as electron backscattered diffraction (EBSD). In-situ-based analysis can also be done using SEMs for nanomaterials. Figure 3.14 shows the schematic diagram comparing the parts in SEM and TEM. In a typical EM, accelerated electrons having a small wavelength are emitted from a filament source into the vacuum chamber. These high potential electrons can penetrate a solid sample up to several microns in depth. These negatively charged potential electrons can be deflected through magnetic or electric fields. When the sample is probed by the accelerating focused beam of an electron, the image is formed of the scanned area called raster scanning. TEM uses an accelerated electron beam as the source of illuminating radiation which has a much smaller wavelength and much higher resolving power than light-based optical microscopes. TEM can provide information on nanoparticle size, grain size, size distribution, and morphology. A typical TEM has a resolving power of about 0.2 nm and a maximum magnification of about 1,000,000×. The resolution limit of SEM is ~0.5 nm and with aberration correction lens-assisted TEMs, a resolution of <50 pm can be achieved. The main difference between TEM and SEM is that imaging in SEM is done when deflected or knocked-off electrons are detected by the detectors, whereas, in a TEM, an electron passes through the sample to create an image.

As seen in Figure 3.14, the main components in an EM are a source of the electron, lenses in the series for deflection and focusing, and also to control the shape and path of the beam, and the electron aperture. All the components are fixed at particular locations in a long cylindrical vacuum chamber. An SEM uses an accelerating voltage of around 30 kV whereas a TEM uses an accelerating voltage in the range of 60–300 kV [39, 40].

3.4.4 X-ray diffraction

XRD is a highly versatile non-destructive testing (NDT) technique frequently employed to investigate the crystal structure of materials. Since the dimension of the X-ray wavelengths is of the same magnitude as the size of the nanostructures, XRD and associated analytical techniques are primary tools for research in nanotechnology. Powder XRD is a common characterization technique for nanostructured materials due to the simplicity of the technique. Analysis of the XRD plot obtained from a sample provides information that is complementary to the information obtained from microscopic and stereoscopic techniques. XRD analysis can be used for phase identification, determination of sample purity, crystallite size, residual stress, and morphology. In the field of surface engineering, XRD is a basic tool to identify the surface phase detection, elemental distribution, compound formation, and size of coating (if any). It can also provide chemical information for elemental analysis as well as for texture analysis. The Debye–Scherrer equation given below in Equation (3.2) provides a relation between the crystal sizes (D) (usually diameter) to the width, diffracting angle, and height. XRD can determine the interlayer distance between the atoms and can detect the orientations of mono- or polycrystal or a single grain. It is effective in finding out the crystal features of any unknown material along with its physical characteristics including internal stresses and the size of a small crystallite region [41–43].

$$D = \frac{k\lambda}{\beta \cos\theta} + 4\varepsilon \tan\theta \qquad\qquad (3.2)$$

where D = crystallite size, E = lattice strain, k = Scherrer constant or shape factor (0.89), λ = wavelength (constant value depends on the type of X-ray used for operation), β = width of the peak at half of its height or full width at half maximum (FWHM), and θ = Bragg angle (radians).

Figure 3.15(a) shows the X-ray diffractometer consisting of an X-ray source and an X-ray detector. A diffractometer instrument analyzes a

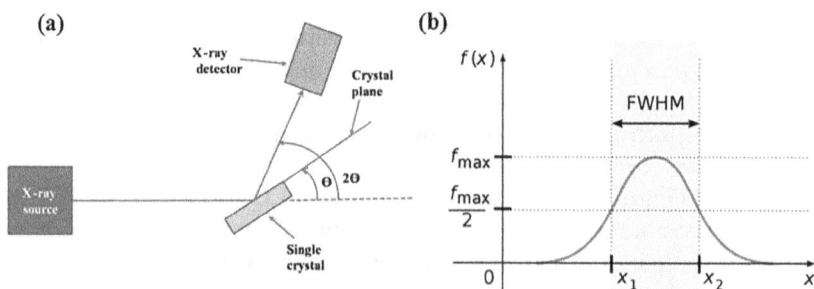

Figure 3.15 (a) Schematic operation diagram of an X-ray diffractometer and (b) FWHM in an X-ray diffraction peak.

material by bombarding X-rays and a scattering pattern is generated when rays interact with the atoms and molecules of the material. As the crystallite size decrease from bulk to nanoscale dimension, the XRD peak broadens. Figure 3.15(b) shows the FWHM for an XRD peak. The advanced tools, small-angle X-ray scattering (SAXS) can determine the size, shape, distribution, orientation, and correlation of nanoparticles properties in solids or solutions. High-resolution XRD can measure layer thickness, roughness, chemical composition, and lattice spacing.

3.5 APPLICATIONS OF SURFACE ENGINEERING

Nowadays, surface coating plays an important role to protect the substrate from wear, corrosion, and damage. To protect the substrate, there are different coating techniques and these coating techniques have different areas of application like aeronautical, sports, chemical, electronics, food, biomedical, and transport industries. In recent times, thermal spray coating has been

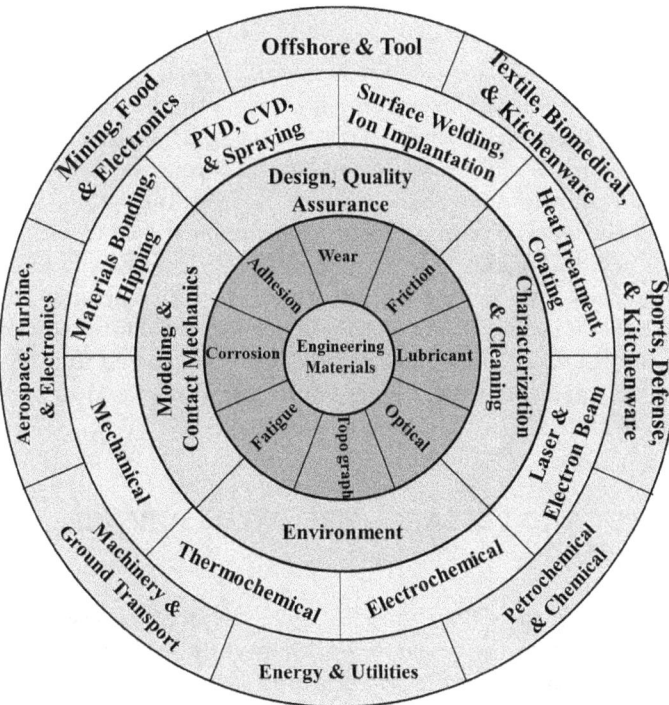

Figure 3.16 Surface engineering application wheel showing different sectors and associative technologies.

utilized in biomedical orthopedic implants [44], cancer therapy, dentistry, shaft, anti-skid decking, propellers, marine anti-fouling coating [45], insulators, electronics RF shielding [46], production equipment of silicon chip [45] transportation and automobile engine and drive train components [47], textile machinery stretch-tow rollers, thread guides [48], cutting tool steel cutting tool, non-steel cutting tool equipment [49], nuclear reactor equipment [50], etc. Nowadays, surface engineering has an important place in sports industries, e.g., to control friction in boats and canoes, in roller sockets of golf club heads, and for resistance from wear. Nanocomposite coatings are used in ice skates and roller skates; in racing cars, for coating engine components, thermal oxidation coating techniques are used [51]. In the automotive industry, surface engineering plays an important role to protect all parts from wear and corrosion. In piston rings, hard chrome material is used for coating. For coating piston rings generally, arc evaporation techniques are used; apart from that, the PVD technique is used [52]. For coating cylinder bores with carbon steel reinforced by ceramic, thermal spray coating is used [53–55]. By using plasma nitriding and oxidation techniques, ball pivots can be coated [56]. Crankshaft bearing can also be coated by thermal spray techniques. Diamond-like carbon coatings are used to coat injector needled which are used in diesel engines [57]. Due to the non-degradable property and low interstitial excellent corrosion resistance, materials such as SS316L, pure stainless steel, chromium cobalt alloy, pure titanium, and titanium alloy Ti_6Al_4V are used in biomedical applications like hip joint replacement and spinal cord replacement. To coat this material, laser surface treatment has been used. Atmosphere plasma spray is used to coat many aerospace engine parts like blade shroud notch, torque tubes, compressor or turbine casing, flow mixing device, turbine blade airfoil, turbine blade snap, and wing main spar. The cutting tool for machining is coated with cobalt or cemented carbide. For this type of coating, the PVD process has been used. The PVD process can be used in cutting tools for dry machining processes where no coolant is used and without reducing the tool life and productivity. In European countries, most of their food items are coated to protect from air, water, etc., and thin-walled glass bottles are also coated to prevent the formation of cracks.

3.6 MARKET AND RESEARCH TRENDS IN SURFACE ENGINEERING

In the modern era, the global surface engineering market has a substantial growth rate. Most of the research work currently being done in the area of surface engineering has the potential for sustainable growth and most of these surface treatment techniques are also eco-friendly. Due to the rapidly growing demand for highly durable and wear-resistant products, the international market for surface engineering is continuously growing. It is also predicted that the demand for manufacturing new materials and the need for better industrial activity will also boost the surface engineering market by 2025. Surface engineering is a crucial technology implemented to increase the corrosion resistance and wear resistance of any particular material. This process is majorly used for metals and alloys and the treatment of plastic and metal-made PCBs. Moreover, the coating industry with a wide range of applications also plays a crucial role in the full-fledged implementation of surface engineering. As per a review done by a group of researchers, it is found that the global market of surface engineering can be segmented based on the chemical type, the base materials, the end-use industry, and the region. Researchers have further segmented it into cleaners, plating chemicals, and conversion coatings based on the chemical type. The global market estimate suggests that it will be dominated by the plating chemical segment during the forecast period. This owes mostly to the extensive use of plating chemicals in end-use industries like transportation, construction, and industrial machinery. Now, coming to the base materials, global surface treatment is categorized into metals, plastic, and others. Out of them, the plastic category is expected to undergo huge growth in the coming decade as it provides better chemical and corrosion resistance over metals. Due to the lightweight of plastic materials, the use of these materials has increased extensively in recent years. Surface engineering is a critical technology, and it has a lot to do with the markets. Moreover, surface engineering plays a vital role in material science and the coating industry. Recent trends have shown that coating technology is expected to grow by up to five times in the upcoming 2–3 years. One of the prime reasons that led to the inefficiency of industrial plants is the progressive deterioration of metallic surfaces that are in use in these plants. If these problems are not checked and controlled, it might even lead to the shutdown of the industry. In the United States alone, there is an expenditure of hundreds of billions of dollars annually that is directly or indirectly related to corrosion and wear failure in materials. For example, in terms of the current prices, corrosion of metals itself costs the U.S. economy almost $300 billion every year, which is about 4.2% of the gross national product [58]. In 1995, it was estimated that the U.K. market for surface engineering accounted for about £10 billion. Out of this amount, about £4.5 billion was spent on 'engineering' coatings to improve corrosion or wear resistance. Implementing these treatments severely affected the value of the manufactured products, which accounted for £95.5 billion (about 7% of U.K. GDP). The aerospace, agriculture, automotive, electrical consumer goods, and electronic sectors were particularly the ones that were predicted to have a steep growth. The report concluded: 'surface engineering provides one of the most important means of engineering product differentiation in terms of performance, quality and life-cycle cost' [59].

3.7 FUTURE OF SURFACE ENGINEERING

In the present growing scenario, surface engineering has become an integral part of all industries due to its technique, process, and quality. PVD, CVD, thermal spray, electrodeposition, sol-gel, and laser cladding techniques are used to coat metal surfaces to resist wear and corrosion. Surface engineering processes have put a large impact on several industries and are an integral part of these industries. Some of the areas where surface modification/engineering is being widely used and extensive research is being done are as follows:

- Surface modification/engineering for non-ferrous alloys/metals
- Surface modification/engineering for composites and polymers
- Surface modification/engineering for ceramics materials
- Surface modification/engineering technologies for the second generation
- NDT of surface engineering components
- Mathematical modeling of surface modification/engineering
- Statistical process control for surface engineering techniques

In the future, several new surface engineering techniques will be introduced in the market and commercialized which will provide novel surface engineered products. Some of these techniques and their application in new areas have been listed below.

- Laser treatment of plasma spray coating and thermal spray coating [60, 61]
- PVD of pre-nitride steels [62]
- Thermo-chemical treatment of pre-laser hardened steels and pre-carburized steel [63, 70]
- Ion beam addition and ion-assisted coatings [64, 65]
- Hipping of overlay coating [66–69]
- CVD of pre-carburized steel [71]

REFERENCES

1. *Nanomaterials and Surface Engineering*, J. Takadoum (Ed.), John Wiley & Sons, Inc., 2013.
2. H. H. Berghan, et al., High temperature plasma with a cold gas blanket in a toroidal magnetic field, *Proc. 3rd IAEA Conf. on Plasma Physics and Controlled Nuclear Fusion Research*, Novosibirsk, vol. II, 1969, 113–124.
3. *Surface Engineering of Modern Materials*, K. Gupta (Ed.), Springer, 2020.
4. S. Shi, F. Chen, E. M. Ehlerding, W. Cai, Surface Engineering of Graphene-Based Nanomaterials for Biomedical Applications, *Bioconjugate Chemistry* 25(9), 2014, 1609–1619.
5. S. Mornet, G. Drisko, *Nanoscale Surface Engineering, A special issue of Nanomaterials*, 2020.

6. *Tribology, Friction and Wear of Engineering Materials, Second Edition*, Ian Hutchings and Philip Shipway, Butterworth-Heinemann, 2017.

7. D. K. Dwivedi, *Surface Engineering: Enhancing Life of Tribological Components*, Springer, 2018.

8. L. Mazzola, Commercializing Nanotechnology, *Nature Biotechnology* 21, 2003, 1137–1143.

9. B. Xu, W. Zhang, Progress and Application of Nano-Surface Engineering in China, Novel Materials Processing by Advanced Electromagnetic Energy Sources, *Proceeding of the International Symposium on Novel Materials Processing by Advanced Electromagnetic Energy Sources*, 19–22 March, 2004, Osaka, Japan, 2005, 339–343.

10. D. Mariotti, J. Patel, V. Švrček, P. Maguire, Plasma-liquid Interactions at Atmospheric Pressure for Nanomaterials Synthesis and Surface Engineering, *Plasma Processes and Polymers*, 9(11–12), 2012, 1074–1085.

11. D. S. Rao, *Surface Engineering*, Daya Publishing House, 2010.

12. J. K. Patra, S. Gouda, Application of Nanotechnology in Textile Engineering: An Overview, *Journal of Engineering and Technology Research*, 5(5), 2013, 104–111.

13. B. Xu, W. Zhang, Novel Materials Processing by Advanced Electromagnetic Energy Sources, *Proceedings of the International Symposium on Novel Materials Processing by Advanced Electromagnetic Energy Sources, Progress and Application of Nano-Surface Engineering in China*, March 19–22, 2004, Osaka, Japan, 2005, 339–343.

14. B. Fotovvati, N. Namdari, A. Dehghanghadikolaei, On Coating Techniques for Surface Protection: A Review, *Journal of Manufacturing and Materials Processing* 3, 2019, 28.

15. Y. Sun, T. Bell, Plasma Surface Engineering of Low Alloy Steel, *Materials Science and Engineering A*, 140, 1991, 419–434.

16. R. Suchentrunk, H. J. Fuesser, G. Staudigl, D. Jonke, M. Meyer, Plasma Surface Engineering-Innovative Processes and Coating Systems for High-Quality Products, *Surface and Coatings Technology*, 112(I1–3), 1999, 351–357.

17. S. C. Behera Mishra, Dependence of Adhesion Strength of Plasma Spray on Coating Surface Properties, *Journal of Materials and Metallurgical Engineering*, 2(1), 2012, 23–30.

18. S. Sivaram, *Chemical Vapour Deposition: Thermal and Plasma Deposition of Electronic Materials*, Springer, USA, 1995.

19. A. Kempster, The Principles and Applications of Chemical Vapour Deposition, *The International Journal of Surface Engineering and Coatings*, 70(2), 1992, 68–75.

20. Z. Li, K. A. Khor, Preparation and Properties of Coatings and Thin Films on Metal Implants, in *Encyclopedia of Biomedical Engineering, Biomaterials: Science and Engineering*, R. Narayan (Editor-in-Chief), Elsevier Inc., vol. 1, 2019, 203–212.

21. K. Reichelt, X. Jiang, The Preparation of Thin Films by Physical Vapour Deposition Methods, *Thin Solid Films*, 191(1), 1990, 91–126.

22. G. Casalino, Computational Intelligence for Smart Laser Materials Processing, *Optics & Laser Technology*, 100, 2018, 165–175.

23. J. D. Majumdar, I. Manna, Laser Surface Engineering, in *Handbook of Manufacturing Engineering and Technology*, A. Nee (Ed.), Springer, London, 2014.

24. P. P. Shukla, P. T. Swanson, C. J. Page, Laser Shock Peening and Mechanical Shot Peening Processes Applicable for the Surface Treatment of Technical Grade Ceramics: A Review, *Proceedings of the Institution of Mechanical Engineers, Part B: Journal of Engineering Manufacture*, 228(5), 2014, 639–652.
25. *Shot Peening Comprehensive Materials Finishing, Reference Work*, M. S. J. Hashmi (Editor-in-Chief), 2017, Elsevier Inc.
26. K. Sikder, D. S. Misra, D. Singhbal, S. Chakravorty, Surface Engineering of Metal-Diamond Composite Coatings on Steel Substrates using Chemical Vapour Deposition and Electroplating Routes, *Surface and Coating Technology*, 1114(2–3), 1999, 230–234.
27. N. Kanani, *Electroplating-Basic Principles, Processes and Practice*, Elsevier, 2004.
28. M. Knez, K. Nielsch, L. Niinistö, Synthesis and Surface Engineering of Complex Nanostructures by Atomic Layer Deposition, *Advanced Materials*, 19(21), 2007, 3425–3438.
29. S. Murty, P. Shankar, B. B. Baldev Raj, J. M. Rath, *Textbook of Nanoscience and Nanotechnology*, University Press-IIM Series in Metallurgy and Materials Science, Springer-University Press, 2012.
30. Y. K. Kim, Nanotechnology-Based Advanced Coatings and Functional Finishes for Textiles, in *Smart Textile Coatings and Laminates*, 2nd Ed., The Textile Institute Book Series, 2019, 189–203.
31. P. Erkoc, F. Ulucan-Karnak, Nanotechnology-Based Antimicrobial and Antiviral Surface Coating Strategies, *Prosthesis*, 3(1), 2021, 25–52.
32. K. Kurosaki, Y. Saito, H. Muta, M. Uno, S. Yamanaka, Nanoindentation Studies of UO_2 and $(U,Ce)O_2$, *Journal of Alloys and Compounds*, 381(1–2), 2004, 240–244.
33. S. Sasmal, M. B. Anoop, Nanoindentation for Evaluation of Properties of Cement Hydration Products, in *Nanotechnology in Eco-efficient Construction, Materials, Processes and Applications, Woodhead Publishing Series in Civil and Structural Engineering*, 2nd Ed., 2019, 141–161.
34. W. C. Oliver, G. M. Pharr, Pharr Improved Technique for Determining Hardness and Elastic Modulus Using Load and Displacement Sensing Indentation Experiments, *Journal of Materials Research*, 7(6), 1992, 1564–1580.
35. L. W. McKeen, Introduction to the Tribology of Plastics and Elastomers, in *Fatigue and Tribological Properties of Plastics and Elastomers*, 3rd Ed., 2016, 27–44.
36. N. Gobind, K. Grover, J. Parshad, Techniques of Measuring Wear for Bulk Materials and Advanced Surface Coatings, *IOSR Journal of Mechanical and Civil Engineering (IOSR-JMCE)* 12(2), 2015, 101–106.
37. M. Kennedy, M. S. J. Hashmi, Methods of Wear Testing for Advanced Surface Coatings and Bulk Materials, *Journal of Materials Processing Technology*, 77(1–3), 1998, 246–253.
38. M. Chandran, *Synthesis, Characterization and Application of Diamond Films, Carbon-Based Nanofillers and their Rubber Nanocomposites, Carbon Nano-Objects*, 2019, 183–224.
39. *Transmission Electron Microscopy Characterization of Nanomaterials*, C. S. S. R. Kumar (Ed.), Springer-Verlag, Berlin, Heidelberg, 2014.
40. B. Williams, C. B. Carter, *TEM Transmission Electron Microscopy, A Textbook for Materials Science*, Springer Nature Switzerland AG.

41. D. Cullity, S. R. Stock, *Elements of X-Ray Diffraction*, 3rd Ed., Pearson, 2001.
42. F. Holder, R. E. Schaak, Tutorial on Powder X-ray Diffraction for Characterizing Nanoscale Materials, *ACS Nano*, 13(7), 2019, 7359–7365.
43. P. Whitfield, L. Mitchell, X-Ray Diffraction Analysis of Nanoparticles: Recent Developments, Potential Problems and Some Solutions, *International Journal of Nanoscience*, 3(6), 2004, 757–763.
44. P. F. Kurunczi, Cerionx, Inc., Assignee. Atmospheric pressure non-thermal plasma device to clean and sterilize the surfaces of probes, cannulas, pin tools, pipettes and spray heads. United States patent US 7, 094, 314, 2006.
45. E. Park, R. A. Condrate, Graded Coating of Hydroxyapatite and Titanium by Atmospheric Plasma Spraying. *Materials Letters*, 40(5), 1999, 228–34.
46. R. Hui, Z. Wang, O. Kesler, L. Rose, J. Jankovic, S. Yick, R. Maric, D. Ghosh, Thermal Plasma Spraying for SOFCs: Applications, Potential Advantages, and Challenges. *Journal of Power Sources*, 170(2), 2007, 308–23.
47. U. Kogelschatz, Atmospheric-Pressure Plasma Technology. *Plasma Physics and Controlled Fusion*, 46(12B), 2004, B63.
48. J. R. Roth, *Industrial Plasma Engineering: Volume 2-Applications to Non thermal Plasma Processing*, CRC Press, 2001.
49. C. Tendero, C. Tixier, P. Tristant, J. Desmaison, P. Leprince. Atmospheric Pressure Plasmas: A Review. *Spectrochimica Acta Part B: Atomic Spectroscopy*, 61(1), 2006, 2–30.
50. S. Sampath Thermal Spray Applications in Electronics and Sensors: Past, Present, and Future. *Journal of Thermal Spray Technology*, 19(5), 2010, 921–49.
51. H. Dong, T. Bell and A. Mynott, *Sports Engineering*, 1999, 2, 213–219.
52. H. E. Jones, Morgan: Carnegie Schol. *Mem. Iron Steel Inst.*, 21, 1932, 39.
53. G. Barbezat, Thermal Spray Coatings for Tribological Applications in the Automotive Industry. *Advanced Engineering Materials*, 8, 2006, 678–681.
54. T. Lampe, S. Eisenberg, E. Rodriguez Cabeo, Plasma Surface Engineering in the Automotive Industry-Trends and Future Prospectives. *Surface and Coatings Technology*, 174, 2003, 1–7.
55. K. Bobzin, F. Ernst, K. Richardt, T. Schlaefer, C. Verpoort, G. Flores, Thermal Spraying of Cylinder Bores with the Plasma Transferred Wire Arc Process. *Surface and Coatings Technology*, 202, 2008, 4438–4443.
56. J. Vetter, G. Barbezat, J. Crummenauer, J. Avissar, Surface Treatment Selections for Automotive Applications. *Surface and Coatings Technology*, 200, 2005, 1962–1968.
57. C. M. Cotell, J. A. Sprague, *Preface, Surface Engineering*, vol 5, ASM Handbook, ASM International, 1994.
58. R. A. Matthews, P. Holiday, *2005 Revisited: The UK Surface Engineering Industry to 2010*, NASURF, Farnborough, 1998.
59. C. P. O. Treutler, Industrial Use of Plasma-deposited Coatings for Components of Automotive Fuel Injection Systems. *Surface and Coatings Technology*, 200, 2005, 1969–1975.
60. A. K. Lugscheider, J. Wilden, *Proc. Materials '87*, London, The Institute of Metals, 1988, 26.1.
61. R. Sivakumar, B. L. Mordike: *Surface Engineering*, 1987, 3, (4), 299–309.
62. P. Ernst, G. Barbezat, Thermal Spray Applications in Powertrain Contribute to the Saving of Energy and Material Resources. *Surface and Coatings Technology*, 202, 2008, 4428–4431.

63. T. Bell, A. Bloyce, *Heat Treatment '84', Paper 36*, The Metals Society, London, 1984.

64. G. Carter, *Ion Assisted Surface Treatments, Techniques and Processes*, vol. 8, The Institute of Metals, London, 1982.

65. G. Dearnaley, *Surface Engineering*, 1986, 2(3), 213–222.

66. L. E. Tidbury, B. A. Rickinson, *Proc. Materials '87*, London, The Institute of Metals, 1988, 44.1.

67. Rickinson, *Proc. East Midlands Electricity Board Seminar*, Solihull, September 1986.

68. M. Anderson, B. A. Rickinson, *Proc. 1st Int. Conf. on 'Surface Engineering'*, vol. 16, The Welding Institute, Abington, 1985.

69. H. D. Steffens, R. Dansmer, U. Fisher, *Surface Engineering*, 1988, 4(1), 39–43.

70. H. Yasbandha, MSc(Eng) Thesis, University of Birmingham, 1987.

71. S. McLean, PhD Thesis, University of Birmingham.

Chapter 4

Surface engineering of nanomaterials

Processing and applications

Mainak Saha

Indian Institute of Technology Madras, Chennai, Tamil Nadu, India

Manab Mallik

National Institute of Technology, Durgapur, Durgapur, West Bengal, India

CONTENTS

4.1 INTRODUCTION

As correctly pointed out by Richard Feynman in his book *There's Plenty of Room at the Bottom*, nanoengineered systems have the potential to revolutionize the field of materials research [1–3]. From a historical viewpoint, this may be considered the origin of a new avenue of materials research for heterogeneous material systems which helps to separate the surface from the bulk properties [3–7]. Besides, this approach has been utilized to discover materials with excellent bulk and surface/interface properties [8–17]. Synthesis techniques for nanomaterials (NMs) may be primarily split into two groups: (i) top-down and (ii) bottom-up approaches. Methods based on the top-down approach involve comminution or dissociation of bulk materials into finer components whereas those for the bottom-up approach are used for assembling atoms together [18–22]. The former was originally developed for the microelectronics industries [23–25]. However, with the progress of technology for nanopatterning and deposition of thin films, this approach made its way into the fabrication of NMs [18]. Methods based on this approach commonly include solid-state milling (such as ball milling), photolithography, laser machining, soft lithography, nanosphere lithography, colloidal lithography, ion implantation, and deposition [19]. Common examples of synthesis techniques involving a bottom-up approach include plasma arcing, chemical vapor deposition (CVD), atomic layer deposition (ALD), metal-organic decomposition, laser pyrolysis, molecular beam epitaxy, sol-gel method, and wet synthesis [19].

One of the areas where surface engineering of NMs is being explored (at present) is the design of electrode materials for energy storage applications [26]. This avenue utilizes the excellent charge storage capacity and high rechargeability of NMs [27]. At present, commercial-grade batteries (for energy storage applications) mainly use microscale materials for electrodes [28]. This may be attributed to (i) fabrication challenges for NMs and (ii) the high reactivity (due to high surface area to volume ratio) NMs which leads to the formation of solid electrolyte interphase (SEI) [29–31]. The two aforementioned factors may be held responsible for an increased consumption of electrolyte during the formation of the electrode–electrolyte interface, consequently, degrading the charge storage capacity [32]. This is one of the challenges which may be overcome through surface engineering [33, 34].

SEI must facilitate charge transfer reactions for the purpose of storing and releasing energy without any degradation in charge storage and release capacity for battery applications [34]. From a surface engineering viewpoint, it is extremely important to engineer electrode materials for energy storage applications [32, 35, 36]. This necessitates the isolation of electrode materials with enhanced bulk and surface properties for high electrochemical performance (including high storage capacity, internal resistance, and so on [23, 24, 37, 38]. The present chapter is aimed at addressing the present state of research in the context of surface-engineering-based processing techniques and applications of NMs. This has been followed by a discussion on the challenges and future perspectives in the avenue of surface-engineered NMs.

4.2 SURFACE ENGINEERING APPROACH FOR NM FABRICATION

4.2.1 CVD and ALD

To the best of the authors' knowledge, the synthesis of materials in gas phase (to date) is the most widely used procedure for surface engineering of NMs [39]. These kinds of processing approaches primarily include a mix of precursor gases which flow across a chamber (maintained at high temperature) to react with the substrate surface [40]. In the context of CVD, reactions between the substrate surface and gas phase precursor are influenced by a number of factors which include (i) precursor partial pressure, (ii) temperature, and (iii) reactivity of the electrode [41]. In the case of ALD (Figure 4.1), gas phase precursors react in a self-limiting process. As a result, ALD involves the pulsing of a number of precursors (generally two or more) in an alternating manner [42, 43]. In this context, it is worthwhile mentioning that although CVD is a favorable technique to tailor the properties of substrate surfaces, however, it is highly challenging to control the thickness [43]. For instance, the deposition of carbon-based NMs on a variety of substrates (using CVD) has been reported to enhance mechanical properties, corrosion resistance, and electrical conductivity [44]. Another interesting characteristic of CVD is the deposition of high-purity films [44]. However, the high temperatures prevalent during CVD may lead to the decomposition of a number of substrates limiting the choice of gas phase precursors for a given substrate during the process [45, 46]. Moreover, conventional CVD techniques have a number of limitations which include (i) high temperature (of reaction), (ii) low deposition rates, and (iii) operation at low pressures. This has been the reason for the development of enhanced CVD techniques such as lasers, hot filaments, and plasma [46]. This has given way to the

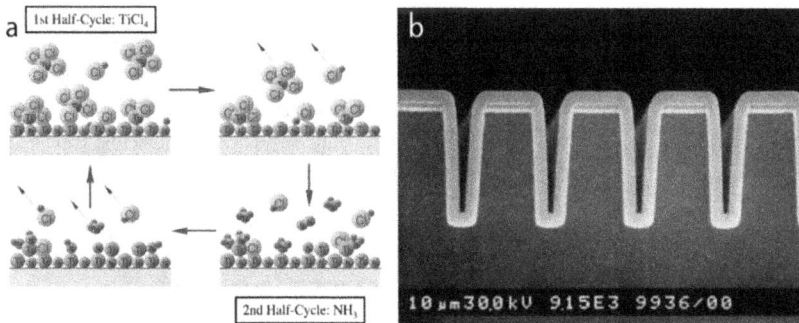

Figure 4.1 ALD: (a) schematic and (b) scanning electron microscope (SEM) image of a cross section of 300-nm-thick Al_2O_3 ALD film on a trench-structured Si wafer [42, 43].

development of metal-organic chemical vapor deposition (MOCVD) [39, 46]. Instead, ALD includes a two-step process that enables processing for a number of depositions [41–43]. An inert gas, for example, Ar or N_2, is used to prevent a CVD reaction between precursors [45, 47]. The major advantages of ALD over CVD are (i) thickness control with high precision, (ii) lower operating temperatures, and (iii) coatings with a high level of design complexity that may be deposited on substrate surfaces [45]. However, ALD suffers from the high requirements of the precursor, viz. the precursor must be (i) volatile and (ii) able to undergo a self-limiting reaction [43].

4.2.2 Wet-chemistry-based surface modification

This approach uses reagents or solvents for the chemical modification or functionalization of the material surfaces [48]. In addition, this approach provides an additional control of the engineered interface. Either non-covalent interactions [40, 48] or covalent functionalization [41, 46] may be utilized to produce a coating using the aforementioned approach. An example of the former is an ionic surfactant-coated carbon nanotube (CNT) [49]. As well as providing the integrity of the CNT, the presence of an ionic surfactant prevents bundling and enables dispersion in liquid media [39]. On the other hand, chemical grafting of reactive functional groups (such as hydroxyl and carboxyl groups) to the surface (using strong oxidizers such as nitric acid and hydrogen peroxide) may be used to achieve covalent interactions. This leads to the formation of highly reactive defect sites through an increase in wettability. Although wet-chemistry-based methods are user-friendly, however, chemical selection for a particular reaction is most challenging [50]. In addition, the wet chemistry route may affect the formation of functionalized surfaces which may undergo partial corrosion or dissolution in the solution [51]. Besides, strong oxidizers used for achieving covalent interactions are highly dangerous, especially for high-temperature reactions [50]. As a result, several rinsing steps (time-consuming in nature) are essential to prevent the accumulation of harmful chemical residue after functionalization [46]. One of the dry chemical processes is reactive ion etching (RIE) which facilitates adjustment of surface properties such as the hydrophobicity and reflectivity obtained using wet chemical modification techniques [52]. Besides, RIE may be used to fabricate nanostructures which are difficult to obtain by the wet chemical method [27, 47, 48, 52].

4.2.3 Electrochemical or electrophoretic surface modification

Electrodeposition (ED) is a chemical precipitation technique in which ionic species are attracted in solution to the electrode using electric current. As a result, reduction/oxidation occurs followed by subsequent deposition. A number of materials including metals, polymers and ceramics may be deposited using ED [49, 50]. A variety of properties such as adhesion, hardness, and crystallinity may be controlled during ED [49]. Besides, a number of advantages are offered by ED, which include fast deposition rates with deposition over a large area [53], no post-deposition treatment [51], and most importantly, low cost. However, only a few materials may be electrodeposited from aqueous solutions [53, 54].

Electrophoretic deposition (EPD) is a colloidal route for forming coatings from a stable suspension. However, EPD does not involve chemical reactions [55]. A DC electric field is applied to a colloidal suspension during EPD. This leads to the migration and subsequent deposition of charged particles on the oppositely charged electrode [52]. Stabilization of dispersed particles in the solvent is ensured by the addition of surfactants [56–60]. This, in addition, also provides charged positions to enable migration [48]. Recent EPD-based investigations by Carter et al. [6] and Oakes et al. [61] have determined surfactant-free EPD of nanostructures (NSs) by the use of polar solvents. Teranishi et al. [62] have reported that a number of materials (such as metals, nitrides, and oxides) may be deposited using EPD. Carter et al. [57] have demonstrated the fabrication of 3D foam materials with enhanced mechanical properties. In contrast to ED, EPD is highly scalable and economic [54]. Besides, EPD enables homogeneous microstructure [58], and densely packed coatings [53].

4.2.4 Glancing angle deposition

The glancing angle deposition (GLAD) technique involves the condensation of sputtered atoms on the substrate surface (Figure 4.2) [63]. A variety of physical and chemical phenomena influence the thin film deposition process, especially during the early stages of film growth [63]. The growth mode of the coating is influenced by a number of factors, viz. (i) temperature, (ii) crystallography, (iii) surface conditions of the substrate, (iv) energy of the condensed particles, and (v) nature of interactions between sputtered atoms and the substrate [27]. After the initial stages of film deposition, film growth becomes dominant. During thin film growth, two phenomena: (i) atomic-level shadowing and (ii) surface atom distribution [63]. Several microstructural factors like crystallinity, size, and density of nucleation sites influence film growth [63]. For the case of low temperatures on the surface of the substrate ($\sim 0.1 T_m$, where T_m is the substrate melting temperature), surface diffusion of incident (sputtered) atoms is reduced [27]. This is followed by the formation of a columnar structure by the condensation of the incident atoms at the nearest nucleation sites [27].

When the flux of the sputtered atoms is incident at a non-perpendicular angle to the surface of a substrate, the nucleation sites intersect the incident atoms [27, 63]. This leads to a shadowing effect subsequently causing a tilted columnar grain growth (Figure 4.3) [63]. It may be noted that during room temperature GLAD deposition, there occurs a competition for the

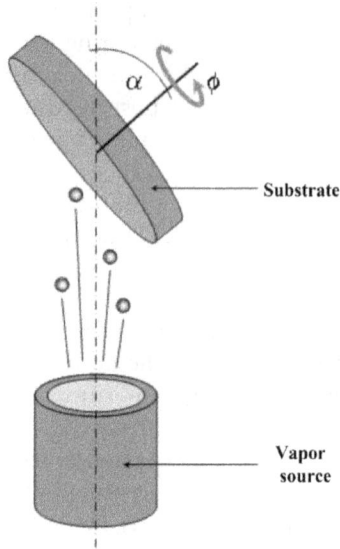

Figure 4.2 Schematic of the experimental setup used for the GLAD technique. α is the angle between the direction of the incident vapor and the substrate normal. ϕ is the angle of rotation along the axis centered on the substrate [63].

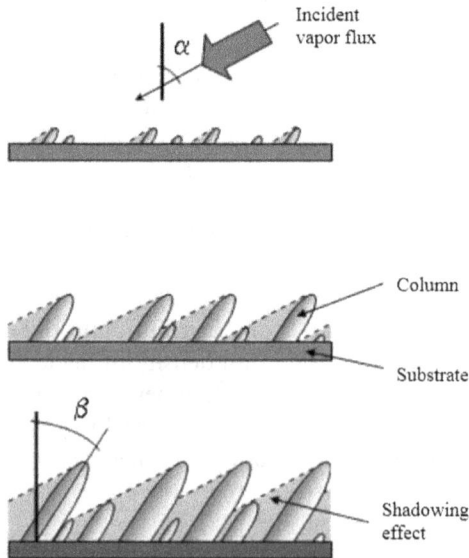

Figure 4.3 Columnar grains with a particular orientation on a substrate. α is the angle of incidence of the sputtered atoms on the substrate relative to the normal from the substrate. β is the angle of growth of the columnar grain with reference to the normal from the substrate [63].

growth of columnar grains [27]. This leads to the evolution of fractal micro-structure columns, resulting in the shadowing effect [27]. Considering that the sputtered atoms are incident on the substrate surface at an angle α with respect to the normal (to the substrate surface), columnar grains may be considered to grow at an angle $\beta < \alpha$ [60]. α and β may be related to each other using the "tangent rule" which is given by [60]

$$\tan\alpha = 2\tan\beta \tag{4.1}$$

For $\alpha < 60°$, this rule shows a clear view of the columnar structure [63]. A modified version of Equation 4.1 has also been proposed in this context. This is given as [64]

$$\beta = \alpha - \sin^{-1}\frac{(1-\cos\alpha)}{2} \tag{4.2}$$

The latter (Equation 4.2) is much more accurate than the tangent rule (Equation 4.1). This is because the latter calculates β with much higher accuracy than that of the former, especially for high values of α ($\alpha\sim90°$). Besides, the latter is based on a 2D geometric analysis. It considers that the structure of the films resembles columns with inclined hemispheric summits. Besides, surface diffusion and the minimum distance between the growth site of the particle flux and the top of columnar grains are also taken into account in Equation 4.2. Nevertheless, unavoidable factors such as substrate polarization and surface contamination may give rise to discrepancies between the proposed theory [64] and experimental results. Hara et al. [65] have reported that the atoms incident on the film during its growth, undergo preferential diffusion. Beydaghya et al. [66] have calculated the inclination of the columns (β) by measuring the refractive index of evaporated Si films. Lintymer et al. [67] have reported that Equations 4.1 and 4.2 are valid even for the case of sputter-deposited thin films of Cr. It has been reported that sputtering pressure leads to significant deviations in theoretical inclination (β) predicted by the tangent rule (Equation 4.1) [67]. Moreover, these deviations have been attributed to the reduction of the shadowing effect is reduced with decreasing mean free path of sputtered atoms [67].

Using GLAD, Nieuwenhuizen and Haanstra [68] have reported a chevron-like architecture of Al films deposited on a mobile substrate. The aforementioned architecture was reported to be obtained through an alternate tilting of the substrate by $+\alpha$ and $-\alpha$ with an equal deposition period [67]. The microstructure obtained was reported to comprise columnar grains grown in a zigzag manner. Moreover, the morphology and size of the grains have been reported to be highly influenced by two different factors, viz. (i) angle of deposition and (ii) rate of deposition [66]. This has been observed in the case of sputtered Cr films. In addition, the directional nature of columnar grain growth may also be utilized to obtain a wide variety of architectures [63]. This may be achieved by changing the position of the source (of sputtered atoms) with respect to the columnar grains during film growth [64]. During such an operation, it is only the substrate which rotates through the center of the substrate [66]. The source of sputtered atoms remains stationary while the substrate is rotated [60, 63]. Such an instrument offers: (i) a rotation axis for the purpose of varying the angle of incidence of the particle flux (α) and (ii) a rotary axis at an azimuthal angle (φ) for an indirect modification of the sputtered source position [66]. Overall, the GLAD technique significantly reduces the chances of shadowing which otherwise may be created by tilting the substrate and a change in the position of the substrate during the deposition process [65].

4.3 NMS SURROUND OUR DAILY LIVES

In addition to the naturally existing NMs, it is possible to artificially fabricate NMs which are designed to match the different commercial requirements [45]. Synthetic (or artificial) NMs presently find extensive applications in a number of areas such as clothing, cosmetics, electronic goods, and many other items of daily use [47]. In the context of medical applications, these materials find extensive applications in areas such as drug delivery, diagnosis, and imaging. The unique properties of NMs (especially high room temperature mechanical strength and chemical reactivity) may be attributed to a high surface area to volume ratio and the onset of quantum effects in these materials [45]. There are a number of physical and chemical techniques for the synthesis of nanoparticles (NPs) [47]. Moreover, there exist a number of natural sources of NMs such as ash (volcanic), fine sand, dust, and biological matter [19, 45, 47]. Practically, it is hard to control the shape and size of synthetic NPs [47]. This is because these NPs are obtained as by-products of human activities (such as combustion of automobile engines, and mining) and are comprised of a number of elements [45]. The size, shape, and chemical composition of synthetic NPs may be engineered by selecting an appropriate fabrication technique [47]. At present, a number of synthetic NPs are being investigated which include semiconductors (nanostructured), nanoceramics, nanosized magnetic particles, and metallic powders [45].

4.3.1 Energy storage: an emerging trend in surface engineering of nanomaterials

In the context of energy storage devices, pseudocapacitors (based on a Faradaic storage mechanism) and hybrid supercapacitors involve faster charging and discharging rates when compared with those of typical batteries [31, 67]. A schematic representing the role of surface-engineered NMs in the context of energy storage applications is shown in Figure 4.4. Moreover,

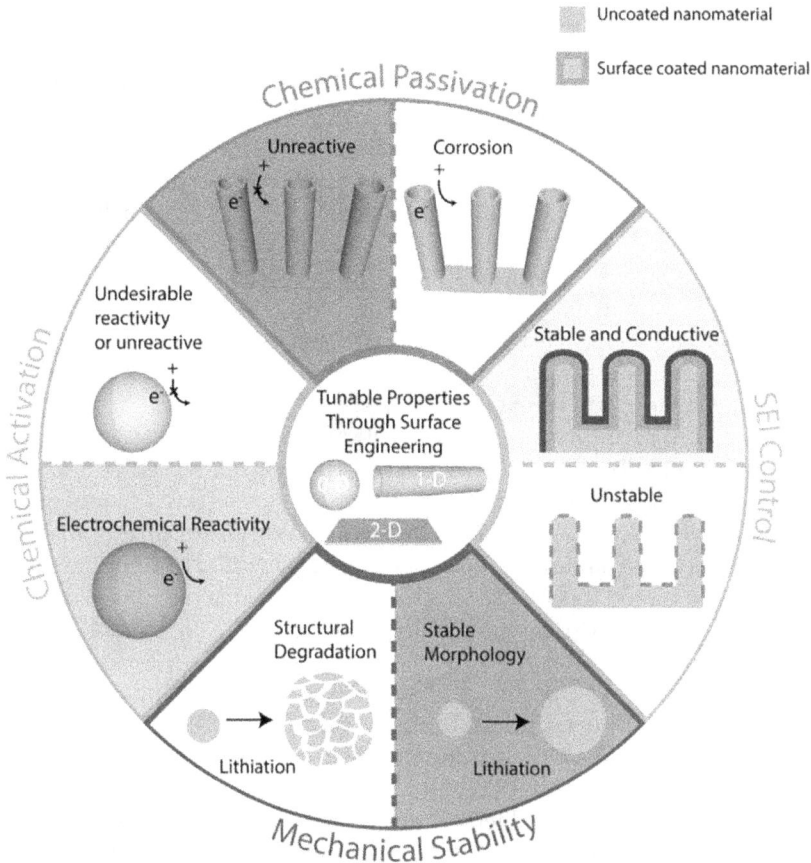

Figure 4.4 Control of different properties of NMs (for energy storage applications) using surface engineering [31].

the energy densities of these devices are higher than those of non-Faradaic supercapacitors [69]. This enables a pseudocapacitor to store charge over a wide voltage range. Common examples of pseudocapacitor electrodes include RuO_2 and MnO_2 [70–77]. Additional cycles lead to the lowering of charge storage performance [78, 79]. This is because increased thickness may lead to an impediment to the mobility of charge carriers toward the surface and ionic diffusion through the formation of vanadium oxide [80].

Surface engineering may be used to overcome the limitations of pseudocapacitors and hybrid supercapacitors through an optimization of the bulk and surface [81, 82]. This enhances the rate of ionic diffusion in addition to providing a high surface area [83]. These devices make use of metal oxides with high electrical resistance. Conductive pathways with short diffusion lengths must be provided by the engineered scaffolds [84]. These pathways are required for enhancing charge transfer with no irreversible reaction [85].

These shortcomings may be overcome during active metal deposition through (i) increasing surface area and (ii) prevention of undesirable reactions by completely covering the substrate [85–87]. One of the most common ways to achieve these goals includes nanostructured conductive carbon templates or nanowire core/shell [88, 89]. Surface-engineering-based processing of scaffolds and active materials involves a two-step process for optimizing both surface area and chemical reactivity [90]. The high electrical conductivity of the substrate leads to an improvement in capacitance [91, 92]. In addition, the high electrical conductivity leads to a decrease in the infrared (IR) drop in galvanostatic testing.

Dubal et al. [94] have used chemical bath deposition to deposit MnO_2 on Si nanowires (NWs) with a complete coating on the array of NW. This prevents the reaction between Si NWs with the electrolyte. This gave them the opportunity to optimize the ionic liquid electrolyte each having a 2.2 V range [94]. As reported by Ghodbane et al. [95], the ionic conductivity and reactivity may be controlled through an appropriate optimization of active material crystal structure. Besides, a separate processing of the bulk and substrate surface (enabled by surface engineering) results in an optimization of the active material during deposition [95]. Aqueous ammonia has been reported to optimize the crystallinity, morphology, and surface area of $Ni(OH)_2$ during the chemical bath deposition process [96]. Fewer amounts of ammonia were reported to result in thin nanowalls with a high surface area resulting in a low resistance to charge transfer [96]. On the other hand, $Ni(OH)_2$ was observed to act as an active material [96]. Carbon-based materials such as graphene [94] and CNTs [70] have been reported to provide high surface area conductive substrates for faradaic metal oxides by acting as supercapacitors themselves. However, the collective effects of carbon and metal oxides lead to an enhancement in the overall capacitance [86, 87, 97]. For instance, the deposition of a thin layer of carbon on a pseudocapacitor provides a conductive wrapping leading to an increase in the overall capacitance by ~20% [89, 98].

4.3.2 SEI control through surface engineering

As briefly described in the Introduction section (Section 4.1), SEI forms from the electrolytic reaction on the electrode surface [98]. Moreover, a number of parameters such as the type and morphology of the electrode, and testing parameters influence the composition and function of the SEI [98]. Although this layer provides a stable interface, however, it has also been reported to impede ionic diffusion [99]. In addition, exposure of additional material may be caused due to a volumetric expansion of the electrode [92, 100, 101]. This in turn may lead to the formation of high amounts of SEI with subsequent irreversible consumption of ions and electrolytes [102]. It has been observed that additives such as vinylene carbonate [65, 103] and CO_2 lead to the formation of SEI with a much lower interfacial resistance as compared to that in the case of additive-free electrolytes [103]. The final aim for SEI-based surface engineering is to form a surface which is stable, ionically conductive surface and most importantly shows minimal degradation over a prolonged period of time [93, 105]. Hence, appropriate control of SEI is essential for separate engineering of the bulk and surface of NMs which tend to be highly reactive in an electrochemical environment, owing to their high surface area to volume ratio [106].

ALD has been reported to emerge as an important deposition technique toward manipulating SEI [106, 107]. In the context of ALD coatings, the most effective deposition occurs on the assembled electrode in order to ensure the rapid transfer of electrons [104]. As briefly mentioned in Section 4.1, ALD materials such as Al_2O_3 and ZnO have the ability to control SEI formation through protection of electrode on account of possessing low ionic conductivity. As highlighted by Cheng et al. [108], ALD-deposited $LiAlO_2$ has a much higher ionic conductivity as compared to that of Li_2O–Al_2O_3 [108]. It was reported that the application of ALD-based surface coating to the anode in a full cell leads to a significant enhancement in the number of cycles for charge storage [108]. Besides, the anode is protected from chemical attack by the cathode comprising of transition metals [108]. In this context, Al has been reported to prevent mechanical failure through Li intercalation, compression of Si, and a change in SEI composition [109]. The surface coating with the highest coulombic efficiency (over 100 cycles) was achieved for the case of 3 wt.% Al [109]. Even annealing may be used to modify the electrode surface to promote enhanced SEI formation [109]. Han et al. [110] have reported that the removal of chemically adsorbed water and hydroxyl groups from annealed TiO_2 surface results in the formation of SEI with low ionic resistivity.

4.3.2.1 SEI: characterization

One of the approaches toward developing surface engineering approaches for controlling SEI formation is through understanding the reactions occurring in the SEI [111–113]. This further necessitates an understanding of the composition of SEI. A number of characterization techniques such as Atomic Force Microscopy (AFM), Electrochemical Impedance Spectroscopy (EIS), and Fourier Transform Infrared Spectroscopy (FTIR) [110] have been used for investigating surface-engineered microstructures. Spectroscopic investigations on the SEI have revealed the presence of ionic insulators such as Li_2CO_3 and LiF [108]. The use of Secondary Ion Mass Spectroscopy (SIMS) and X-Ray Photoelectron Spectroscopy (XPS) has revealed that SEI on Al_2O_3 coated electrode is much thinner as compared to that of $LiAlO_2$ which has been shown to possess excellent ionic conductivity of Li^+ [109]. Lipson et al. [114] have investigated SEI formation for ALD-deposited Al_2O_3 coating on MnO electrode using Scanning Ion Conductance Microscopy (SICM) topography. It was shown that 3Å of Al_2O_3 leads to partial prevention of thick

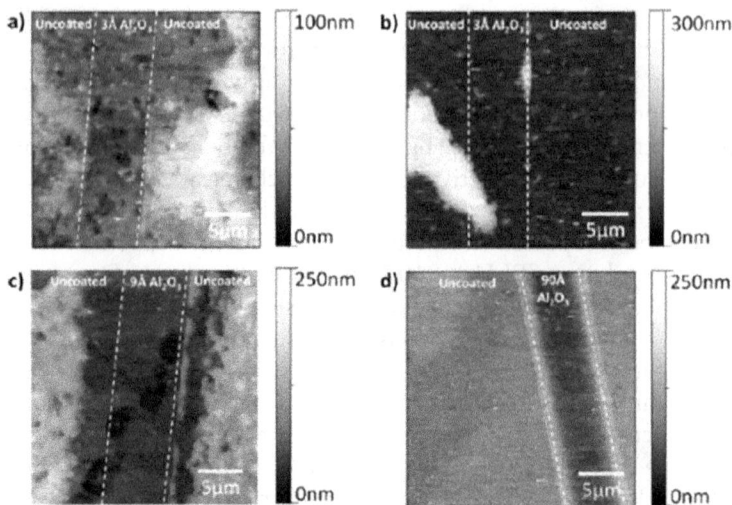

Figure 4.5 SICM tomography images showing a striped ALD Al$_2$O$_3$ with thicknesses of (a, b) 3Å ((a) and (b) represent two different regions), (c) 9Å, and (d) 90Å on MnO substrate surface [114].

SEI layer formation [114]. On the other hand, 9Å of Al$_2$O$_3$ was reported to completely prevent the formation of a thick SEI layer (Figure 4.5) [114]. The formation of SEI on Si NWs with hydride and methyl terminated surfaces was investigated with FTIR and XPS [111]. The termination of methyl group on the surface of Si renders the surface relatively unreactive [107].

4.3.3 Chemical passivation

Owing to the reduced dimensions (typically less than 100 nm), NMs possess a high surface area to volume ratio. This leads to high surface energy, which leads to a high reactivity in many environments [111]. Supercapacitors are composed of an electrolyte between two electrodes and store energy through electrostatic attraction between opposite charges [112]. In other words, ions in the electrolyte are attracted to the oppositely charged electrodes leading to the formation of an electric double layer on the electrode surface [113]. This is unlike most of the other energy storage devices which are based on charge storage using Faradaic electron transfer reactions [110]. The electrostatic nature of the supercapacitors leads to very fast charge and discharge [109]. In addition, chemical passivation of NMs may lead to the use of a whole new range of chemically compatible materials with high charge storage durability for applications in supercapacitors [107]. This, in principle, may also be used to increase the capacitance of highly porous nanoelectrodes [104]. Surface engineering with high precision (at the atomic scale) may lead to partial dissolution of ions in solvents [114]. This leads to an enhancement in the overall capacitance [116]. Chmiola et al. [115] have demonstrated the aforementioned approach in carbide-derived carbon. Porous Si has been demonstrated to provide the high surface area required for a supercapacitor electrode [115]. Electrochemical etching has been reported to optimize electrodes for ionic conductivity, surface area, and penetration of electrolytes in porous Si [114]. It has been shown that porous Si reacts with ionic liquid electrolytes. This makes it a poor electrode material for supercapacitors [114]. In order to overcome the aforementioned problem, CVD-based thin coating of graphene on the surface of porous Si has been shown to extend the voltage window to ~2.7 V, with an increase in discharge times by over 20 times [114]. Besides, nanostructures in aqueous and organic media have also been reported to be stabilized by carbon coating [111, 114]. Passivation of porous Si by TiN using ALD has been reported in Ref. [115]. Mesoporous Si may also act as a support for porous carbon with subnanometer-sized pores [108]. A number of coatings on Si NWs may lead to the prevention of undesirable reactions, leading to an increase in the voltage window [107].

4.4 INFLUENCE OF SURFACE ENGINEERING ON THE MECHANICAL PROPERTIES

In the context of energy storage systems with high charge storage capacities, mechanical properties play a very important role. This is because volumetric changes often promote cracking. This leads to the loss of active [108, 114]. Surface engineering may be used as a potential tool to provide mechanical support for the purpose of minimizing active material loss [107]. In the context of Li-ion batteries (LIBs), it has been reported that alloying materials such as Ge, Si, and Sn and metal oxides have high Li ion storage capacities [104]. This leads to high volume expansion (~300–400%). In the case of conventional bulk counterparts, volumetric expansions lead to mechanical fracture and subsequent degradation of active material (electrode) [95]. This ultimately leads to battery failure [93]. At present, two strategies have been devised to overcome these difficulties. The first strategy involves designing NMs for the purpose of accommodating large volume expansions while the second one involves the use of thin surface coatings as an adhesive during the storage of Li [70]. Both these strategies are highly essential for the purpose of overcoming these mechanical challenges [72]. A number of surface engineering approaches have been devised to minimize the amount of mechanical expansion and degradation [73]. One of these approaches is to use an external bulk coating on the entire electrode. The other approach As reported by Lu et al. [116], involves using conductive carbon coating on an electrode made of Si NP. Besides, it was also highlighted that the use of individual NPs as coating materials leads to mechanical fracture of the coating. The other strategy, as shown by Nguyen et al. [117], involves the use of individual NM or NW as a protective coating. $NiSi_x$ NW was initially grown using a silane CVD technique, subsequently followed by the growth of an amorphous Si layer using an additional silane CVD step [117]. Finally, the ALD-based coating of Al_2O_3 was deposited. A high cyclability with 150 cycles at a full discharge capacity as high as up to ~ 3000 mAh/g and with 700 cycles at a partial discharge capacity of ~1200 mAh/g, was reported [117]. As highlighted in Ref. [116], Al_2O_3 coating undergoes fracture even after repeated cycles of charging and discharging, and shows high cyclability. Similar observations have been observed for Si core–shell structured NPs with in-situ polymer coatings [112] and Fe_3O_4 with carbon coatings [114]. Moreover, it was also highlighted that although the aforementioned coatings improved cyclability, however, mechanical fracture during the cycles could not be prevented [107]. This led to the development of a yolk/shell structure [106]. The main difference between a yolk/shell and a core/shell structure is based on the difference in the structure of the inner core [94]. In other words, the inner core completely fills the outer shell in a core–shell structure whereas, there is a void space (between the inner core and outer shell) for a yolk/shell structure [69]. The accommodation of volume expansion occurs with the help of the void space (in the yolk–shell structure). This is subsequently supported by the outer shell [74]. The application of external loads on a multi-layer NP followed by the removal of the outer layer of the NP has been reported for the fabrication of yolk/shell structures [1, 75, 120, 121]. Liu et al. [119] have demonstrated that a pomegranate-like design leads to enhanced performance (total capacity: ~3000 mAh/g with a small amount of degradation even after 1000 cycles). A similar approach has been used to enhance the overall performance in S/TiO_2, S/Carbon, and Sn/C yolk–shells [119]. In addition, there have been a number of recent experimental investigations highlighting the efficiency of yolk–shell and core–shell structures toward improving the overall cyclability of Na-ion batteries (SIBs) [1].

4.5 SURFACE ENGINEERING: FUTURE PERSPECTIVES

4.5.1 Prevention of microcracks

Typical magnitudes of Li intercalation potentials (with respect to Li/Li⁺) used for most of the cathodes are ~4V [114]. Most of these are highly stable and presently find applications in commercial batteries [121]. However, Li metal oxides ($LiMO_x$) are brittle ceramics. Hence, Li intercalation, leading to lattice distortions, may often lead to microcrack formation thereby resulting in a degradation of the electrodes' ability to intercalate Li [106].

4.5.2 Prevention of polysulfide shuttling

In terms of the charge storage capacity, Li-S batteries pose a major challenge to conventional LIBs [1, 119]. However, these batteries suffer from a number of limitations, rendering them unsuitable for commercial applications

[1]. A conventional Li-S battery makes use of a Li anode and S cathode and typically involves C networks for the immobilization of S for the purpose of providing electrical connectivity across the electrode [118]. Similar to the other high-performance electrodes, S cathodes also show large volume expansions [2]. However, there are a number of challenges with these electrodes. During the reaction of S cathode with Li, two things may occur: (i) active material may undergo huge large volume expansions, and (ii) intermediate phases may be formed (in the Li/S reaction chain) [112]. The former leads to large volume expansions resulting in an agglomeration of active S and consequently leading to losses in ionic conductivity [113]. The latter is unique to S cathodes. Seh et al. [123] have used wet chemical processing to fabricate S/TiO_2 yolk/shell electrodes with a cyclability of ~800mAh/g for 1000 cycles. Moreover, pairing of these electrodes with pre-lithiated Si/C yolk–shell particles has been reported to enhance energy density in LIB [123]. In addition, the other challenge to be overcome in the context of Li-S batteries is the shuttling of polysulfide, i.e., reduction in the active S content in the cathode through the dissolution of polysulfide into the electrolyte [122, 124]. Recent experimental investigations have highlighted that thin metal oxide coatings, such as Al_2O_3, indium-tin oxide (ITO), and vanadium oxide result in strong interface (surface) binding of polysulfides, minimizing the detrimental effect of polysulfide shuttling [125]. In the context of Li-S battery research, this is a recently emerging research avenue with a high potential toward designing cathodes with cyclability as high as ~100 cycles with more than 90% of the initial S capacity [121–126].

4.5.3 Prevention of dendritic growth for metal anodes

In the context of rechargeable LIBs, Li metal was initially used as the anode. However, dendritic growth in these anodes were reported to lead to a quick mechanical. The coating techniques devised to prevent dendritic growth led to a low cyclability of LIBs. Drop-cast-based deposition of polystyrene NPs on a Cu foil followed subsequently by flash evaporation of carbon fibers has been reported in Ref. [125]. A conductive carbon yolk–shell structure for enhanced cycling performance in Li metal anodes was designed [125]. Most importantly, this electrode was shown to possess high cyclability with a coulombic efficiency of ~ 99.9% for 150 cycles [125]. The other strategy involves the use of ALD for the purpose of coating thin layers of Al_2O_3 on a metallic foil of Li [114, 125]. In both approaches, the performance of Li metal anode was improved. Similar to the Li-ion systems, Na ion systems have also been shown to suffer from the same problem of dendritic growth in the metal anodes (Na anode in this context). Kazyak et al. [127] have demonstrated that a 2–3 nm Al_2O_3 coating on the Li metal anode led to the high stability of the device (for ~1200 cycles). Nevertheless, the two aforementioned approaches have been successfully demonstrated to prevent dendritic growth in Li metal anode [126, 127]. Besides, dendritic growth in Na ion systems and K ion systems in Potassium-Ion Batteries (PIBs) is an avenue which is largely unexplored at present. This avenue is expected to offer tremendous potential toward the exploration of Ca and Mg in rechargeable battery systems.

4.6 CONCLUSIONS

The present review highlights that microstructural evolution in surface-engineered NMs is highly influenced by fabrication (deposition) techniques. In the context of energy research (especially high-performance energy storage devices), this has opened up a number of new avenues which include enhanced chargeability, high durability (for electrodes), and high energy density. A decoupling of the space characteristics from the bulk properties enables one to separately optimize the properties of the surface coatings with those of the underlying substrates. Besides, surface engineering may also be used to overcome a number of challenges, primarily associated with the utilization of nanostructures in the aforementioned devices.

REFERENCES

1. A. S. Aricò, P. Bruce, B. Scrosati, J.-M. Tarascon, and W. van Schalkwijk, "Nanostructured materials for advanced energy conversion and storage devices," *Nature Mater.*, vol. 4, no. 5, pp. 366–377, May 2005, doi: 10.1038/nmat1368.
2. M. Armand, and J. M. Tarascon, "Building better batteries," *Nature*, vol. 451, no. 7179, pp. 652–657, Feb. 2008, doi: 10.1038/451652a.
3. B. S. Murty, P. Shankar, B. Raj, B. B. Rath, and J. Murday, "Unique properties of nanomaterials," *Textbook of Nanoscience and Nanotechnology*, pp. 29–65, 2013, doi: 10.1007/978-3-642-28030-6_2.
4. T. Brousse, D. Bélanger, and J. W. Long, "To be or not to be pseudocapacitive?," *Journal of The Electrochemical Society*, vol. 162, no. 5, pp. A5185–A5189, 2015, doi: 10.1149/2.0201505JES.
5. C. K. Chan et al., "High-performance lithium battery anodes using silicon nanowires," *Nature Nanotechnology*, vol. 3, no. 1, pp. 31–35, Jan. 2008, doi: 10.1038/NNANO.2007.411.
6. R. Carter et al., "Solution assembled single-walled carbon nanotube foams: Superior performance in supercapacitors, lithium-ion, and lithium-air batteries," *Journal of Physical Chemistry C*, vol. 118, no. 35, pp. 20137–20151, Sep. 2014, doi: 10.1021/JP5054502.
7. A. D. Maynard, "Don't define nanomaterials," *Nature 2011 475:7354*, vol. 475, no. 7354, pp. 31–31, Jul. 2011, doi: 10.1038/475031a.
8. J. Lahann, "Nanomaterials clean up," *Nature Nanotechnology 2008 3:6*, vol. 3, no. 6, pp. 320–321, May 2008, doi: 10.1038/nnano.2008.143.

9. Y. Tang et al., "Unravelling the correlation between the aspect ratio of nano-tubular structures and their electrochemical performance to achieve high-rate and long-life lithium-ion batteries," *Angewandte Chemie*, vol. 126, no. 49, pp. 13706–13710, Dec. 2014, doi: 10.1002/ANGE.201406719.

10. M. D. Slater, D. Kim, E. Lee, and C. S. Johnson, "Sodium-ion batteries," *Advanced Functional Materials*, vol. 23, no. 8, pp. 947–958, Feb. 2013, doi: 10.1002/ADFM.201200691.

11. C. Y. Su et al., "Direct formation of wafer scale graphene thin layers on insu-lating substrates by chemical vapor deposition," *Nano Letters*, vol. 11, no. 9, pp. 3612–3616, Sep. 2011, doi: 10.1021/NL201362N.

12. P. Simon and Y. Gogotsi, "Materials for electrochemical capacitors," *Nature Materials*, vol. 7, no. 11, pp. 845–854, Nov. 2008, doi: 10.1038/NMAT2297.

13. I. Shterenberg, M. Salama, Y. Gofer, E. Levi, and D. Aurbach, "The challenge of developing rechargeable magnesium batteries," *MRS Bulletin*, vol. 39, no. 5, pp. 453–460, 2014, doi: 10.1557/MRS.2014.61.

14. M. J. Sailor, "Fundamentals of porous silicon preparation," *Porous Silicon in Practice*, pp. 1–42, Jan. 2012, doi: 10.1002/9783527641901.CH1.

15. M. V. Reddy, G. V. Subba Rao, and B. V. R. Chowdari, "Metal oxides and oxy-salts as anode materials for Li ion batteries," *Chemical Reviews*, vol. 113, no. 7, pp. 5364–5457, Jul. 2013, doi: 10.1021/CR3001884.

16. A. Ponrouch, C. Frontera, F. Bardé, and M. R. Palacín, "Towards a calcium-based rechargeable battery," *Nature Materials*, vol. 15, no. 2, pp. 169–172, Feb. 2016, doi: 10.1038/NMAT4462.

17. Y. Qiu et al., "High-rate, ultralong cycle-life lithium/sulfur batteries enabled by nitrogen-doped graphene," *Nano Letters*, vol. 14, no. 8, pp. 4821–4827, Aug. 2014, doi: 10.1021/NL5020475.

18. M. Y. Naz, S. Shukrullah, A. Ghaffar, K. Ali, and S. K. Sharma, "Synthesis and processing of nanomaterials," *Solar Cells*, pp. 1–23, 2020, doi: 10.1007/978-3-030-36354-3_1.

19. S. Kumar, P. Bhushan, and S. Bhattacharya, "Fabrication of nanostructures with bottom-up approach and their utility in diagnostics, therapeutics, and others," *Energy, Environment, and Sustainability*, pp. 167–198, 2018, doi: 10.1007/978-981-10-7751-7_8.

20. M. Saha, "γ-TiAl alloy: Revisiting tensile creep deformation behaviour and creep life at 832 °C," 2021, doi: 10.1080/2374068X.2021.1949175.

21. M. Saha, "A brief discussion on the tensile creep deformation behaviour of wrought single-phase γ-TiAl," *Materials Today: Proceedings*, Jan. 2021, doi: 10.1016/j.matpr.2020.11.189.

22. M. Saha, "Understanding the role of Al_2O_3 formed during isothermal oxida-tion in a dual phase AlCoCrFeNi2.1 eutectic high-entropy alloy," *Journal of Materials NanoScience*, vol. 7, no. 2, pp. 68–72, Nov. 2020. Available: http://thesciencein.org/journal/index.php/jmns/article/view/119

23. B. Yao, J. Zhang, X. Fan, J. He, and Y. Li, "Surface engineering of nanomateri-als for photo-electrochemical water splitting," *Small*, vol. 15, no. 1, Jan. 2019, doi: 10.1002/SMLL.201803746.

24. Y. Jung, Y. Huh, and D. Kim, "Recent advances in surface engineering of porous silicon nanomaterials for biomedical applications," *Microporous and Mesoporous Materials*, vol. 310, p. 110673, Jan. 2021, doi: 10.1016/J.MICROMESO.2020.110673.

25. M. Saha and M. Mallik, "Additive manufacturing of ceramics and cermets: Present status and future perspectives," *Sādhanā 2021 46:3*, vol. 46, no. 3, pp. 1–35, Aug. 2021, doi: 10.1007/S12046-021-01685-2.
26. H. Gleiter, "Nanostructured materials: Basic concepts and microstructure," *Acta Materialia*, vol. 48, no. 1, pp. 1–29, Jan. 2000, doi: 10.1016/S1359-6454(99)00285-2.
27. J. Takadoum, "Nanomaterials and surface engineering," *Nanomaterials and Surface Engineering*, Mar. 2013.
28. G. Baret and P. P. Jobert, "Nanostructured coatings," *Nanomaterials and Surface Engineering*, pp. 271–292, Mar. 2013, doi: 10.1002/9781118618523.CH10.
29. R. Bhattacharyya, B. Key, H. Chen, A. S. Best, A. F. Hollenkamp, and C. P. Grey, "In situ NMR observation of the formation of metallic lithium microstructures in lithium batteries," *Nature Materials*, vol. 9, no. 6, pp. 504–510, 2010, doi: 10.1038/NMAT2764.
30. M. Armand and J. M. Tarascon, "Building better batteries," *Nature*, vol. 451, no. 7179, pp. 652–657, Feb. 2008, doi: 10.1038/451652A.
31. K. Share, A. Westover, M. Li, and C. L. Pint, "Surface engineering of nanomaterials for improved energy storage – A review," *Chemical Engineering Science*, vol. 154, pp. 3–19, Nov. 2016, doi: 10.1016/J.CES.2016.05.034.
32. F. Sanchette, C. Ducros, and A. Billard, "Aluminum-based nanostructured coatings deposited by magnetron sputtering for corrosion protection of steels," *Nanomaterials and Surface Engineering*, pp. 207–226, Mar. 2013, doi: 10.1002/9781118618523.CH7.
33. A. Christmann, C. Longuet, and J. M. L. Cuesta, "Transparent polymer nanocomposites: A new class of functional materials," *Nanomaterials and Surface Engineering*, pp. 31–52, Mar. 2013, doi: 10.1002/9781118618523.CH2.
34. N. Martin, K. Robbie, and L. Carpentier, "Architecture of thin solid films by the GLAD technique," *Nanomaterials and Surface Engineering*, pp. 1–30, Mar. 2013, doi: 10.1002/9781118618523.CH1.
35. R. Constantin, P. A. Steinmann, and C. Manasterski, "Decorative PVD coatings," *Nanomaterials and Surface Engineering*, pp. 109–161, Mar. 2013, doi: 10.1002/9781118618523.CH5.
36. D. Stuerga and T. Caillot, "Microwave chemistry and nanomaterials: From laboratory to pilot plant," *Nanomaterials and Surface Engineering*, pp. 163–206, Mar. 2013, doi: 10.1002/9781118618523.CH6.
37. D. Pilloud and J. F. Pierson, "High temperature oxidation resistance of nanocomposite coatings," *Nanomaterials and Surface Engineering*, pp. 329–347, Mar. 2013, doi: 10.1002/9781118618523.CH12.
38. J. Shen, M. Shafiq, M. Ma, and H. Chen, "Synthesis and surface engineering of inorganic nanomaterials based on microfluidic technology," *Nanomaterials 2020, Vol. 10, Page 1177*, vol. 10, no. 6, p. 1177, Jun. 2020, doi: 10.3390/NANO10061177.
39. H. Tiznado, M. Bouman, B. C. Kang, I. Lee, and F. Zaera, "Mechanistic details of atomic layer deposition (ALD) processes for metal nitride film growth," *Journal of Molecular Catalysis A: Chemical*, vol. 281, no. 1–2, pp. 35–43, Feb. 2008, doi: 10.1016/J.MOLCATA.2007.06.010.

40. M. D. Groner, F. H. Fabreguette, and J. W. Elam, and S. M. George, "Low-temperature Al_2O_3 atomic layer deposition," *Chemistry of Materials*, vol. 16, no. 4, pp. 639–645, Feb. 2004, doi: 10.1021/CM0304546.

41. N. Liu, H. Wu, M. T. McDowell, Y. Yao, C. Wang, and Y. Cui, "A yolk-shell design for stabilized and scalable li-ion battery alloy anodes," *Nano Letters*, vol. 12, no. 6, pp. 3315–3321, Jun. 2012, doi: 10.1021/NL3014814.

42. M. Leskelä and M. Ritala, "Atomic layer deposition chemistry: Recent developments and future challenges," *Angewandte Chemie International Edition*, vol. 42, no. 45, pp. 5548–5554, Nov. 2003, doi: 10.1002/ANIE. 200301652.

43. H. Tiznado, M. Bouman, B. C. Kang, I. Lee, and F. Zaera, "Mechanistic details of atomic layer deposition (ALD) processes for metal nitride film growth," *Journal of Molecular Catalysis A: Chemical*, vol. 281, no. 1–2, pp. 35–43, Feb. 2008, doi: 10.1016/J.MOLCATA.2007.06.010.

44. Y. Talyosef et al., "Comparing the behavior of nano- and microsized particles of LiMn1.5Ni0.5O4 spinel as cathode materials for li-ion batteries," *Journal of The Electrochemical Society*, vol. 154, no. 7, p. A682, May 2007, doi: 10.1149/1.2736657.

45. C.-Y. Su et al., "Direct formation of wafer scale graphene thin layers on insulating substrates by chemical vapor deposition," *Nano Letters*, vol. 11, no. 9, pp. 3612–3616, Sep. 2011, doi: 10.1021/NL201362N.

46. S. Lal, U. Jana, P. K. Manna, G. P. Mohanta, R. Manavalan, and S. L. Pal, "Nanoparticle: An overview of preparation and characterization," *Journal of Applied Pharmaceutical Science*, vol. 2011, no. 6, pp. 228–234, 2011.

47. J. W. Elam, D. Routkevitch, P. P. Mardilovich, and S. M. George, "Conformal coating on ultrahigh-aspect-ratio nanopores of anodic alumina by atomic layer deposition," *Chemistry of Materials*, vol. 15, no. 18, pp. 3507–3517, Sep. 2003, doi: 10.1021/CM0303080.

48. G. Luka et al., "Transparent and conductive undoped zinc oxide thin films grown by atomic layer deposition," *Physica Status Solidi (A) Applications and Materials Science*, vol. 207, no. 7, pp. 1568–1571, Jul. 2010, doi: 10.1002/ PSSA.200983709.

49. S. J. Klaine et al., "Nanomaterials in the environment: Behavior, fate, bioavailability, and effects," *Environmental Toxicology and Chemistry*, vol. 27, no. 9, pp. 1825–1851, Sep. 2008, doi: 10.1897/08-090.1.

50. B. O. Park, C. D. Lokhande, H. S. Park, K. D. Jung, and O. S. Joo, "Performance of supercapacitor with electrodeposited ruthenium oxide film electrodes—effect of film thickness," *Journal of Power Sources*, vol. 134, no. 1, pp. 148–152, Jul. 2004, doi: 10.1016/J.JPOWSOUR.2004.02.027.

51. V. C. Moore et al., "Individually suspended single-walled carbon nanotubes in various surfactants," *Nano Letters*, vol. 3, no. 10, pp. 1379–1382, Oct. 2003, doi: 10.1021/NL034524J.

52. J. M. Englert et al., "Covalent bulk functionalization of graphene," *Nature Chemistry 2011 3:4*, vol. 3, no. 4, pp. 279–286, Mar. 2011, doi: 10.1038/ nchem.1010.

53. K. Novoselov, "Electric field effect in atomically thin carbon films," *Science*, vol. 306, no. 5696, pp. 666–669, Oct. 2004, doi: 10.1126/science.1102896.

54. Y. He, C. Jiang, H. Yin, J. Chen, and W. Yuan, "Superhydrophobic silicon surfaces with micro–nano hierarchical structures via deep reactive ion etching and galvanic etching," *Journal of Colloid and Interface Science*, vol. 364, no. 1, pp. 219–229, Dec. 2011, doi: 10.1016/J.JCIS.2011.07.030.

55. J. Takadoum, "Characterization of coatings: Hardness, adherence and internal stresses," *Nanomaterials and Surface Engineering*, pp. 293–327, Mar. 2013, doi: 10.1002/9781118618523.CH11.

56. A. J. Yin, J. Li, W. Jian, A. J. Bennett, and J. M. Xu, "Fabrication of highly ordered metallic nanowire arrays by electrodeposition," *Applied Physics Letters*, vol. 79, no. 7, p. 1039, Aug. 2001, doi: 10.1063/1.1389765.

57. A. Bumm et al., "Giant molecules here, there every-where, academic," *Macromolecules*, vol. 12, no. 8, p. 200, 2000, doi: 10.1002/(SICI)1521-4095 (200004)12:8.

58. V. C. Moore et al., "Individually suspended single-walled carbon nanotubes in various surfactants," *Nano Letters*, vol. 3, no. 10, pp. 1379–1382, Oct. 2003, doi: 10.1021/NL034524J.

59. "Effect of solid electrolyte interface (SEI) film on cyclic performance of Li4Ti5O12 anodes for Li ion batteries," *Journal of Power Sources*, vol. 239, pp. 269–276, Oct. 2013, doi: 10.1016/J.JPOWSOUR.2013.03.141.

60. R Carter, L Oakes, N Muralidharan, AP Cohn, A Douglas, CL Pint. Polysulfide anchoring mechanism revealed by atomic layer deposition of V2O5 and sulfur-filled carbon nanotubes for lithium–sulfur batteries. *ACS Applied Materials & Interfaces*. 2017 Mar 1;9(8):7185*#x2013;92.

61. L. Oakes et al., "Surface engineered porous silicon for stable, high performance electrochemical supercapacitors," *Scientific Reports 2013 3:1*, vol. 3, no. 1, pp. 1–7, Oct. 2013, doi: 10.1038/srep03020.

62. T. Teranishi, M. Hosoe, and T. Tanaka, and M. Miyake, "Size control of monodispersed pt nanoparticles and their 2D organization by electrophoretic deposition," *Journal of Physical Chemistry B*, vol. 103, no. 19, pp. 3818–3827, May 1999, doi: 10.1021/JP983478M.

63. N. Martin, K. Robbie, and L. Carpentier, "Architecture of thin solid films by the GLAD technique," *Nanomaterials and Surface Engineering*, pp. 1–30, Mar. 2013, doi: 10.1002/9781118618523.CH1.

64. R. N. Tait, T. Smy, and M. J. Brett, "Structural anisotropy in oblique incidence thin metal films," *Journal of Vacuum Science & Technology A: Vacuum, Surfaces, and Films*, vol. 10, no. 4, p. 1518, Jun. 1998, doi: 10.1116/1.578037.

65. K. Hara, K. Itoh, M. Kamiya, H. Fujiwara, K. Okamoto, and T. Hashimoto, "Alignment of crystallites in obliquely deposited cobalt films," *Japanese Journal of Applied Physics*, vol. 33, no. 6R, p. 3448, Jun. 1994, doi: 10.1143/JJAP.33.3448.

66. G. Beydaghyan, C. Buzea, Y. Cui, C. Elliott, and K. Robbie, "Ex situ ellipsometric investigation of nanocolumns inclination angle of obliquely evaporated silicon thin films," *Applied Physics Letters*, vol. 87, no. 15, p. 153103, Oct. 2005, doi: 10.1063/1.2084329.

67. J. Lintymer, J. Gavoille, N. Martin, and J. Takadoum, "Glancing angle deposition to modify microstructure and properties of sputter deposited chromium thin films," *Surface and Coatings Technology*, vol. 174–175, pp. 316–323, Sep. 2003, doi: 10.1016/S0257-8972(03)00413-4.

68. J. M. Nieuwenhuizen and H. B. Haanstra, "Microfractography of thin films".

69. A. P. Cohn et al., "All silicon electrode photocapacitor for integrated energy storage and conversion," *Nano Letters*, vol. 15, no. 4, pp. 2727–2731, Apr. 2015, doi: 10.1021/ACS.NANOLETT.5B00563.

70. C. Portet, P. L. Taberna, P. Simon, and C. Laberty-Robert, "Modification of Al current collector surface by sol-gel deposit for carbon-carbon supercapacitor applications," *Electrochimica Acta*, vol. 49, no. 6, pp. 905–912, Mar. 2004, doi: 10.1016/J.ELECTACTA.2003.09.043.

71. P. L. Taberna, P. Simon, and J. F. Fauvarque, "Electrochemical characteristics and impedance spectroscopy studies of carbon-carbon supercapacitors," *Journal of The Electrochemical Society*, vol. 150, no. 3, p. A292, 2003, doi: 10.1149/1.1543948.

72. C. Portet, P. L. Taberna, P. Simon, and E. Flahaut, "Modification of Al current collector/active material interface for power improvement of electrochemical capacitor electrodes," *Journal of The Electrochemical Society*, vol. 153, no. 4, p. A649, 2006, doi: 10.1149/1.2168298.

73. C. Portet, G. Yushin, and Y. Gogotsi, "Effect of carbon particle size on electrochemical performance of EDLC," *Journal of the Electrochemical Society*, vol. 155, no. 7, p. A531, 2008, doi: 10.1149/1.2918304.

74. I. Kovalenko, D. G. Bucknall, and G. Yushin, "Detonation nanodiamond and onion-like-carbon-embedded polyaniline for supercapacitors," *Advanced Functional Materials*, vol. 20, no. 22, pp. 3979–3986, Nov. 2010, doi: 10.1002/ADFM.201000906.

75. L. Wei and G. Yushin, "Electrical double layer capacitors with sucrose derived carbon electrodes in ionic liquid electrolytes," *Journal of Power Sources*, vol. 196, no. 8, pp. 4072–4079, Apr. 2011, doi: 10.1016/J.JPOWSOUR.2010.12.085.

76. A. E. Fischer, K. A. Pettigrew, D. R. Rolison, R. M. Stroud, and J. W. Long, "Incorporation of homogeneous, nanoscale MnO$_2$ within ultraporous carbon structures via self-limiting electroless deposition: Implications for electrochemical capacitors," *Nano Letters*, vol. 7, no. 2, pp. 281–286, Feb. 2007, doi: 10.1021/NL062263I.

77. J. Chmiola, G. Yushin, R. Dash, and Y. Gogotsi, "Effect of pore size and surface area of carbide derived carbons on specific capacitance," *Journal of Power Sources*, vol. 158, no. 1, pp. 765–772, Jul. 2006, doi: 10.1016/J.JPOWSOUR.2005.09.008.

78. J. Chmiola, G. Yushin, Y. Gogotsi, C. Portet, P. Simon, and P. L. Taberna, "Anomalous increase in carbon at pore sizes less than 1 nanometer," *Science*, vol. 313, no. 5794, pp. 1760–1763, Sep. 2006, doi: 10.1126/SCIENCE.1132195.

79. A. Kajdos, A. Kvit, F. Jones, J. Jagiello, and G. Yushin, "Tailoring the pore alignment for rapid ion transport in microporous carbons," *Journal of the American Chemical Society*, vol. 132, no. 10, pp. 3252–3253, Mar. 2010, doi: 10.1021/JA910307X.

80. Y. Korenblit et al., "High-rate electrochemical capacitors based on ordered mesoporous silicon carbide-derived carbon," *ACS Nano*, vol. 4, no. 3, pp. 1337–1344, Mar. 2010, doi: 10.1021/NN901825Y.

81. L. Wei, M. Sevilla, A. B. Fuertes, R. Mokaya, and G. Yushin, "Hydrothermal carbonization of abundant renewable natural organic chemicals for high-performance supercapacitor electrodes," *Advanced Energy Materials*, vol. 1, no. 3, pp. 356–361, May 2011, doi: 10.1002/AENM.201100019.

82. C. Thomsen and S. Reich, "Double resonant raman scattering in graphite," *Physical Review Letters*, vol. 85, no. 24, pp. 5214–5217, 2000, doi: 10.1103/PHYSREVLETT.85.5214.

83. P. H. Tan, S. Dimovski, and Y. Gogotsi, "Raman scattering of non-planar graphite: Arched edges, polyhedral crystals, whiskers and cones," *Philosophical Transactions of the Royal Society A: Mathematical, Physical and Engineering Sciences*, vol. 362, no. 1824, pp. 2289–2310, Nov. 2004, doi: 10.1098/RSTA.2004.1442.

84. T. D. Nguyen and T. O. Do, "Solvo-hydrothermal approach for the shape-selective synthesis of vanadium oxide nanocrystals and their characterization," *Langmuir*, vol. 25, no. 9, pp. 5322–5332, May 2009, doi: 10.1021/LA804073A.

85. J. S. Anderson and A. S. Khan, "Phase equilibria in the vanadium-oxygen system," *Journal of The Less-Common Metals*, vol. 22, no. 2, pp. 209–218, 1970, doi: 10.1016/0022-5088(70)90021-4.

86. L. Hu et al., "Symmetrical MnO_2-carbon nanotube-textile nanostructures for wearable pseudocapacitors with high mass loading," *ACS Nano*, vol. 5, no. 11, pp. 8904–8913, Nov. 2011, doi: 10.1021/NN203085J.

87. M. Zhang et al., "Materials science: Strong, transparent, multifunctional, carbon nanotube sheets," *Science*, vol. 309, no. 5738, pp. 1215–1219, Aug. 2005, doi: 10.1126/SCIENCE.1115311.

88. K. Evanoff et al., "Towards ultrathick battery electrodes: Aligned carbon nanotube-enabled architecture," *Advanced Materials*, vol. 24, no. 4, pp. 533–537, Jan. 2012, doi: 10.1002/ADMA.201103044.

89. M. G. Willinger, G. Neri, E. Rauwel, A. Bonavita, G. Micali, and N. Pinna, "Vanadium oxide sensing layer grown on carbon nanotubes by a new atomic layer deposition process," *Nano Letters*, vol. 8, no. 12, pp. 4201–4204, Dec. 2008, doi: 10.1021/NL801785B.

90. K. le Van et al., "Amorphous vanadium oxide films synthesised by ALCVD for lithium rechargeable batteries," *Journal of Power Sources*, vol. 160, no. 1, pp. 592–601, Sep. 2006, doi: 10.1016/J.JPOWSOUR.2006.01.049.

91. R. N. Reddy and R. G. Reddy, "Porous structured vanadium oxide electrode material for electrochemical capacitors," *Journal of Power Sources*, vol. 156, no. 2, pp. 700–704, Jun. 2006, doi: 10.1016/J.JPOWSOUR.2005.05.071.

92. Z. J. Lao, K. Konstantinov, Y. Tournaire, S. H. Ng, G. X. Wang, and H. K. Liu, "Synthesis of vanadium pentoxide powders with enhanced surface-area for electrochemical capacitors," *Journal of Power Sources*, vol. 162, no. 2 SPEC. ISS., pp. 1451–1454, Nov. 2006, doi: 10.1016/J.JPOWSOUR.2006.07.060.

93. B. Wang, K. Konstantinov, D. Wexler, H. Liu, and G. X. Wang, "Synthesis of nanosized vanadium pentoxide/carbon composites by spray pyrolysis for electrochemical capacitor application," *Electrochimica Acta*, vol. 54, no. 5, pp. 1420–1425, Feb. 2009, doi: 10.1016/J.ELECTACTA.2008.09.028.

94. D. P. Dubal et al., "3D hierarchical assembly of ultrathin MnO_2 nanoflakes on silicon nanowires for high performance micro-supercapacitors in Li- doped ionic liquid," *Scientific Reports 2015 5:1*, vol. 5, no. 1, pp. 1–10, May 2015, doi: 10.1038/srep09771.

95. O. Ghodbane, J.-L. Pascal, and F. Favier, "Microstructural effects on charge-storage properties in MnO_2-based electrochemical supercapacitors," *ACS Applied Materials and Interfaces*, vol. 1, no. 5, pp. 1130–1139, May 2009, doi: 10.1021/AM900094E.

96. Q. Ke, M. Zheng, H. Liu, C. Guan, L. Mao, and J. Wang, "3D TiO_2@Ni(OH)$_2$ core-shell arrays with tunable nanostructure for hybrid supercapacitor application," *Scientific Reports 2015 5:1*, vol. 5, no. 1, pp. 1–11, Sep. 2015, doi: 10.1038/srep13940.

97. J. N. Barisci, G. G. Wallace, and R. H. Baughman, "Electrochemical studies of single-wall carbon nanotubes in aqueous solutions," *Journal of Electroanalytical Chemistry*, vol. 488, no. 2, pp. 92–98, Jul. 2000, doi: 10.1016/S0022-0728(00)00179-0.

98. Y. S. Jung et al., "Ultrathin direct atomic layer deposition on composite electrodes for highly durable and safe Li-Ion batteries," *Advanced Materials*, vol. 22, no. 19, pp. 2172–2176, May 2010, doi: 10.1002/ADMA.200903951.

99. S. D. Perera et al., "Vanadium oxide nanowire-carbon nanotube binder-free flexible electrodes for supercapacitors," *Advanced Energy Materials*, vol. 1, no. 5, pp. 936–945, Oct. 2011, doi: 10.1002/AENM.201100221.

100. V. Khomenko, E. Frackowiak, and F. Béguin, "Determination of the specific capacitance of conducting polymer/nanotubes composite electrodes using different cell configurations," *Electrochimica Acta*, vol. 50, no. 12, pp. 2499–2506, Apr. 2005, doi: 10.1016/J.ELECTACTA.2004.10.078.

101. D. Choi, G. E. Blomgren, and P. N. Kumta, "Fast and reversible surface redox reaction in nanocrystalline vanadium nitride supercapacitors," *Advanced Materials*, vol. 18, no. 9, pp. 1178–1182, May 2006, doi: 10.1002/ADMA.200502471.

102. I.-H. Kim, J.-H. Kim, B.-W. Cho, Y.-H. Lee, and K.-B. Kim, "Synthesis and electrochemical characterization of vanadium oxide on carbon nanotube film substrate for pseudocapacitor applications," *Journal of the Electrochemical Society*, vol. 153, no. 6, p. A989, 2006, doi: 10.1149/1.2188307.

103. P. Simon and Y. Gogotsi, "Materials for electrochemical capacitors," *Nature Materials*, vol. 7, no. 11, pp. 845–854, Nov. 2008, doi: 10.1038/NMAT2297.

104. D. R. Rolison and B. Dunn, "Electrically conductive oxide aerogels: New materials in electrochemistry," *Journal of Materials Chemistry*, vol. 11, no. 4, pp. 963–980, 2001, doi: 10.1039/B007591O.

105. D. Aurbach, K. Gamolsky, B. Markovsky, Y. Gofer, M. Schmidt, and U. Heider, "On the use of vinylene carbonate (VC) as an additive to electrolyte solutions for Li-ion batteries," *Electrochimica Acta*, vol. 47, no. 9, pp. 1423–1439, Feb. 2002, doi: 10.1016/S0013-4686(01)00858-1.

106. Z. Chen et al., "High-performance supercapacitors based on intertwined CNT/V_2O_5 nanowire nanocomposites," *Advanced Materials*, vol. 23, no. 6, pp. 791–795, Feb. 2011, doi: 10.1002/ADMA.201003658.

107. C. Hungru, J. A. Dawson, and J. H. Harding, "Effects of cationic substitution on structural defects in layered cathode materials LiNiO 2," *Journal of Materials Chemistry A*, vol. 2, no. 21, pp. 7988–7996, May 2014, doi: 10.1039/C4TA00637B.

108. H. Liu, F. Lin, J. Zhai, and L. Jiang, and D. Zhu, "Reversible wettability of a chemical vapor deposition prepared ZnO film between superhydrophobicity and superhydrophilicity," *Langmuir*, vol. 20, no. 14, pp. 5659–5661, Jul. 2004, doi: 10.1021/LA036280O.

109. F. Cheng, Y. Xin, Y. Huang, J. Chen, H. Zhou, and X. Zhang, "Enhanced electrochemical performances of 5 V spinel LiMn1.58Ni0.42O4 cathode materials by coating with $LiAlO_2$," *Journal of Power Sources*, vol. 239, pp. 181–188, Oct. 2013, doi: 10.1016/J.JPOWSOUR.2013.03.143.

110. E. L. Memarzadeh, W. Peter Kalisvaart, A. Kohandehghan, Z. Beniamin, C. M. B. Holt, and D. Mitlin, "Silicon nanowire core aluminum shell coaxial nanocomposites for lithium ion battery anodes grown with and without a TiN interlayer," *Journal of Materials Chemistry*, vol. 22, no. 14, pp. 6655–6668, Mar. 2012, doi: 10.1039/C2JM16167B.

111. L. Hu et al., "Facile synthesis of amorphous Ni(OH)2 for high-performance supercapacitors via electrochemical assembly in a reverse micelle," *Electrochimica Acta*, vol. 174, pp. 273–281, Aug. 2015, doi: 10.1016/J.ELECTACTA.2015.05.170.

112. W. Xu, S. S. S. Vegunta, and J. C. Flake, "Surface-modified silicon nanowire anodes for lithium-ion batteries," *Journal of Power Sources*, vol. 196, no. 20, pp. 8583–8589, Oct. 2011, doi: 10.1016/J.JPOWSOUR.2011.05.059.

113. C. Q. Zhang, J. P. Tu, Y. F. Yuan, X. H. Huang, X. T. Chen, and F. Mao, "Electrochemical performances of Ni-coated ZnO as an anode material for lithium-ion batteries," *Journal of The Electrochemical Society*, vol. 154, no. 2, p. A65, Dec. 2006, doi: 10.1149/1.2400609.

114. A. L. Lipson et al., "Nanoscale investigation of solid electrolyte interphase inhibition on li-ion battery MnO electrodes via atomic layer deposition of Al_2O_3," *Chemistry of Materials*, vol. 26, no. 2, pp. 935–940, Jan. 2014, doi: 10.1021/CM402451H.

115. J. Chmiola, C. Largeot, P.-L. Taberna, P. Simon, and Y. Gogotsi, "Desolvation of ions in subnanometer pores and its effect on capacitance and double-layer theory," *Angewandte Chemie International Edition*, vol. 47, no. 18, pp. 3392–3395, Apr. 2008, doi: 10.1002/ANIE.200704894.

116. B. Scrosati and J. Garche, "Lithium batteries: Status, prospects and future," *Journal of Power Sources*, vol. 195, no. 9, pp. 2419–2430, May 2010, doi: 10.1016/J.JPOWSOUR.2009.11.048.

117. V. Georgakilas et al., "Functionalization of graphene: Covalent and non-covalent approaches, derivatives and applications," *Chemical Reviews*, vol. 112, no. 11, pp. 6156–6214, Nov. 2012, doi: 10.1021/CR3000412.

118. X. Lu et al., "Improving the cycling stability of metal–nitride supercapacitor electrodes with a thin carbon shell," *Advanced Energy Materials*, vol. 4, no. 4, p. 1300994, Mar. 2014, doi: 10.1002/AENM.201300994.

119. H. Tran Nguyen, M. Robert Zamfir, L. Dinh Duong, Y. Hee Lee, P. Bondavalli, and D. Pribat, "Alumina-coated silicon-based nanowire arrays for high quality Li-ion battery anodes," *Journal of Materials Chemistry*, vol. 22, no. 47, pp. 24618–24626, Nov. 2012, doi: 10.1039/C2JM35125K.

120. P. G. Bruce, S. A. Freunberger, L. J. Hardwick, and J. M. Tarascon, "Li–O_2 and Li–S batteries with high energy storage," *Nature Mater.*, vol. 11, no. 1, pp. 19–29, 2012, doi: 10.1038/nmat3191.

121. N. Liu et al., "A pomegranate-inspired nanoscale design for large-volume-change lithium battery anodes," *Nature Nanotechnology 2014 9:3*, vol. 9, no. 3, pp. 187–192, Feb. 2014, doi: 10.1038/nnano.2014.6.

122. J. T. PD Yang, "Towards systems materials engineering," *Nat. Mater.*, vol. 11, no. 7, pp. 560–563, 2012, doi: 10.1038/nmat3367.

123. Z. Wei Seh et al., "Sulphur–TiO_2 yolk–shell nanoarchitecture with internal void space for long-cycle lithium–sulphur batteries," *Nature Communications 2012 4:1*, vol. 4, no. 1, pp. 1–6, Jan. 2013, doi: 10.1038/ncomms2327.

124. P. G. Bruce, S. A. Freunberger, L. J. Hardwick, and J. M. Tarascon, "Li-O$_2$ and Li-S batteries with high energy storage," *Nat. Mater.*, vol. 11, no. 1, pp. 19–29, 2012, doi: 10.1038/nmat3191.

125. A. S. Aricò, P. Bruce, B. Scrosati, J.-M. Tarascon, and W. van Schalkwijk, "Nanostructured materials for advanced energy conversion and storage devices," *Nat. Mater.*, vol. 4, no. 5, pp. 366–377, May 2005, doi: 10.1038/nmat1368.

126. "Interconnected hollow carbon nanospheres for stable lithium metal anodes," *nature.com*, Accessed: Aug. 21, 2021. Available: https://idp.nature.com/authorize/casa?redirect_uri=https://www.nature.com/articles/nnano.2014.152&casa_token=CHTj2W9iuwoAAAAA:KLGctQAtdIDYPC2A-Cet3T5Zi4cx3ppfumNGss2vRrjD6_RXXSOUz3N2-YLF4ji4EMnkOmin5oKVGyCrig

127. E. Kazyak, K. N. Wood, and N. P. Dasgupta, "Improved cycle life and stability of lithium metal anodes through ultrathin atomic layer deposition surface treatments," *Chemistry of Materials*, vol. 27, no. 18, pp. 6457–6462, Sep. 2015, doi: 10.1021/ACS.CHEMMATER.5B02789.

Chapter 5

Laser surface modification of metal additive manufactured parts

A case study of ex-situ and in-situ methodology

Abhishek Kumar, Bijaya Bikram Samal,
Ashish Kumar Nath, and Cheruvu Siva Kumar
Indian Institute of Technology Kharagpur, Kharagpur, West Bengal, India

CONTENTS

DOI: 10.1201/9781003319375-5

5.1 INTRODUCTION TO ADDITIVE MANUFACTURING

From time immemorial, humans have strived hard to progress as much as possible with the help of science, engineering, and technology. The progress of human civilization demanded the fabrication of new tools, resulting in the manufacturing processes development. In recent times, the demands for customization, complex designs, and superior mechanical properties have driven additive manufacturing (AM) [1]. Additive manufacturing technology produces the required form by layering material, ideally by stacking contoured layers on top of each other. As a result, it is also known as layer (or layered) technology. The layer technology concept is founded on the notion that every object, theoretically, may be divided into layers and reconstructed using these levels, regardless of its geometry's complexity. AM is a layer-based automated fabrication technique [2]. AM combines two major subprocesses: the actual fabrication of each layer and the sequential combination of successive layers to create the component. Both procedures take place at the same time [3]. The 3D data of the part, also known as the virtual product model, is required for the AM construction process. It is a feature of additive manufacturing that the geometry and material characteristics of the component are created during the build process [4]. Additive manufacturing is the umbrella term for all manufacturing methods that automatically generate and connect volume pieces known as voxels. The volume components are typically even-thickness layers. As a result, additive manufacturing/3D printing directly converts 3D computer-aided design (CAD) data (the virtual product model) into a physical or actual component. Because scaling is simple in the CAD file, parts of various sizes and materials may be produced from the same data set.

The phrase additive manufacturing refers to a variety of methods, including 3D printing, rapid prototyping (RP), direct digital manufacturing (DDM), layered manufacturing, and additive fabrication. The possibilities for using AM are endless. Early applications of AM in the form of RP focused on preproduction visualization models. Recently, AM has been utilized to create end-use goods like airplane parts, dental restorations, medical implants, food items, car components, and fashion items [5]. While the layer-upon-layer method is primary, there are numerous uses of additive manufacturing technology with varying grades of complexity to suit a variety of requirements, such as:

- A design visualization tool that enables to build highly personalized goods for both consumers and experts
- Industrial tooling
- To fabricate small quantities of manufacturing parts
- Design-based robust structures
- Small batch production of customized products

5.2 DIFFERENT TYPES OF ADDITIVE MANUFACTURING PROCESSES

According to the ASTM Committee F42, AM process can be classified into seven types [6]. They are as follows.

5.2.1 Binder jetting

The terms inkjet powder printer and 3D printing are also known for binding jetting. Under the term "3D printing", MIT has designated this particular technique, subsequently often used to characterize most additive production technologies against binder jetting. Binder jetting technique has specific characteristics with document printers, which led to the term 3D printing. The jetting process is the most remarkable of these parallels. When a document printer discharges ink selectively on a page, the binder throws adhesive into a powder bed selectively. A jet binder may hold up to five print heads with up to 300 jets per print head. Like any additive manufacturing process, the digital geometric model translation to finite-height layers begins. The binder jetting binds one powder layer at a time, and each layer is above the previous one. The final product is a power cube with lots of solid components within. To eliminate the remaining powder, these pieces must be dug from the powder cube and then treated through an air compressor. During certain instances, the post-processing additionally uses heat treatment. Any unused powder may be again utilized. The most prevalent substance is a powder based on gypsum. There are several materials, from plastic to metal [7].

5.2.2 Directed energy deposition

The welding and cladding process is modified to construct a part layer-wise using metal powders/rods/filaments, known as Directed Energy Deposition (DED). A melting pool is produced when concentrated energy in the form of laser, electron, or plasma melts the input feedstock provided. The feedstock can be metal powder, metal wires, or rods. When the melting pool is moved (guided), the molten material left over quickly solidifies as a supplementary substance. Arc, laser, and electron beams are the most popular energy sources. In contrast to many other additive technologies, DED typically adds material to components that already exist, preforms, etc. DED has had less frequent usage than some other additive methods, but in the future, new hybrid methods merging them with CNC machining guarantee that this technology will be used more widely [7].

5.2.3 Material extrusion

FFF utilizes a polymer material as input. The feedstock can be of plastic pallets, polymer filaments, metal wires, composites, etc. Considering the prominently used form, i.e., filament, it is supplied on the expansion valve that regulates the flow. A numerical control process controlled directly by CAM software directs it to heat the nozzle to melt the material and, therefore, can move horizontally and vertically. Due to the force provided by the extruder motor, the molten polymeric material extruded/comes out of the nozzle. The removed thermoplastic material falls onto a heated bed/ platform in layers as the material hardens immediately due to cooling by the envelope temperature. For moving the extrusion head, stepper or servo motors are typically utilized. The filament for FFF is made of polymers such as poly-lactic acid (PLA), acrylonitrile butadiene styrene (ABS), thermoplastic polyurethane (TPU), polycarbonate (PC), and polymer composites. In most situations, the polymer is formed of virgin resin in the form of a filament. There are several open-source efforts aimed at turning plastic waste into filaments after use.

5.2.4 Material jetting

Material jetting additive manufacturing is inspired by the inkjet printing of papers. It consists of several heads which are used to deposit the material in the form of liquid selectively on the two planar axes and then on the vertical Z-axis, to create the part layer-wise. Typically, post-processing is needed, which involves removing the support material using a jet of water. A liquid photopolymer is the most often utilized substance in material jetting. This substance will stay as a liquid until exposed to ultraviolet radiation, upon which there will be solidification. UV lamps are usually connected to each side/portion of the print head to ensure that the material solidifies immediately after it is put down. Material jetting's primary benefit above other techniques is its capability to distribute selectively and combine components. However, the vat in the photopolymerization process only contains single material. The material jetting may selectively put various materials in various model areas and mix these photopolymers to produce composites. Material properties vary from stiff to flexible, from high temperature to biocompatibility [7].

5.2.5 Powder bed fusion

One of the most promising technology for both metals and polymers is the selective fusion of powders by an energy source in a bed. Basically, there are three main stages in this process, namely, the built platform or bed, the powder dispenser platform, and the powder collector platform. The process starts by layering the powder finely on the built platform by a recoater. This is done by moving the dispenser platform up. Once the powder layer is layered up, the laser/electron beam fuses the powder in a predefined path provided by the CAD geometry. Then, the recoater moves again to dispense a fresh layer of powder. The remaining powder gets collected in the collector platform. This process continues to provide an excellent dimensional accurate part. Once the process is completed, the part is removed from the powders, and the excess powder is used again. This process uses an unfused powder medium to provide support during the printing, due to which there is a low need for support, and overhang structures can be prepared with ease. Different types of technologies used are selective laser sintering (SLS), selective laser melting (SLM), electron beam melting (EBM), etc. SLM utilizes a laser to directly blend the metal powder to form a very dense component with improved mechanical properties almost equivalent to conventional metals. SLS uses the laser to sinter the powder material creating a melt pool. EBM produces the part in a very high vacuum using electron beams. Due to the interaction of the electron beam, the powder completely melts and then solidifies. EBM manufactured components are thick, non-porous, and have high strength. A wide range of materials can be processed using this technology, such as titanium, stainless steel, copper, nickel-based superalloys, PEEK, PA, and elastomers [7].

5.2.6 Sheet lamination

Layers of adhesive-coated paper, plastic, or metal laminates are bonded together and shaped using a conventional cutter or laser. Objects produced using this method may be further changed after printing by machining or drilling. The material feedstock determines the typical layer resolution for this process, which varies in thickness from one to a few sheets of copy paper.

5.2.7 Vat photopolymerization

Stereolithography (SLA) primarily uses photopolymerization to produce a substantial part of a liquid. It involves the vat of photopolymers onto which there is a printing stage for going down in the Z-axis. The X and Y pattern is made by the interaction of laser on the photopolymer, due to which crosslinking takes place and the liquid solidifies into solid. Similarly, this process is repeated to make the part layer by layer. In another method, the digital light processing (DLP) projector is exposed to a liquid polymer vat. Due to the exposure of the liquid polymer resin, polymerization takes place, and thus it hardens. Once a single layer is exposed and solidifies, the built platform moves down in Z-axis. The movement of the platform is made as per the predefined sliced thickness. Due to this movement, fresh polymer liquid is available at the top for exposure again. This complete process is repeated until the entire component is fabricated. Once the process is completed, the polymer liquid is drained out to get the manufactured part.

5.3 POST-PROCESSING IN METAL AM

Post-processing activities are typically needed after manufacture using AM methods to create the proper functionality and shape. Post-processing methods are utilized to improve AM components manufactured to overcome additive manufacturing restrictions [8]. This includes the elimination of support, improvements in the texture of the surface, cosmetic enhancements, and enhancements in properties via thermal methods. Post-processing plays a significant part in price marketing, and competitive businesses may thus ignore post-processing to cut expenses [9].

5.3.1 Support removal

Support removal is a crucial step in AM methods. In general, AM components are supported by two types: natural supports (i.e., materials that surround the part throughout the manufacturing process) and synthetic supports (i.e., extra rigid structures that connect the central part to the substrate to support and restrain the main part) [10]. In powder bed fusion (PBF), AM methods involve separating the component from the surrounding powder. The manufactured member must be allowed to cool in PBF methods before removing the powders; otherwise, the non-uniform cooling causes structural deformation. The time it takes to cool down depends on the material and the size of the component. There are various ways to remove the parts from the surrounding powder after cooling down, including brushes, compressed air, mild bead blasting, and dental cleaning equipment. Not all AM processes need synthetic support, although nearly all PBF AM techniques need such supports. These supports may be created from either the material constructed or a supplementary substance [11]. These materials are typically manufactured using lower processing settings. They are still too powerful to be manually removed. Hammering, strip sawing, blade cutting, EDM wire, and other metal cutting techniques are thus widely utilized.

5.3.2 Surface texture improvements

AM components may have unwanted surface texture characteristics to remove. These include stair steps, adherence of powder, fill patterns from the constructed material, and signs of support removal. To flatten the stair-stepping problem, a thin layer thickness on the component may be considered to avoid this mistake during fabrication time.

5.3.3 Aesthetic improvements

For aesthetic, creative, or marketing use of AM components, the aesthetic of the finished part may be regarded as a critical element to be fully appreciated. In most instances, improvements in the surface texture are sufficient to offer aesthetics of the final product. Some situations may need the component to be dipped into suitable color containers. This method works because of the inherent porosity in PBF AM techniques due to the accompanying high absorption. Before painting, the component may need to be sealed. The manufactured AM component may also be plugged in with metals such as chromium and nickel, strengthening the part and improving wear resistance.

5.3.4 Property enhancements

Thermal treatments improve the finished AM component's characteristics. Traditional thermal treatment is a typical thermal method to alleviate residual stresses and to develop the required microstructure. Recent methods for heat treatment have also been used to maintain the fine grain microstructure in AM components, alleviate stress, and improve ductility. Before introducing AM methods capable of direct metal production, many approaches for producing metal green components using AM were devised. Furnace post-processing was thus necessary to produce dense, useable metal components. Due to many process factors that need to be adjusted and numerous stages, controls over shrinkage and dimensional accuracy in post-processing furnaces are complicated. A variety of additional methods were developed over the years, and conventional heat treatment so combines AM with furnace processing to create metal components [12]. One of them is known as SLS/HIP. As stated previously, the laser only scans outside the outlines of the element throughout the SLS process and finally produces a metal "can" filled with loose powder. These components are subsequently treated with HIP to their total density. The SLS/HIP method has been effective for manufacturing complicated 3D components for aerospace applications in Inconel 625 and Ti–6Al–4V.

5.4 LASER SURFACE FINISHING FOR METAL ADDITIVE MANUFACTURED PARTS

Grinding and polishing methods are often employed to decrease surface roughness. Polishing using laser radiation is a novel technique for achieving such high-quality surfaces. There are three distinct process variations in concept.

5.5 TYPES OF LASER SURFACE MODIFICATION/POLISHING

There are mainly three kinds of laser polishing processes as shown in Figure 5.1. They are as follows.

5.5.1 Surface modification/polishing by the mechanism of large area ablation

Material is flattened over the whole surface. This results in greater ablation of the surface peaks and reduced depression in the valleys [13]. It is mainly utilized in CVD diamond films and plates. Lasers (excimer) are utilised (ArF, KrF, XeCl). Laser polishing is done at an angle of incidence of up to 85° as normal as the target surface to guarantee enhanced material removal at the profile peaks. By turning the sample during processing, an additional decrease in roughness may be obtained. The processing duration ranges from a few minutes to many hours a square centimeter depending on the laser source, one or two processing stages, and beginning roughness.

5.5.2 Surface modification/polishing by localized ablation

Localized ablation polishing is based on a controlled removal of profile peaks using pulsed laser radiation [14]. A sophisticated and expensive profile

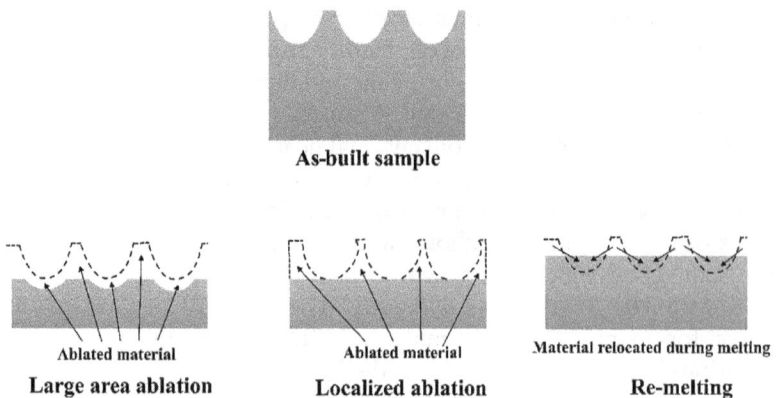

As-built sample

Ablated material
Large area ablation

Ablated material
Localized ablation

Material relocated during melting
Re-melting

Figure 5.1 Various types of laser surface polishing.

measuring equipment is needed to identify the location of the profile peaks. It needs an accurate measuring instrument to measure the initial surface profile. After comparing the nominal/actual value, only the profile peaks are linked to a regulated laser pattern [15].

5.5.3 Surface modification/polishing by re-melting

A thin layer of surface melts, and the surface tension leads to a material flow from the peaks into the valleys. No material is eliminated but relocated throughout the molten process. It is a novel way to automatically polish 3D surfaces. There are two types of polishing by re-melting, depending upon the process variation, i.e., macropolishing and micropolishing [16].

5.5.3.1 Macropolishing

It is performed using continuous laser wave (CW) radiation. Depending on the material and surface roughness initially, the beam width and re-melting depth need to be determined. Fiber-coupled Nd: YAG lasers usually are utilized. The roughness achieved relies on many contributing variables:

- Initial surface roughness and, in particular, lateral surface structural dimensions
- Thermophysical properties include heat conductivity and capacity, absorption coefficient, viscosity, surface tension, temperature melting, and evaporation
- Homogeneity of material: segregations, including decreasing the quality of the surface
- The tiny grain is recommended to have a medium grain size and a statistic distribution

5.5.3.2 Micropolishing

Micropolishing using pulsed laser radiation can be achieved. The length of the pulse is usually between 20 and 1000 ns, and the re-melting depth is 0.5–5 μm. Only fine pre-produced surfaces (e.g., micro-milled and grinding) may be polished using a micropolishing version. Due to the limited depth of re-melting, more prominent surface structures are unimpacted and thus cannot be removed. Pulse length and intensity are the most critical process factors. Longer pulses may remove more significant surface features laterally. The intensity must be selected based on the pulse duration and the polished material. To provide a homogeneous re-melting depth, a top-hat intensity distribution is ideal. The Nd: YAG and Excimer lasers are generally used for micropolishing. Micropolishing is perfect for tribological and medicinal applications due to its tiny microroughness. The targeted polishing and modification of the gloss level also produce design surfaces [14]. The micropolishing technique frequently results in more acceptable microroughness and greater gloss levels. As a result, a mixture of the two types is utilized for specific applications: first macropolishing to remove tracks from milling or to turn, then micropolishing to increase the gloss level. However, most research in laser polishing focuses mainly on re-melting mechanisms because of many benefits over ablation-based techniques, including a more extensive scope for automation, shorter processing times, decreased environmental impacts, improved surface roughness control, and localized processing skills.

5.6 MECHANISM OF LASER POLISHING: THE STATE OF THE ART

The polishing of additive manufactured parts/components using lasers is an established method working on the process of fusion taking place in the layer of material. The layer being in the order of a few micrometers gets modified due to the melting as a result of heat from a high-intensity laser. Laser polishing is sometimes referred to as laser surface re-melting. It involves melting a thin layer of the material, which allows the material to flow from peaks into valleys because of the surface tension. Laser polishing is a process in which the material being worked on is not removed; instead, it is repositioned as a molten pool. To fuse a tiny layer of substrate rapidly, a laser beam of appropriate energy density is used. The surface roughness imperfections are bridged by this [17]. Because of the fluidity of the molten metal, it easily flows into the surrounding valley. Once it is done, the molten metal re-solidifies and enables the solidified layer to be attached to the surface of the substrate or the part. This, therefore, lowers the asperity peaks and valleys, which makes the surface smoother [18]. As the surface roughness value of additive manufacturing components is one of the most critical constraints, many studies try to reduce their roughness. Various beam power and scan speed factors are studied to see how laser polishing (or laser surface re-melting) capabilities are impacted [19].

Researchers [20] provided thorough research on CO_2 laser finishing. They carried out operations with laser power changes but a constant scanning speed of 3000 mm/min in different alloys. When the beam intensity was too high, and the focus distance was too short, the surface quality was found to have deteriorated. The performance of laser polishing was also shown to improve with the initial surface roughness. The ability to polish a chosen small patch (0.1 mm²) on its substrate area SLP (selective laser polishing) was achieved by another group [21]. Besides enhancing the surface quality, SLP was demonstrated to generate a homogeneous surface (uniform alloy element composition) and decrease surface roughness by up to 50%. It is needed to have an appropriate selection of process parameters

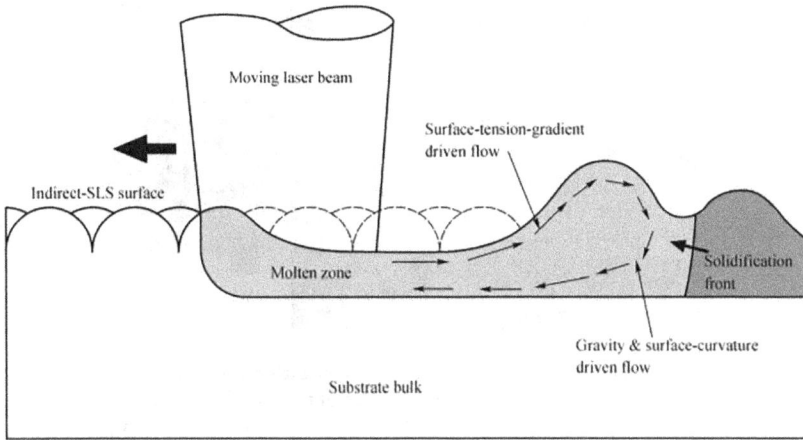

Figure 5.2 Laser polishing mechanisms by the formation of a periodic surface structure during surface over-melting [23].

like laser power to get an aerial surface roughness of the polished surface, more minor than the obtained. Another group of researchers [18] reduced the surface roughness (Ra) from 7.5 μm to 1.2–1.3 μm of the AISI 420 stainless steel sintered with a bronze substrate. Another group [22] studied the potential for the surface quality of titanium alloys to enhance a nanosecond pulsed fiber laser was also examined. At 1.21 W/cm², the laser power density was constant, and the scan speed was 200 mm/s. The surface roughness was reduced from 5 mm to 1 mm, but along with it, there was an enhancement of microhardness and wear resistance. Figure 5.2 shows laser polishing mechanisms: (a) formation of a periodic surface structure during surface over melting, (b) rough/coarse laser polishing based on the surface over melting, and (c) gentle laser polishing based on shallow surface melting [23].

5.7 FIRST CASE STUDY: EX-SITU LASER SURFACE RE-MELTING OF METAL ADDITIVE MANUFACTURED PARTS

In this type, Post-processing is done in different machines/systems. It is most commonly used due to flexibility in choosing the kind of processes and process parameters in different ranges. Limitations of the process depend upon the restriction of the individual machine selected with an increase in manufacturing lead time compared to in-situ re-melting due to changes in the machine and job loading/unloading process. Figure 5.3 shows (a) 3-axis CNC-based laser surface re-melting setup and (b) ex-situ laser polished samples.

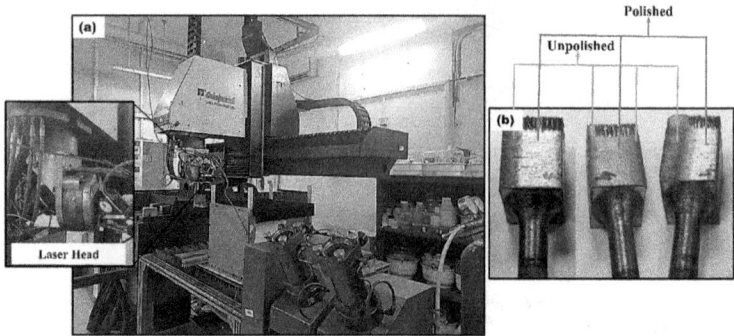

Figure 5.3 (a) 3-axis CNC-based laser surface re-melting setup and (b) ex-situ laser polished samples. (Copyright and Courtesy: LASER Lab, Department of Mechanical Engineering, IIT Kharagpur, West Bengal, India.)

5.7.1 Materials and methods

The material used for the ex-situ laser re-melting experimentation is pre-alloy stainless steel "EOS Stainless Steel PH1". The chemical compositions of the proprietary material include "DIN 1.4540" and "UNS S15500". It possesses excellent resistance to corrosion and high mechanical properties, mainly if it is precipitation strengthened. It is widely used with superior hardness, strength, and corrosion resistance in a broad spectrum of medical, aeronautical, and other technological applications. The material is also suited for partial construction applications, including prototypes for functional metal, small series items, customized products, and replacement components. The standard process parameters were used, with the complete geometry melting and a 20 mm layer thickness which in all directions were relatively uniform. After that, surface roughness using a "Taylor and Hobson" profilometer was assessed in two orthogonal directions. Data was gathered perpendicular to the laser beam's scanning direction in both the longitudinal and transverse orientations (the laser beam scanning path).

5.7.2 Results and discussions

For various samples, the initial average sample surface roughness was varied. The decrease in surface roughness together with the ultimate surface roughness must thus be taken into consideration. The best surface finish (i.e., the lowest surface roughness) can be obtained from the sample following a variety of process parameters (laser power = 1800 W; scanning speed of 2250 mm in one step; laser power density is 179.1 W/mm^2, and line power is 2.4 J/mm) after laser area recasting by various process parameters, such as laser power, scanning speed and number of passes. Table 5.1 shows different input process parameters and their corresponding percentage reduction in surface roughness.

When starting roughness value was 11.6 μm, the lowest attainable finishing is 1.0 μm. The laser-polished surface finishing roughness relies on the original sample roughness. The greater the roughness at first, the better

Table 5.1 Different input process parameters and their corresponding percentage reduction in surface roughness

Sl. No.	Laser power (watt)	Laser scan speed (mm/s)	No. of pass	Laser power density (W/mm²)	Laser line energy (J/mm)	Percentage reduction in roughness (%)
1	1000	1250	1	99.5	2.4	81.6
2	1400	1000	1	139.3	4.2	83.78
3	1400	1250	1	99.5	3.0	86.48
4	1400	500	1	139.3	8.4	87.38
5	1400	750	1	139.3	5.6	90
6	1500	1875	1	149.2	2.4	66.96
7	1800	2250	1	179.0	2.4	88.18
8	1800	2250	2	179.0	2.4	91.05
9	2000	1000	1	198.9	6.0	88.11
10	2000	2000	1	198.9	3.0	83.78
11	2000	2500	1	198.9	2.4	84.53
12	2000	3000	1	198.9	2.0	85.57
13	2150	3000	1	213.9	2.1	84.35
14	2150	3000	2	213.9	2.1	86.61

the laser's capacity to reduce roughness. Comparing the results obtained, the lowest surface roughness, i.e., best surface finish, is attained at a laser power of about 1800 W, a scan speed of 2250 mm/min, laser power density is 213.9 W/mm², and line energy is 2.15 J/mm for a single pass. Maximum surface roughness is achieved in a single pass at an output of 2150 W; scan speeds of 3000 mm/min, laser density of 213.9 W/mm², and power lines of 2.15 J/mm. The roughness of the original area was 14.7 μm, while the polished roughness of the surface was 2.3 μm (Figures 5.4–5.6).

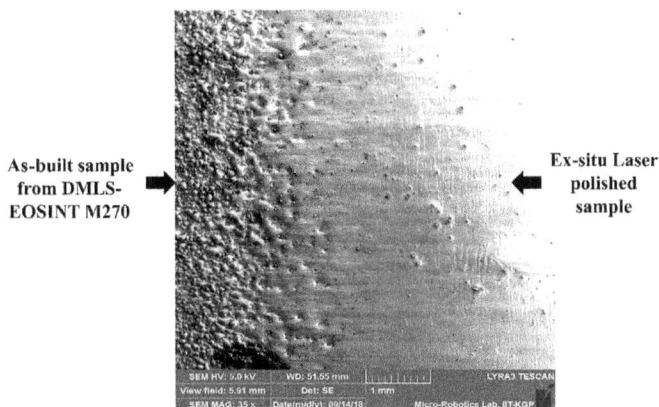

As-built sample from DMLS-EOSINT M270

Ex-situ Laser polished sample

Figure 5.4 SEM micrograph of as-built and ex-situ laser polished (re-melted) sample.

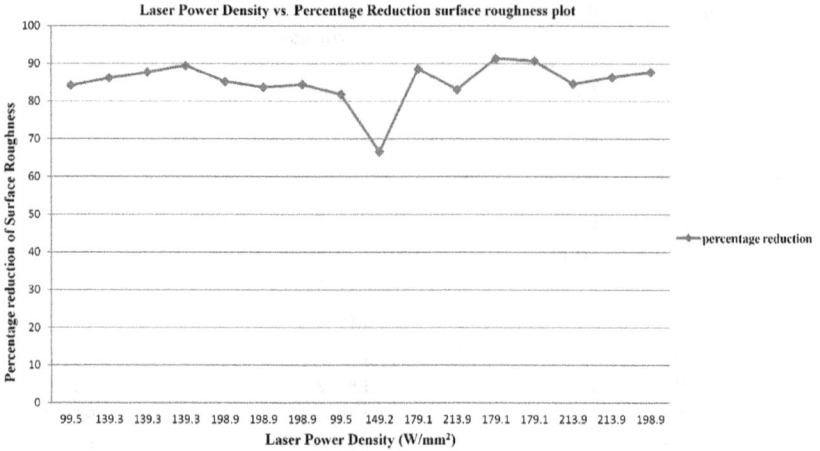

Figure 5.5 Graph showing the percent change in surface roughness versus laser power density.

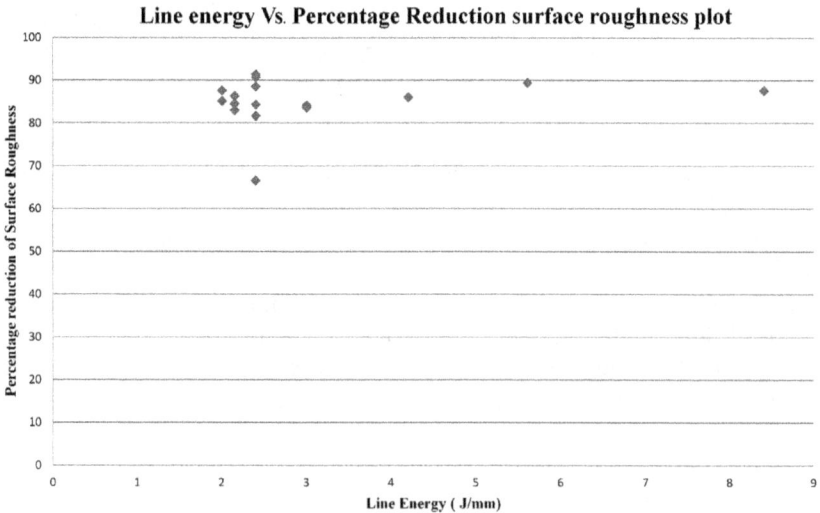

Figure 5.6 Graph showing the percent change in surface roughness versus line energy.

5.8 SECOND CASE STUDY: IN-SITU LASER SURFACE RE-MELTING BY DMLS PROCESS

In situ, laser surface polishing is a method of polishing the parts during the additive manufacturing process. In this type of process, in the same machine with different process parameters. This ultimately results in fast processing time than that of ex-situ polishing methods. Limitation includes a narrow range of process parameters that is available for that machine (Figure 5.7).

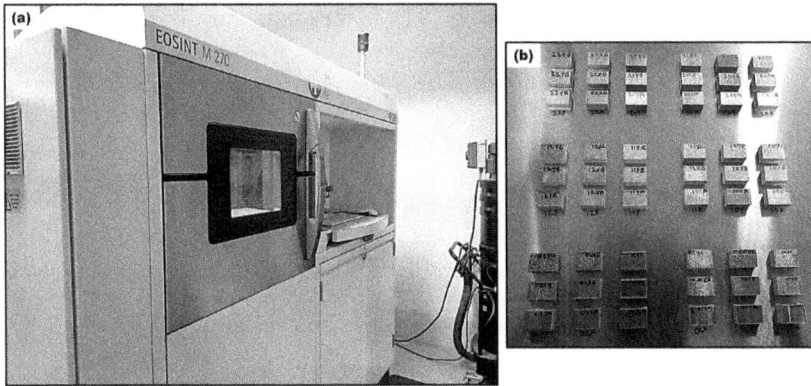

Figure 5.7 (a) DMLS-based EOSINT M270 machine (IIT Kharagpur) and (b) built specimen. (Copyright and Courtesy: Direct Digital Manufacturing Laboratory, Department of Mechanical Engineering, IIT Kharagpur, West Bengal, India.)

5.8.1 Materials and methods

Cobalt–chromium (CoCr) alloy is taken as material for this study. It is popularly used for fabricating surgical and dental implants due to its high weight-to-strength, bio-compatibility, and toxicity.

In this experiment, laser surface re-melting of CoCr samples was performed in the same machine where the sample was built. For this process, the bulk part of the sample was built using a standard EOS process parameter. On the top surface, process parameters (i) scan speed, (ii) hatching distance, and (iii) hatching direction were varied. Initial process parameters were first set before doing the experiment in the software and then, the final samples were built using EOS M 270 machine up to 5 mm in height. Three replicates were built with sample sets (A, B, C) for each process parameter. For the final layer, sample sets A, B, and C were given different process conditions. In the first set of the sample (A), a single exposure (no powder layer recoated at the last layer) was given. In the second set (B), only one layer thickness of 80 microns of powder was spread by the recoater blade, and then single exposure of the specified process parameter was given. In the third sample (C), two layers, each having a thickness of 80 microns of powder, were spread by the recoater blade, and then single exposure of the specified process parameter was given.

5.8.2 Results and discussions

After experimenting with different process parameters as discussed above, output was taken as surface roughness analysis. In surface roughness analysis, average surface roughness (Ra) in both longitudinal and transverse directions is considered for analysis. The longitudinal direction indicates that the surface roughness is measured perpendicular to the scan direction of the laser beam that is the breadth of the sample. The transverse direction means that surface roughness is measured along the path of the scan direction of the laser beam that is the length of the sample.

Built sample process parameters and output surface roughness are displayed in the following observation table.

Calculation of energy density can be done by the following equation:

$$\text{Energy Density} = \frac{\text{Laser Power}}{\text{Scan Speed} \times \text{Hatch Distance} \times \text{Layer Thickness}} \quad (5.1)$$

Thus, the lowest surface finish in the longitudinal direction is obtained as 1.02 microns at a scan speed of 800 and a hatch distance of 0.04 mm. The lowest surface finish in the transverse direction is 1.02 microns at a scan speed of 800 and a hatch distance of 0.08 mm. There was a 94% reduction in surface roughness by laser surface re-melting carried out in the same machine (EOS M 270) in which the sample was built. The minimum average surface roughness obtained by measuring it with Taylor and Hobson contact profilometer was 1.43 microns. The average surface roughness (A, B, C) with different scan strategies are as given in Figures 5.8–5.11 and Table 5.2.

Figure 5.8 SEM micrographs: (a) as built sample (b) in-situ laser polished (re-melted) sample.

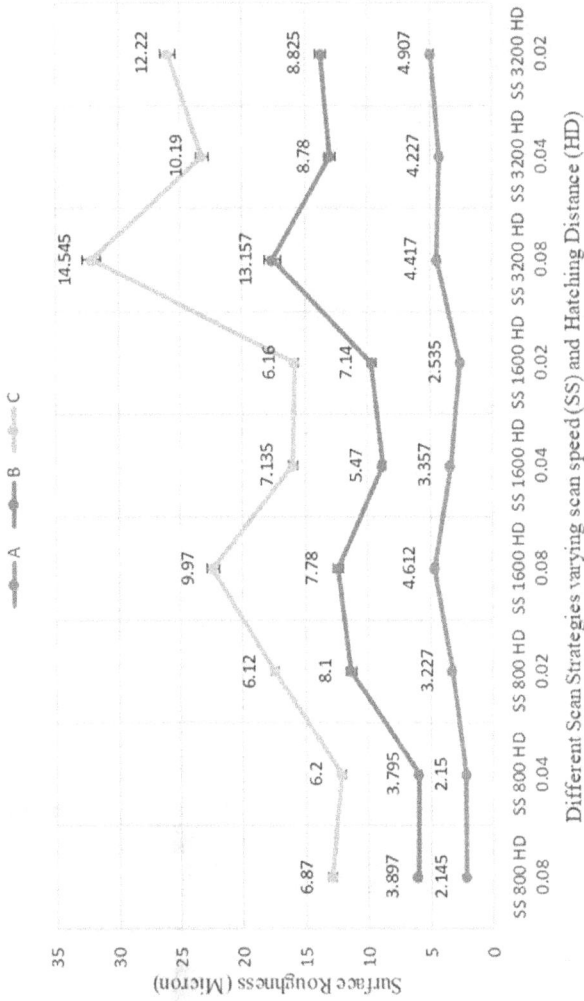

Figure 5.9 Graph for three different process variations and their corresponding surface roughnesses with change in various scan strategies (i.e., scan speed and hatch distance).

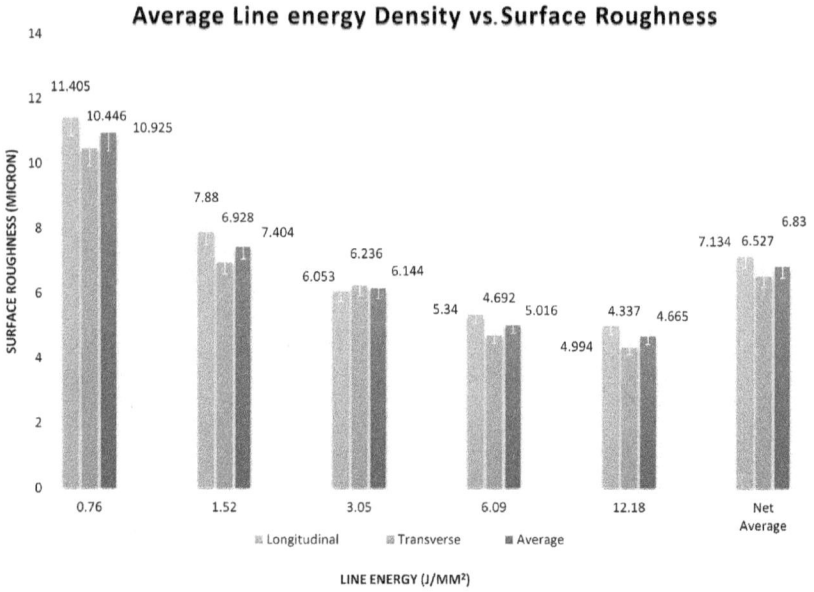

Figure 5.10 Graph showing the change in surface roughness in two directions and its average for different values of average line energy.

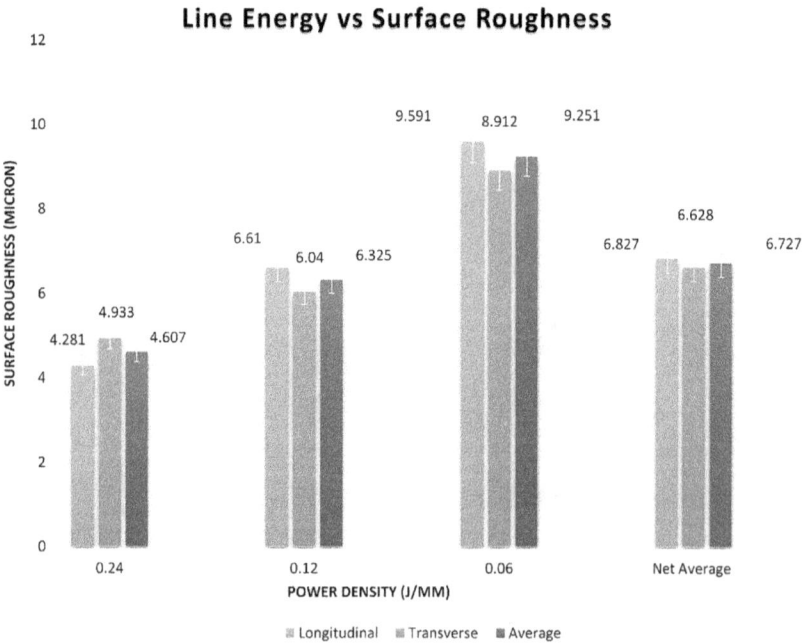

Figure 5.11 Graph showing the change in surface roughness in two directions and its average with the variation in power density.

Table 5.2 Observation table showing different process parameters and the corresponding surface roughness values in two directions as well as its average

Sr no.	Laser power (watt)	Scan speed (mm/s)	Hatch distance (mm)	Hatch direction	Sample set	Energy density (J/mm³)	Surface roughness, longitudinal (micron)	Surface roughness, transverse (micron)	Average surface roughness (micron)
1	195	800	0.08	X	A	38.08	2.14	1.03	1.585
2	195	800	0.08	X	B	38.08	2.91	1.08	1.995
3	195	800	0.08	X	C	38.08	3.27	1.69	2.48
4	195	800	0.08	Y	A	38.08	1.83	3.58	2.705
5	195	800	0.08	Y	B	38.08	5.27	6.33	5.8
6	195	800	0.08	Y	C	38.08	7.39	12.05	9.72
7	195	800	0.04	X	A	76.17	3.73	2.01	2.87
8	195	800	0.04	X	B	76.17	6.52	4.33	5.425
9	195	800	0.04	X	C	76.17	10.83	7.47	9.15
10	195	800	0.04	Y	A	76.17	1.02	1.84	1.43
11	195	800	0.04	Y	B	76.17	1.50	2.83	2.165
12	195	800	0.04	Y	C	76.17	1.41	4.93	3.17
13	195	800	0.02	X	A	152.34	2.01	1.48	1.745
14	195	800	0.02	X	B	152.34	4.08	4.07	4.075
15	195	800	0.02	X	C	152.34	8.24	6.12	7.18
16	195	800	0.02	Y	A	152.34	3.74	5.68	4.71

(Continued)

Table 5.2 (Continued) Observation table showing different process parameters and the corresponding surface roughness values in two directions as well as its average

Sr no.	Laser power (watt)	Scan speed (mm/s)	Hatch distance (mm)	Hatch direction	Sample set	Energy density (J/mm³)	Surface roughness, longitudinal (micron)	Surface roughness, transverse (micron)	Average surface roughness (micron)
17	195	800	0.02	Y	B	152.34	6.90	17.35	12.125
18	195	800	0.02	Y	C	152.34	—	—	—
19	195	1600	0.08	X	A	19.04	7.35	3.33	5.34
20	195	1600	0.08	X	B	19.04	9.02	5.22	7.12
21	195	1600	0.08	X	C	19.04	8.80	7.64	8.22
22	195	1600	0.08	Y	A	19.04	3.58	4.19	3.885
23	195	1600	0.08	Y	B	19.04	6.96	9.92	8.44
24	195	1600	0.08	Y	C	19.04	13.33	12.30	12.815
25	195	1600	0.04	X	A	38.08	2.35	1.81	2.08
26	195	1600	0.04	X	B	38.08	5.81	4.14	4.975
27	195	1600	0.04	X	C	38.08	10.30	6.96	8.63
28	195	1600	0.04	Y	A	38.08	4.14	5.13	4.635
29	195	1600	0.04	Y	B	38.08	4.02	7.91	5.965
30	195	1600	0.04	Y	C	38.08	4.26	7.31	5.785
31	195	1600	0.02	X	A	76.17	3.07	2.90	2.985
32	195	1600	0.02	X	B	76.17	11.63	12.51	12.07
33	195	1600	0.02	X	C	76.17	15.22	5.85	10.535
34	195	1600	0.02	Y	A	76.17	2.15	2.02	2.085
35	195	1600	0.02	Y	B	76.17	1.27	3.15	2.21
36	195	1600	0.02	Y	C	76.17	5.73	6.47	6.10

									6.1
37	195	3200	0.08	X	A	9.52	5.28	4.03	4.655
38	195	3200	0.08	X	B	9.52	11.63	9.24	10.435
39	195	3200	0.08	X	C	9.52	14.02	12.56	13.29
40	195	3200	0.08	Y	A	9.52	3.20	5.16	4.18
41	195	3200	0.08	Y	B	9.52	16.60	15.16	15.88
42	195	3200	0.08	Y	C	9.52	17.70	16.53	17.115
43	195	3200	0.04	X	A	19.04	5.86	3.54	4.7
44	195	3200	0.04	X	B	19.04	9.22	8.34	8.78
45	195	3200	0.04	X	C	19.04	11.78	10.19	10.985
46	195	3200	0.04	Y	A	19.04	2.90	4.61	3.755
47	195	3200	0.04	Y	B	19.04	—	—	—
48	195	3200	0.04	Y	C	19.04	—	—	—
49	195	3200	0.02	X	A	38.08	4.11	5.10	4.605
50	195	3200	0.02	X	B	38.08	8.58	8.34	8.46
51	195	3200	0.02	X	C	38.08	14.50	12.09	13.295
52	195	3200	0.02	Y	A	38.08	4.82	5.60	5.21
53	195	3200	0.02	Y	B	38.08	8.62	9.76	9.19
54	195	3200	0.02	Y	C	38.08	14.64	12.35	13.495

5.9 CONCLUSION

Ex-situ laser polishing by ablation requires additional processing. It increases cost, but the improvement of surface quality is relatively tremendous, and it can be utilized for manufacturing parts that need very high surface quality. The effect of process parameters on the surface roughness was studied, and it was found that the higher the initial roughness, the smoother the surface can be generated by re-melting. Among all, the best sample whose roughness was reduced is 85% better than the as-built part.

In-situ laser surface re-melting can be very beneficial for improving the surface roughness during the printing process hence reducing the requirement for expensive post-processing and reducing the manufacturing time. It can be achieved by wisely choosing the optimum process parameters and modifying the laser interaction on the surface of the additively produced part. It can be concluded that by increasing scan speed, surface finish rises in general. Increasing the hatching distance, initially surface roughness decreases but then improves.

Since the surface roughness of the additively produced components is a significant constraint, this chapter attempts to provide a detailed study of the laser surface re-melting methodology and the polishing mechanism in great detail. This chapter also provided two critical case studies to provide insight into how process parameters can affect the surface quality in both in-situ and ex-situ laser polishing. The case studies show the potential of laser surface modification as a vital tool for improving surface quality to a great extent.

REFERENCES

1. T. D. Ngo, A. Kashani, G. Imbalzano, K. T. Q. Nguyen, and D. Hui, "Additive manufacturing (3D printing): A review of materials, methods, applications and challenges," *Composites Part B: Engineering*, vol. 143, pp. 172–196, 2018, doi: 10.1016/j.compositesb.2018.02.012.
2. W. Gao et al., "The status, challenges, and future of additive manufacturing in engineering," *Computer-Aided Design*, vol. 69, pp. 65–89, 2015, doi: 10.1016/j.cad.2015.04.001.
3. H. Lee, C. H. J. Lim, M. J. Low, N. Tham, V. M. Murukeshan, and Y.-J. Kim, "Lasers in additive manufacturing: A review," *International Journal of Precision Engineering and Manufacturing-Green Technology*, vol. 4, no. 3, pp. 307–322, 2017, doi: 10.1007/s40684-017-0037-7.
4. J. Gardan, "Additive manufacturing technologies: State of the art and trends," *International Journal of Production Research*, vol. 54, no. 10, pp. 3118–3132, 2016, doi: 10.1080/00207543.2015.1115909.
5. N. Guo and M. C. Leu, "Additive manufacturing: Technology, applications and research needs," *Frontiers of Mechanical Engineering*, vol. 8, no. 3, pp. 215–243, 2013, doi: 10.1007/s11465-013-0248-8.

6. M. D. Monzón, Z. Ortega, A. Martínez, and F. Ortega, "Standardization in additive manufacturing: Activities carried out by international organizations and projects," *The International Journal of Advanced Manufacturing Technology*, vol. 76, no. 5, pp. 1111–1121, 2015, doi: 10.1007/s00170-014-6334-1.

7. Society of Manufacturing Engineers, "Additive Manufacturing Glossary." https://www.sme.org/technologies/additive-manufacturing-glossary/ (accessed Sep. 06, 2021).

8. J.-Y. Lee, A. P. Nagalingam, and S. H. Yeo, "A review on the state-of-the-art of surface finishing processes and related ISO/ASTM standards for metal additive manufactured components," *Virtual and Physical Prototyping*, vol. 16, no. 1, pp. 68–96, 2021, doi: 10.1080/17452759.2020.1830346.

9. W. Hung, "Post-processing of additively manufactured metal parts," *Journal of Materials Engineering and Performance*, 2021, doi: 10.1007/s11665-021-06037-z.

10. J. Gardan, "Additive manufacturing technologies: State of the art and trends," *International Journal of Production Research*, vol. 54, no. 10, pp. 3118–3132, 2016, doi: 10.1080/00207543.2015.1115909.

11. L. Bian, N. Shamsaei, and J. M. Usher, Eds., *Laser-Based Additive Manufacturing of Metal Parts*. Boca Raton: CRC Press, Taylor & Francis, 2017. doi: 10.1201/9781315151441.

12. X. Peng, L. Kong, J. Y. H. Fuh, and H. Wang, "A review of postprocessing technologies in additive manufacturing," *Journal of Manufacturing and Materials Processing*, vol. 5, no. 2, 2021, doi: 10.3390/jmmp5020038.

13. D. Bhaduri et al., "Laser polishing of 3D printed mesoscale components," *Applied Surface Science*, vol. 405, pp. 29–46, 2017, doi: 10.1016/j.apsusc.2017.01.211.

14. E. Willenborg, "Polishing with Laser Radiation," 2010.

15. K. Wissenbach, "Surface Treatment," in *Tailored Light 2: Laser Application Technology*, R. Poprawe, Ed. Berlin, Heidelberg: Springer, 2011, pp. 173–239. doi: 10.1007/978-3-642-01237-2_11.

16. G. Leuteritz and R. Lachmayer, "Concepts for Integrating Laser Polishing Into an Additive Manufacturing System," in *Laser 3D Manufacturing VI*, 2019, vol. 10909, pp. 32–40. Available: 10.1117/12.2509447

17. S. Marimuthu, A. Triantaphyllou, M. Antar, D. Wimpenny, H. Morton, and M. Beard, "Laser polishing of selective laser melted components," *International Journal of Machine Tools and Manufacture*, vol. 95, pp. 97–104, 2015, doi: 10.1016/j.ijmachtools.2015.05.002.

18. A. Lamikiz, J. A. Sánchez, L. N. López de Lacalle, and J. L. Arana, "Laser polishing of parts built up by selective laser sintering," *International Journal of Machine Tools and Manufacture*, vol. 47, no. 12, pp. 2040–2050, 2007, doi: 10.1016/j.ijmachtools.2007.01.013.

19. A. Kumar, S. Saha, C. S. Kumar, and A. K. Nath, "Laser surface re-melting of additive manufactured samples with a line focused beam," *Materials Today: Proceedings*, vol. 26, pp. 1221–1225, 2020, doi: 10.1016/j.matpr.2020.02.245.

20. A. Lamikiz, J. A. Sánchez, L. N. L. de Lacalle, D. del Pozo, and J. M. Etayo, *Surface Roughness Improvement Using Laser-Polishing Techniques*, vol. 526. 2006. doi: 10.4028/0-87849-417-0.217.

21. A. Temmler, E. Willenborg, and K. Wissenbach, "Laser polishing," in *Laser Applications in Microelectronic and Optoelectronic Manufacturing (LAMOM) XVII*, 2012, vol. 8243, pp. 171–183. Available: 10.1117/12.906001

22. C. P. Ma, Y. C. Guan, and W. Zhou, "Laser polishing of additive manufactured Ti alloys," *Optics and Lasers in Engineering*, vol. 93, pp. 171–177, 2017, doi: 10.1016/j.optlaseng.2017.02.005.

23. A. Krishnan and F. Fang, "Review on mechanism and process of surface polishing using lasers," *Frontiers of Mechanical Engineering 2019 14:3*, vol. 14, no. 3, pp. 299–319, Mar. 2019, doi: 10.1007/S11465-019-0535-0.

Chapter 6

Review of materials and methods in 3D printing

S. Ganeshkumar, V. Naveenprabhu, K. Sathish, and R. Vivek

Sri Eshwar College of Engineering, Coimbatore, Tamil Nadu, India

CONTENTS

6.1 INTRODUCTION

In Industry 4.0 scenario, the integration of the internet of things, cyber techniques, digital manufacturing, and cloud computing have evolved drastically. There are several steps involved before pitching a product into the market such as conceptual design, modeling, and manufacturing. In conventional subtractive manufacturing processes, the design of components strongly influences the ease of the manufacturing process [1]. After the advent of the additive manufacturing process, the complexity of the design doesn't affect the manufacturing process due to the ease of manufacturing layer by layer using computer numerical control techniques. Hence, three-dimensional (3D) printing techniques open the doors to the designers and manufacturers to convert the 3D model directly into the prototype or the finished products even for manufacturing complicated hollow structures, contours, and surfaces. The 3D-printed component also can withstand high temperatures; hence, it is used in turbine blades, rocket nozzles, aero-engine components, etc. The incredible 3D-printed aerospike rocket engine built with liquid cooling channels directly fed to the rocket nozzles interiors can operate with high performance with varying altitudes. In short, outside of rapid prototyping applications, additive manufacturing can be widely used to manufacture the finished components [2, 3]. The conceptual design is

DOI: 10.1201/9781003319375-6

converted into rapid prototyping and the quality and performance of the products are evaluated and the product will be pitched into the market for the end-user. Additive manufacturing allows topology optimization to the designed components to create the model where exactly the material is needed, which offers to reduce the weight of the material and increase the performance. In contrast to subtractive manufacturing, material wastage is drastically reduced. In the machining of titanium alloys costing 30$ a kilogram, 90% of wastages are thrown away, which leads to material wastage. In additive manufacturing, layer manufacturing is executed and takes the material where it is required; hence, it leads to optimal usage of the material. The Fusion 360 from Autodesk enables the platform for integration of topology optimization and additive manufacturing.

The designer can convert the optimized design directly to the B rep model or STL formats. In contrast to subtractive manufacturing, additive manufacturing technology is evolved to reduce the cost of rapid prototyping and increase the productivity and ease of investigation of the performance of prototyping components [4, 5]. In addition to the additive manufacturing of components, digital twins are being evolved to prototype virtually through augmented reality (AR) and virtual reality (VR). In general, additive manufacturing is classified into solid, liquid, and powder-based additive manufacturing processes. In all the additive manufacturing techniques, the formation of 3D-printed components is via the addition of layer by layer. Based on the raw materials such as powders, filaments, polymers, and liquid resins, additive manufacturing is classified. The additive manufacturing process is used in prototyping the conceptual design and manufacturing the finished product. In additive manufacturing of powder metals, by selective LASER sintering (SLS) process, finished components are made and it has wide applications in industries such as medical, aerospace, defense, and civil engineering. Based on the requirement of the user, the mode of additive manufacturing technique is selected and the conceptual idea is converted to the prototype or finished components [6]. Additive manufacturing creates opportunities in the manufacturing of complex geometries and is easier than conventional subtractive manufacturing processes. The 3D objects are created layer by layer using polymer resins, powder metals, papers, etc. Laminated object manufacturing is one of the additive manufacturing techniques which creates 3D objects by laminating the polymer or paper layers by heating. The heating path is controlled using computer numerical control [7]. The G codes are generated from the CAD data and based on the profiles and contours, the layers are sliced. This laminated object manufacturing is used for prototyping and the surface finish is comparatively low with liquid-type stereolithographic printing. Ultimaker, CURA, MakerBot, and Flashforge are some of the commercial fusion deposition modeling printers, and similar to the Fused Deposition Modeling (FDM), liquid and powder-based printers are available with the advent of Industry 4.0 and the digital manufacturing scenario and it takes human manufacturing to the next evolution.

6.2 3D PRINTING MATERIALS

The source of the 3D-printed components is 3D CAD data and the filament or source material. The raw material used in 3D printing is polymer filament in fusion deposition modeling and often liquid resins are used in the stereolithography technique. Several types of researches are being carried out in the 3D printing materials to print the object and supports. Materials such as polyvinyl alcohol (PVA) are often used as sacrificial supports which can be easily removed after 3D printing by dipping into the water. The sacrificial supports can dissolve in water due to their solubility. Poly Lactic acid (PLA) is a biodegradable material often used in FDM techniques. PLA materials are made from the renewable resources starch formed by condensation and polymerization reaction by the loss of water molecules [8]. The conversion of monomer to polymer occurs in the polymerization reaction. PLA is widely used in rapid prototyping due to its biodegradability property. Commercially, PLA polymers were produced in the 1990s. The available starch in corn is fed to the hydrolysis process and is followed by the fermentation process either by bacteria or fungi. The lactic acid monomers are produced after the reaction is polymerized with the help of light. The production cost is PLA is more compared to other plastics. Fusion deposition modeling PLA filaments are biodegradable hence the filament breaks when exposed to atmospheric conditions. ABS (acrylonitrile butadiene styrene) is a material similar to PLA plastics; ABS is the amorphous thermos plastic used widely in 3D printing systems similar to PLA. ABS serves excellent impact, chemical, and abrasive resistance. These materials are used to manufacture automobile instrument panels, food processing components, rapid prototypes, etc. The continuous mass technique is used in the manufacturing of ABS materials. The color of ABS is transparent ivory in nature and can be readily colored with pigments and dyes. This ABS material is poor in weathering resistance and can be scratched easily [9]. Based on the properties and the requirements of the manufacturer, materials for additive manufacturing are selected. PVA material is often used in sacrificial support materials. This enables the platform for manufacturing complicated inclusions, profiles, large overhangs, cavities, and intricate structures and enables easy support removal. PVA is a water-soluble material that offers the solubility property to dilute the supports in water. Software packages like CURA identify the support material and component filament. The support material filaments (PVA) are thermally stable and hence they can be used for longer periods of printing time [10]. After completing the 3D printing process, in post-processing, the 3D-printed components with supports are placed in water to dissolve the sacrificial supports. This restricts the harmful chemicals in the dissolving of supports, and users can easily dilute the material in tap water. PVA is a biodegradable material similar to PLA. After dissolving PVA in water, 3D-printed component is rinsed and fed to the other additional post-processing techniques. 3D printers like Ultimaker and MakerBot enables the features of dual extrusion such as PLA and ABS. The functional prototypes with aesthetics can be manufactured using PLA, ABS, Nylon, and PVA materials. Nylon materials are similar to PLA and ABS used often in 3D printing, which is a chemically resistant material that can be stable for a long period with good functional behavior. Nylon is the commercial name of poly Amide (PA) [11, 12].

6.3 METHODS OF ADDITIVE MANUFACTURING

The additive manufacturing techniques include fusion deposition modeling, vat polymerization, laminated object manufacturing, stereolithographic additive manufacturing, direct light processing (DLP), SLS, binder jetting, ballistic particle manufacturing (BPM), 3D printing, powder bed fusion, and many more. Fusion deposition modeling additive manufacturing process is a widely used rapid prototyping technique in which the manufacturing cost is low [13]. PLA, Nylon, and ABS are the standard materials used in the FDM process. The surface finish of the FDM is comparatively low than stereolithography and powder additive manufacturing. Hence, this technique is used for rapid prototyping applications [4]. The 3D filament PLA/ABS/Nylon material is fed to the extruder through the two rollers and the material is melted up to the melting point. The graphical user interface of the 3D printer applications enables the features of the temperature setting of the extruder. The optimum temperature setting of the extruder and platform temperature lead to increasing the quality of the 3D-printed components. Thermoplastic materials are generally used in the FDM process and various infill structures enhance weight reduction [14, 15]. The infills such as gyroid, linear, circular, square, triangular, and honeycomb structures are some examples of infill structures available in the packages such as CURA, SLIc3R, Fusion 360, etc. The ratio of infill to the total volume can be altered based on the percentage. The Fusion deposition modeling of a component at different layers and 3D-printed components are depicted in Figure 6.1.

More infills lead to an increase the mechanical strength and vice versa. The selection of percentages and types of infills is based on the applications of the 3D-printed components. Commercial 3D printers such as CURA, Ultimaker, and Flashforge are able to extrude the filaments independently for supports and components. In the BPM technique, a jet of powder metals in form of microdroplets is sprayed through the moving nozzle on the fixed platform. The platform lowers for consecutive layers to complete the 3D-printed components. In the SLS process, the powder metals are compacted through the CO_2 LASER source. The voids between the particles are reduced after the fusion of powder metal particles.

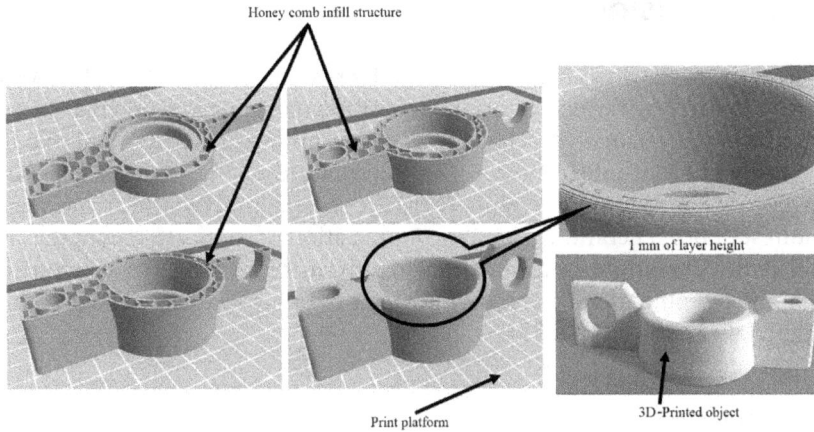

Figure 6.1 3D model with honeycomb infill structure and 3D-printed object.

6.4 STEPS INVOLVED IN 3D PRINTING TECHNIQUES

In all additive manufacturing processes, conversion of the existing component to CAD data. The CAD data model is created using modeling software such as Fusion 360, Creo Elements, etc. In the reverse engineering process, the LASER trigonometric scanners and CMM machines are employed to scan the existing 3D model. Irrespective of the additive manufacturing process such as powder additive manufacturing technique, SLS, and fusion deposition modeling, the preprocessing is identical. The CAD data is fed to the CNC controller to control the tool paths in X- and Y-directions [16, 17]. Whereas the platform is taken as the Z-axis, after completion of one layer, Z-axis is activated to prepare the 3D printing machine for next layer printing. The conversion of CAD data to Standard Tessellation Language or standard triangular language led to the creation and placement of facets of triangles [18, 19]. In additive manufacturing packages, STL formats are imported and the smoothness of the contours and profiles are verified. The model is repaired if data loss occurs and the sliced 3D model in STL format is given to the G code generator for generating the G codes to control the tool paths. In the processing stages, based on the types of 3D printing methods, the parameter is fixed and after completion of the 3D printing, the component is fed to post-processing. Coating, painting processes, and support removal are involved in post-processing [19]. The surface finish and mechanical behavior of the material decide the 3D-printed component usage such as rapid prototyping or finished component [20, 21] In the laminated object manufacturing process, sheet stock is given as feed, whereas in fusion deposition modeling, the filament is given as feed material. The process settings involve print speed, layer height, filament temperature, etc. [22, 23].

6.5 CONCLUSION

The review of the previous survey concludes that materials such as PLA, ABS, and Nylon are used widely in fusion deposition modeling and polymer sheets are used in laminated object manufacturing processes. CURA and Slic3R are firmware that integrate the 3D printer and the CAD data to generate the G codes and support designs. There are several kinds of research carried out in support generation, topology optimization of supports, and optimum tool path generation. In a nutshell, the advent of additive manufacturing technology and materials research leads to creating a drastic change in the industrial revolution [24, 25].

REFERENCES

1. Campbell, T., Williams, C., Ivanova, O. and Garrett, B., 2011. Could 3D printing change the world. *Technologies, Potential, and Implications of Additive Manufacturing*, Atlantic Council, Washington, DC, 3.
2. Lee, J.Y., An, J. and Chua, C.K., 2017. Fundamentals and applications of 3D printing for novel materials. *Applied Materials Today*, 7, pp.120–133.
3. Carneiro, O.S., Silva, A.F. and Gomes, R., 2015. Fused deposition modeling with polypropylene. *Materials & Design*, 83, pp.768–776.
4. Ahn, S., Montero, M., Odell, D., Roundy, S. and Wright, P.K., 2002. Anisotropic material properties of fused deposition modeling ABS. *Rapid Prototyping Journal*, 8(4), pp. 248–257. https://doi.org/10.1108/13552540210441166
5. Wong, K.V. and Hernandez, A., 2012. A review of additive manufacturing. *International Scholarly Research Notices*, 2012, p.208760.
6. Herzog, D., Seyda, V., Wycisk, E. and Emmelmann, C., 2016. Additive manufacturing of metals. *Acta Materialia*, 117, pp.371–392.
7. Guo, N. and Leu, M.C., 2013. Additive manufacturing: Technology, applications and research needs. *Frontiers of Mechanical Engineering*, 8(3), pp.215–243.
8. Lim, S., Buswell, R.A., Le, T.T., Austin, S.A., Gibb, A.G. and Thorpe, T., 2012. Developments in construction-scale additive manufacturing processes. *Automation in Construction*, 21, pp.262–268.
9. Gao, W., Zhang, Y., Ramanujan, D., Ramani, K., Chen, Y., Williams, C.B., Wang, C.C., Shin, Y.C., Zhang, S. and Zavattieri, P.D., 2015. The status, challenges, and future of additive manufacturing in engineering. *Computer-Aided Design*, 69, pp.65–89.
10. Bikas, H., Stavropoulos, P. and Chryssolouris, G., 2016. Additive manufacturing methods and modelling approaches: A critical review. *The International Journal of Advanced Manufacturing Technology*, 83(1–4), pp.389–405.
11. Horn, T.J. and Harrysson, O.L., 2012. Overview of current additive manufacturing technologies and selected applications. *Science Progress*, 95(3), pp.255–282.
12. Ford, S. and Despeisse, M., 2016. Additive manufacturing and sustainability: An exploratory study of the advantages and challenges. *Journal of Cleaner Production*, 137, pp.1573–1587.

13. Drumright, R.E., Gruber, P.R. and Henton, D.E., 2000. Polylactic acid technology. *Advanced Materials*, 12(23), pp.1841–1846.
14. Montero, M., Roundy, S., Odell, D., Ahn, S.H. and Wright, P.K., 2001. Material characterization of fused deposition modeling (FDM) ABS by designed experiments. *Society of Manufacturing Engineers*, 10(13552540210441166), pp.1–21.
15. Hale, W.R., Pessan, L.A., Keskkula, H. and Paul, D.R., 1999. Effect of compatibilization and ABS type on properties of PBT/ABS blends. *Polymer*, 40(15), pp.4237–4250.
16. Wagner, J.J., Shu, H. and Kilambi, R., 2021. *Experimental Investigation of Fluid-particle Interaction in Binder Jet 3D Printing.*
17. Pandey, R. and Ramani, H.B., 2021. A study of metallurgical and processing approach of additive layer manufacturing processes for metal tool components production. *Materials Today: Proceedings*, 47, pp.6642–6646.
18. Ganeshkumar, S., Sureshkumar, R., Sureshbabu, Y. and Balasubramani, S., 2020. A review on cutting tool measurement in turning tools by cloud computing systems in industry 4.0 and IoT. *GIS Science Journal*, 7(8), pp.1–7.
19. Sitthi-Amorn, P., Ramos, J.E., Wangy, Y., Kwan, J., Lan, J., Wang, W. and Matusik, W., 2015. MultiFab: A machine vision assisted platform for multi-material 3D printing. *ACM Transactions on Graphics (Tog)*, 34(4), pp.1–11.
20. Ganeshkumar, S., Thirunavukkarasu, V., Sureshkumar, R., Venkatesh, S. and Ramakrishnan, T., 2019. Investigation of wear behaviour of silicon carbide tool inserts and titanium nitride coated tool inserts in machining of en8 steel. *International Journal of Mechanical Engineering and Technology*, 10(1), pp.1862–1873.
21. Venkatesh, S., Sivapirakasam, S.P., Sakthivel, M., Ganeshkumar, S., Prabhu, M.M. and Naveenkumar, M., 2021. Experimental and numerical investigation in the series arrangement square cyclone separator. *Powder Technology*, 383, pp.93–103.
22. Ganeshkumar, S., Sureshkumar, R., Sureshbabu, Y. and Balasubramani, S., 2019. A numerical approach to cutting tool stress in CNC turning of En8 steel with silicon carbide tool insert. *International Journal of Scientific & Technology Research*, 8(12), pp.3227–3231.
23. Kumar, S.G. and Thirunavukkarasu, V., 2016. Investigation of tool wear and optimization of process parameters in turning of EN8 and EN 36 steels. *Asian Journal of Research in Social Sciences and Humanities*, 6(11), pp.237–243.
24. Gokilakrishnan, G., Ganeshkumar, S., Anandakumar, H. and Vigneshkumar, M., 2021, March. A critical review of production distribution planning models. In *2021 7th International Conference on Advanced Computing and Communication Systems (ICACCS)* (Vol. 1, pp. 2047–2051). IEEE.
25. Ganeshkumar, S., Kumar, S.D., Magarajan, U., Rajkumar, S., Arulmurugan, B., Sharma, S., Li, C., Ilyas, R.A., Badran, M.F., 2022. Investigation of tensileproperties of different infill pattern structures of 3D-printed PLAPolymers: Analysis and validation using finite element analysis in ANSYS. *Materials*, 2022(15), p.5142. https://doi.org/10.3390/ma15155142

Chapter 7

The framework of combining artificial intelligence with additive manufacturing

S. Ganeshkumar
Sri Eshwar College of Engineering, Coimbatore, Tamil Nadu, India,

T. Deepika
Amrita Vishwa Vidyapeetham, Coimbatore, Tamil Nadu, India,

R. Sureshkumar and V. Naveenprabhu
Sri Eshwar College of Engineering, Coimbatore, Tamil Nadu, India

CONTENTS

7.1 INTRODUCTION

Nowadays, buzzwords such as Industry 4.0, Blockchain, the Internet of Things, and digital manufacturing are seen. The integration of these technologies together can strongly influence productivity, data handling, and even a market survey. The pitching of a product to the market with more features and low cost leads to the successful launch and survival of the product. All the potential manufacturing industries are foraging modern techniques and integration of data science, artificial intelligence, machine learning, and digital manufacturing to achieve the finest production and sustainable growth also leads to a huge impact on the global economy. The combining activity of the two different technology leads to attaining the benefits of both. Artificial Intelligence is the technique and discipline which helps to think of the machine from past experiences or real-time data without human intervention. In contrast to machine learning, artificial intelligence is the domain that can able to design the entire data acquisition, data handling, and prediction environment. Artificial intelligence can communicate with real-time machines from the predicted results. The prediction accuracy is improved consistently based on real-time data and fed to digital manufacturing systems such as CNC-operated machines, and factory automated robots. Artificial intelligence is also implemented in robotics such as humanoids to mimic human behaviors. In the manufacturing sector, the role of artificial intelligence strongly influences autonomous manufacturing systems. Digital manufacturing systems include the design and modeling of machine components, analysis, and manufacturing. Artificial intelligence plays a key role in regenerative design in reducing stress concentration regions. Automated topology optimization is the paradigm of integrating artificial intelligence and digital manufacturing.

3D printing and naked eye 3D visualization experiments were carried out by integrating medical image processing, artificial intelligence, and 3D printing. In medical image processing, segmentation, layering, and edge detection techniques are used in capturing the CT and MRI images of osteosarcoma, lumbar spines, local nerves, and vessels. YOLOV3, a real-time object detection algorithm, is used to detect the MRI and CT images. This algorithm can also identify live videos, images, and objects. It was found that the image processing aspirants also utilized the YOLOV3 image detection algorithm in the machine vision system for factory automation. The integration of the segmented images is processed for four-angle view visualization. The 3D images are converted into CAD data for 3D printing. This research exhibits the techniques of conversion of segmented images from image processing algorithms to stereolithographs (SLA) using Standard Triangulation Languages (STLs) [1]. The applications of 3D printing techniques in the pharmaceutical industry were explored by carrying out experiments with ghost tablets. The 3D-printed poly(ethylene glycol) diacrylate (PEGDA)-based 'Ghost tablets' prepared using direct light processing (DLP) type 3D printer drug release kinetics are analyzed to form the dataset. The recorded drug release kinetics is fed to the machine learning algorithm to study the characteristic behavior of drug release. The regression algorithm is generated by a machine learning multiple regression prediction models, and the drug release kinetics is analyzed using binary classification (support vector machine) prediction modes. Moreover, the research states the scope of future 3D-printed ghost tablets, applications of 3D printing technology in the biomedical industry, and paved the way for integrating artificial intelligence and machine learning algorithm in biomedical and additive manufacturing methods [2]. The traditional additive manufacturing industry development with intellectualization is exhibited through the survey of applications of 3D-printed components in various industries such as biomedical, engineering, research and development, etc. The survey states that the precision and accuracy of the 3D-printed component, optimal design, and weight-to-infill ratio can be achieved by machine learning algorithms and paves the way for integrating machine learning and additive manufacturing techniques [3]. The knowledge-based system and product modeling are amalgamated. The dynamic customer requirements and improved product design is achieved by knowledge-based systems (artificial intelligence) from past experiences. The research bridges the gap between product design and additive manufacturing techniques. The CAD tools generate the computations of tool path and infill ratio for the rapid prototyping processes [4]. The 3D printer process parameters are recorded for the intelligent manufacturing systems, such as printer bed temperature, filament melting point, infill density, infill pattern, weight-to-volume ratio, wall thickness, fan speed, roughness, material, and layer height. The data set is utilized to predict the characteristic properties of the 3D-printed components such as mechanical, thermal, and chemical behaviors [4].

7.2 NEED FOR INTELLIGENT SYSTEMS IN ADDITIVE MANUFACTURING TECHNIQUES

The importance of intelligent systems has been explored in additive manufacturing systems and paved the way for researchers to develop intelligent additive manufacturing systems. The conversion of 3D-modeled CAD data into STL formats using slicing algorithms leads to acquiring the degree of surface finish and quality of the 3D-modeled components. The placements of facets of the triangles in the standard triangular language are done using computing algorithms. The codes running behind the real-time-slicing algorithms are exhibited owing to the advent of the Industry 4.0 Revolution. The contour generation process is taken as a case study. The two cases of STL formats are experimented with, such as binary STL and ASCII STL format. The techniques of conversion of lines into segmentation for triangulation are demonstrated and the scope of integration of intelligent techniques such as prediction of the model accuracy and the number of elements and layers using machine learning techniques have been extensively experimented [4, 5]. The fusing of artificial neural networks (ANNs) with 3D printing is explored to solve large datasets. The research explored the various process techniques of additive manufacturing, such as fusion deposition modeling, liquid additive manufacturing using resins and liquid polymers, powder-based additive manufacturing, DLP, and LASER sintering processes. The illustration of various technique characteristics leads to assist in the processing of infill design, tool path generation, etc.; moreover, this research exhibits the color 3D printing technology using six different categories based on the print substrates as follows:

a) Paper-based
b) Organism-based
c) Food-based
d) Plastic-based
e) Metal-based
f) Powder-based

Three classes of ANN models were demonstrated in the research, such as convolution ANNs, recurrent ANNs, and multilayer perceptron. Moreover, the characteristic features of hidden layers, number of neurons activation functions, and loss functions are exhibited in the research. The activation function of the LASER sintering process is sigmoid and Tanh for 4-9-1 and 3-9-1, respectively. In SLA, liquid-type additive manufacturing sigmoid activation function is recorded for 6-20-5 input–output and hidden layers. The ANN techniques are used in 3D printing for inspection for evaluating the performance of 3D-printed components, monitoring of 3D printing process parameters, and optimization of the process parameters such as printing speed, platform temperature, extruder temperature, filament, substrate material, and layer thickness [6]. Real-time defect identification using machine vision and artificial intelligence in 3D printing has experimented with fused filament fabrication techniques. The work described the image-capturing techniques for computer vision using image processing algorithms for defect identification in 3D printing. The object geometries, edge detection, and segmentation are converted to digital images and trained in deep convolution ANNs. The stringing defect was captured by deploying a machine vision camera and frame grabbers. The stringing was recognized bylines with speed and accuracy. The results of the ANN model can terminate the 3D printing process of refining the process parameters to produce errorless 3D-printed components [7]. Prostheses, artificial human bones, and biomedical applications are exhibited. The interaction of 3D printing with artificial intelligence exclusively in SLA 3D printing with photopolymerization is investigated. The deep learning techniques are capable of handling a large number of data sets and lead to acquiring accuracy and precision of the results. The amalgamation includes tool path optimization and printing speed time prediction, infill pattern with strength calculations, and support designs based on past experiences from the datasets. In addition, the recommendations are reviewed for the selection of deep learning algorithms for computing the tool path and support design using ANN [8].

7.3 INTEGRATION OF AI AND AM TECHNIQUES

With CAD data generation using reverse engineering such as coordinate measuring machines and LASER trigonometry under the assistance of range sensors, the existing components are converted to 3D CAD data. The coordinate measuring machine used for measuring the coordinates (contact and non-contact type CMM machines) converts the coordinates to the points. The framework of the integration of additive manufacturing and artificial intelligence is exhibited in Figure 7.1. The point data is converted into lines and lines are converted into the surface using facet features. Software such as Rhino, AutoCAD, and Creo Elements enable the features to convert the point clouds to the surface and solid 3D CAD data. The 3D CAD data is

Additive Manufacturing

| 3D CAD data conversion |
| Conversion of part to stl format |
| Triangle facet placement computation algorithm |
| Repairing the model |
| Slicing the model |
| Generation of supports |
| Generative design / Optimal support design |
| Tool path generation – G codes |
| Object printing |
| Quality inspection using computer vision |

Artificial Intelligence system

| Process parameters | Tool path design | Generative design / Optimal support design |

| Implementation of intelligent system – Artificial intelligence system |
| Retrieval of previous experience using dataset |
| Training the machine learning model / Deep learning model |
| Model prediction accuracy evaluation |
| Hyper parameter tuning of the machine learning model parameters |
| Optimal solution of process parameters, tool path design and generative design of supports |
| Printing the object and data storage to the repository for future retrieval |

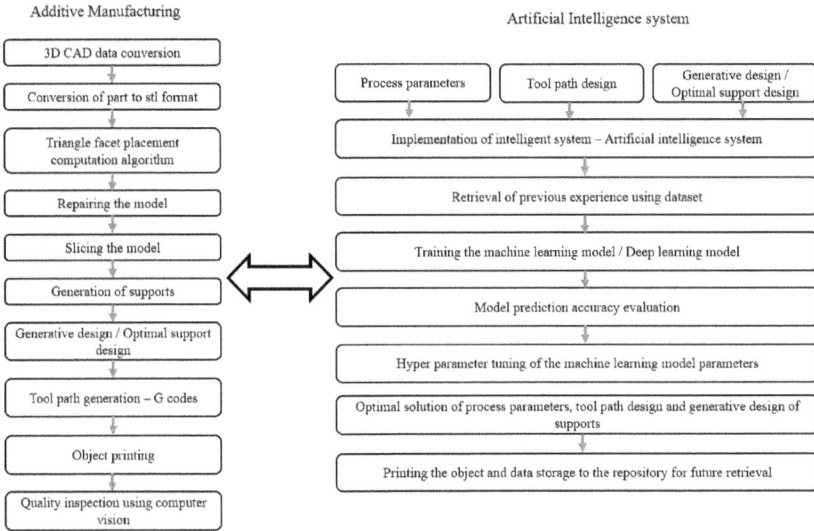

Figure 7.1 Framework of integrating additive manufacturing with artificial intelligence systems.

converted to standard triangle language, in this stage, the placement of facet triangles is computed using facet placing algorithms and packages like slicer and Flashprint applications. The Facet placement options can be achieved by algorithms in JavaScript, Python, R programming languages. The degree of surface smoothness is strongly influenced by the optimum placement of the triangle placement phenomena. The conversion of an object to CAD data with data loss leads to the deviation of accuracy. The mathematical smashing of the lines, floating triangles, shells, holes, edges, and curves to perfect triangles and surfaces causes the restoration of the originality of the existing original component. Mesh mixer, formlabs, magics, blender, netfabb, and many more commercial packages enable the features of repairing the existing model to regain the originality. The repairing of the STL files by manual and automatic modes. In automatic modes, the 3D component is restored by the detection and modification of non-manifold vertices using algorithms. These techniques pave the way to implement artificial intelligence to predict the smooth curves and blending of facets to regain the original 3D components. Supports for the 3D printing components are generated using Auto supports and manual modes using packages like slic3r, CURA, Flashprint, fusion 360, and many more to enable the topology optimized supports. This generates the least structures in an intelligence-based algorithm to avoid more filament usage. Instead of generating the supports on standard triangle languages, directly 3D-printed using B-rep models using support generation algorithm. The algorithm generates the grid points and detects the hanging edges and generates the supporting areas. The two conditions are executed

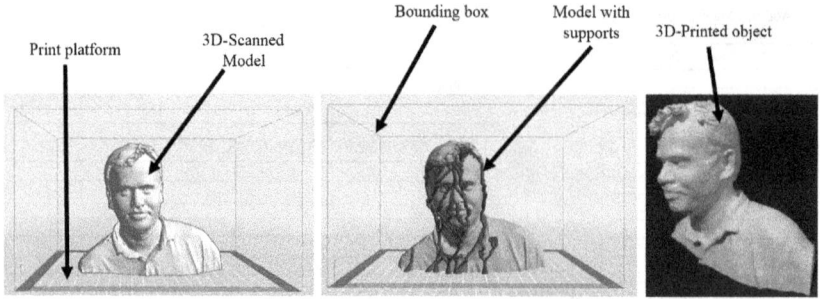

Figure 7.2 3D model with supports in print platform in Flash forge Dreamer.

with programmed B rep models and the accuracy of the 3D-printed models is evaluated [9]. Repairing and support generation of the 3D models are followed by tool path generation. Tool path generation is an essential step in the 3D printing process and, based on the contours of the objects, tool vector and jump vectors are determined using tool path generation algorithms; the tool vector is active on the contours and the extruder fills the filament material on the active regions of the tool vector. The jump vector stands idle at the no-fill regions of the 3D object in the bounding box, which leads to making the extruder go into an idle state, and filament flow is stopped during the no-fill of jump vector regions [10]. The model of a 3D-scanned face and 3D-printed object with supports are exhibited in Figure 7.2. The quality of the 3D-printed components is inspected using machine vision or computer vision techniques and the images acquired from the system are fed to image processing units and frame grabbers [11].

These techniques feed the data to the repository as well as a control system to optimize the 3D printing process parameters. While the machine runs with errors, these control systems assist to refine the process parameters or assist in halting the process to avoid manufacturing wastage. In the integration of artificial intelligence, the errors are taken into account by experience and utilized for the future machine operating parameter setting [12, 13].

7.4 ARTIFICIAL INTELLIGENCE SYSTEM FOR 3D PRINTING

The data set of printing parameters such as extruder temperature, platform temperature, filament material, facet placement techniques, and accuracy of the 3D-printed components are stored in the repository [14, 15]. The machine learning model developed using support vector machines, random forest algorithm, linear or logistic algorithms based on the process nature, and training the models with past experiences or the dataset available in the cloud leads to an increase in the accuracy of the machine learning model. The evaluation of the accuracy using confusion matrices, RMSE, and R-squared values leads to an increase in the performance of prediction accuracy [16, 17]. The accuracy of the machine learning model is refined using hyperparameter tuning such as varying the number of nodes, leaves, and branches of the model the optimal solutions are stored in the cloud repository for future usage. The computing of tool paths, support designs, and process parameter settings in the cloud enables to transfer of the data from one place to another place and it is easy to train the total community similar to the Google Cloud machine learning environment [18, 19].

7.5 CONCLUSION

Researches based on the integration of additive manufacturing with the intelligent environment using artificial intelligence, machine learning, and deep learning enable access to the benefits of both AI and AM [20, 21]. The computing techniques in the cloud environment lead to processes in a common server and the experiences or the dataset from one machine can be shared easily with another machine to achieve the best-trained model leads, which can manufacture with zero defects. The literature exhibits the wide scope of research in integrating artificial intelligence with additive manufacturing in the Industry 4.0 scenario [22, 23].

REFERENCES

1. Jia, G., Huang, X., Tao, S. et al., Artificial intelligence-based medical image segmentation for 3D printing and naked eye 3D visualization, *Intelligent Medicine*, doi: 10.1016/j.imed.2021.04.001.
2. Tagami, T., Morimura, C. and Ozeki, T., 2021. Effective and simple prediction model of drug release from 'ghost tablets' fabricated using a digital light projection-type 3D printer. *International Journal of Pharmaceutics*, 604, p. 120721.
3. Zhou, Y., 2019. Research on development and problems of 3D printing technology under intelligent background. In *2019 12th International Conference on Intelligent Computation Technology and Automation (ICICTA)*, pp.682–685, doi: 10.1109/ICICTA49267.2019.00150.
4. Chang, D. and Chen, C.H., 2014. Integration of knowledge engineering and 3D printing for digital design and manufacturing—A case study. In *Conference on Progress in Additive Manufacturing (Pro-AM 2014)*, Vol. 47, p. 52.
5. Adnan, F.A., Romlay, F.R.M. and Shafiq, M., 2018, April. Real-time slicing algorithm for Stereolithography (STL) CAD model applied in additive manufacturing industry. In *IOP Conference Series: Materials Science and Engineering*, Vol. 342, No. 1, p. 012016. IOP Publishing.
6. Mahmood, M.A., Visan, A.I., Ristoscu, C. and Mihailescu, I.N., 2021. Artificial neural network algorithms for 3D printing. *Materials*, 14(1), p. 163.

7. Paraskevoudis, K., Karayannis, P. and Koumoulos, E.P., 2020. Real-time 3D printing remote defect detection (stringing) with computer vision and artificial intelligence. *Processes*, 8(11), p. 1464.

8. Talaat, F.M. and Hassan, E. (2021) Artificial Intelligence in 3D Printing. In Hassanien, A.E., Darwish, A., Abd El-Kader, S.M., Alboaneen, D.A. (eds) *Enabling Machine Learning Applications in Data Science. Algorithms for Intelligent Systems.* Springer, Singapore, doi: 10.1007/978-981-33-6129-4_6.

9. Li, W., Ye, W. and Chen, W., 2011. Algorithms for mesh repairing to represent automobile parts. *Journal of Modern Transportation*, 19(4), pp.252–260.

10. Shi, K., Cai, C., Wu, Z. and Yong, J., 2019. Slicing and support structure generation for 3D printing directly on B-rep models. *Visual Computing for Industry, Biomedicine, and Art*, 2(1), pp.1–10.

11. Liu, J., Fan, Y., Lu, Q. and Yang, Y., 2014, October. Design of extendable tool path generation software for 3D printing. In *2014 International Conference on Manipulation, Manufacturing and Measurement on the Nanoscale (3M-NANO)*, pp.130–133. IEEE.

12. Ganeshkumar, S., Sureshkumar, R., Sureshbabu, Y. and Balasubramani, S., 2020. A review on cutting tool measurement in turning tools by cloud computing systems in industry 4.0 and IoT. *GIS Science Journal*, 7(8), pp.1–7.

13. Sitthi-Amorn, P., Ramos, J.E., Wangy, Y., Kwan, J., Lan, J., Wang, W. and Matusik, W., 2015. MultiFab: A machine vision assisted platform for multi-material 3D printing. *ACM Transactions on Graphics (Tog)*, 34(4), pp.1–11.

14. Ganeshkumar, S., Thirunavukkarasu, V., Sureshkumar, R., Venkatesh, S. and Ramakrishnan, T., 2019. Investigation of wear behaviour of silicon carbide tool inserts and titanium nitride coated tool inserts in machining of en8 steel. *International Journal of Mechanical Engineering and Technology*, 10(1), pp.1862–1873.

15. Ganeshkumar, S., Thirunavukkarasu, V., Sureshkumar, R., Venkatesh, S. and Ramakrishnan, T., 2019. Investigation of wear behaviour of silicon carbide tool inserts and titanium nitride coated tool inserts in machining of en8 steel. *International Journal of Mechanical Engineering and Technology*, 10(1), pp.1862–1873.

16. Venkatesh, S., Sivapirakasam, S.P., Sakthivel, M., Ganeshkumar, S., Prabhu, M.M. and Naveenkumar, M., 2021. Experimental and numerical investigation in the series arrangement square cyclone separator. *Powder Technology*, 383, pp.93–103.

17. Venkatesh, S., Sivapirakasam, S.P., Sakthivel, M., Ganeshkumar, S., Prabhu, M.M. and Naveenkumar, M., 2021. Experimental and numerical investigation in the series arrangement square cyclone separator. *Powder Technology*, 383, pp.93–103.

18. Ganeshkumar, S., Sureshkumar, R., Sureshbabu, Y. and Balasubramani, S., 2019. A numerical approach to cutting tool stress in CNC turning of En8 steel with silicon carbide tool insert. *International Journal of Scientific & Technology Research*, 8(12), pp.3227–3231.

19. Kumar, S.G. and Thirunavukkarasu, V., 2016. Investigation of tool wear and optimization of process parameters in turning of EN8 and EN 36 steels. *Asian Journal of Research in Social Sciences and Humanities*, 6(11), pp.237–243.

20. Gokilakrishnan, G., Ganeshkumar, S., Anandakumar, H. and Vigneshkumar, M., 2021, March. A critical review of production distribution planning

models. In *2021 7th International Conference on Advanced Computing and Communication Systems (ICACCS)*, Vol. 1, pp. 2047–2051. IEEE.

21. Deepika, T. and Prakash, P., 2020. Power consumption prediction in cloud data center using machine learning. *International Journal of Electrical and Computer Engineering (IJECE)*, 10(2), pp.1524–1532.

22. Deepika, T., Prakash, P. and Dhanya, N.M., 2020. Efficient resource prediction model for small and medium scale cloud data centers. *Journal of Intelligent & Fuzzy Systems*, 39(3), pp.4731–4747.

23. Ganeshkumar, S., Kumar, S.D., Magarajan, U., Rajkumar, S., Arulmurugan, B., Sharma, S., Li, C., Ilyas, R.A., Badran, M.F., 2022. Investigation of tensileproperties of different infill pattern structures of 3D-printed PLAPolymers: Analysis and validation using finite element analysis in ANSYS. *Materials*, 15, p.5142. https://doi.org/10.3390/ma15155142

Chapter 8

A review of metallic deposition in polymer substrate using cold spray additive manufacturing approach

Abdul Faheem
University Polytechnic, Aligarh Muslim University, Aligarh,
Uttar Pradesh, India

Faisal Hasan
Zakir Husain College of Engineering and Technology,
Aligarh Muslim University, Aligarh, Uttar Pradesh, India

Shubhangi Chourasia
Global Institute of Technology and Management, Gurugram, Haryana, India
Delhi Technological University, Delhi, India

Ankit Tyagi
SGT University, Gurgaon, Haryana, India

Shailesh Mani Pandey
National Institute of Technology Patna, Patna, Bihar, India

Qasim Murtaza
Delhi Technological University, Delhi, India

CONTENTS

8.1 INTRODUCTION

In the 20th century, there is a high requirement for the metallization of the polymeric substrate in different industrial sectors. The coating improves the surface and mechanical aspects of various components furthermore guards

DOI: 10.1201/9781003319375-8

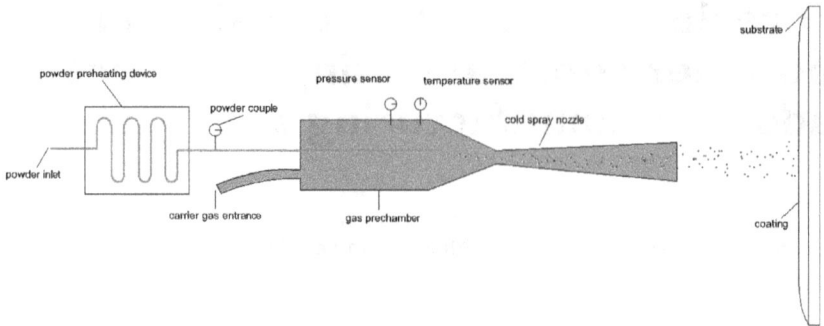

Figure 8.1 Setup of cold gas dynamic spray.

against the hostile environment. Cold spray is a new additive manufacturing approach for micro-/nano-size metallic, non-metallic, composite, and polymeric particles deposition on the substrate under the mechanism of high velocity of the particle. The feedstock particles are accelerated with the help of a convergent/ divergent D-Laval nozzle. The velocity of the particles varies in the range of 300–1200 m/s [1–6]. This coating concept was first introduced by the Russian Institute of Applied Mechanics nearly three decades ago. Recently, the metallization of polymer matrix composite and fiber-reinforced polymer composite substrate has gained attention due to many industrial applications. It provides better high specific strength and enhances the life span of the products. However, polymeric materials have low operational/working temperatures and lower electric conductivity. The life span of the polymeric material may be enhanced using the cold spray additive manufacturing (CSAM) approach; meanwhile, surface conductivity can also be increased. The metallic coating process on polymer substrate is complex; however, preheating micro-sized particles helped in the deposition mechanism [7–9]. Metals such as aluminium (Al), copper (Cu), titanium (Ti), and tin (Sn) have been sprayed at high velocity onto substrates. Generally, the substrate is blended with polystyrene, polycarbonate, polypropylene, acrylonitrile butadiene styrene, and glass-fiber materials [10–13]. Lee et al. [14] investigated that aluminium oxide helped in the deposition of ceramic coating onto the substrate. Sturgeon et al. [15] successfully achieved the metallic coating on the carbon-fiber-reinforced polymer substrate. The review article focuses on the metallic coating on the polymer substrate under cold gas dynamic spray technology. Figure 8.1 shows the setup for the CSAM.

8.2 CSAM APPROACH IN POLYMERIC SUBSTRATE

The metallization on the polymer surface has gained attention, and in recent times, several researches have been done in this domain of coating. The process parameters that helped in the deposition approach in the cold spray are feedstock powder, carrier gas pressure, stand-off distance, gas temperature, nozzle geometry, and powder feed rate. Lupoi et al. [16] investigated the compatibility of the cold spray by spraying aluminium, copper, and tin onto a polymer surface; furthermore, it has been found that approximately 0.02 mJ of energy has been developed by the impact of a single copper particle. At lower carrier gas temperatures, it has been found that a better deposition rate has been achieved in the thermoplastic surface, i.e., deposition of Al micro-sized particles on polyamide-66 substrates [17]. In addition, Al coating polycarbonate, Sn and Cu coating on polyvinyl chloride substrate, and Ti onto polyetheretherketone have been successfully achieved [18–20]. Figures 8.2 and 8.3 show the micrograph of tin coating on ABS, polyamide, polypropylene, and polystyrene substrate, and the deformation of metallic particles on the polymeric substrate, respectively. The low-pressure CS allowed particle velocities in the range of 250–650 m/s.

Generally, the low-pressure CS is in the range of 1 MPa; however, it gives a low deposition rate of particles. Meanwhile, high-pressure CS provides a better deposition rate with velocities in the range of 700–1200 m/s [21–22].

Figure 8.2 Micrograph of copper coating on different substrates [16].

Figure 8.3 Deformed particles on the substrate [16].

8.3 CONCLUSIONS AND FUTURE WORK

The CSAM approach of metallization to deposition on polymer surfaces has gained momentum. CSAM gives highly conductive and well-deposited metallic coatings on various polymers and their fiber-reinforced composites. However, CSAM on polymers has several restrictions on process parameters and coating along with deposition efficiency. Therefore, these aspects need to be enhanced using several optimization approaches. Meanwhile, a huge research area in the metallization of polymer surfaces is still open.

REFERENCES

1. P. Fauchais and G. Montavon, Thermal and Cold Spray: Recent Developments, *Key. Eng. Mater.*, 2008, 384, pp. 1–59.
2. M. McDonald, C. Lamontagne, and S. Moreau, Chandra, Impact of Plasma-Sprayed Metal Particles on Hot and Cold Glass Surfaces, *Thin Solid Films*, 2006, 514(1–2), pp. 212–222.
3. R.W. Smith and R. Knight, Thermal Spraying I: Powder Consolidation- from Coating to Forming, *J. Minerals Metals Mater. Soc.*, 1995, 47(8), pp. 32–39.
4. T. Schmidt, F. Gartner, H. Assadi, and H. Kreye, Development of a Generalized Parameter Window for Cold Spray Deposition, *Acta Mater.*, 2006, 54(3), pp. 729–742.
5. J.R. Davis, *Handbook of Thermal Spray Technology*, ASM International, Materials Park, OH, 2004, pp. 1–36.
6. M.P. Dewar, A.G. McDonald, and A.P. Gerlich, Interfacial Heating During Low-Pressure Cold-Gas Dynamic Spraying of Aluminum Coatings, *J. Mater. Sci.*, 2012, 47(1), pp. 184–198.
7. P. Fauchais, M. Fukumoto, A. Vardelle, and M. Vardelle, Knowledge Concerning Splat Formation: An Invited Review, *J. Therm. Spray Technol.*, 2004, 13(3), pp. 337–360.
8. A. Moridi, S. Hassani-Gangaraj, M. Guagliano, and M. Dao, Cold Spray Coating: Review of Material Systems and Future Perspectives, *Surf. Eng.*, 2014, 30(6), pp. 369–395.
9. A. Faheem, F. Hasan, and Q. Murtaza Material Perspective and Deformation Pattern of Micro-sized Metallic Particle Using Cold Gas Dynamic Spray. In: Narayanan, R., Joshi, S., Dixit, U. (eds) *Advances in Computational Methods in Manufacturing. Lecture Notes on Multidisciplinary Industrial Engineering.* Springer, Singapore, 2019. https://doi.org/10.1007/978-981-32-9072-3_34
10. P. Alkhimov, V. F. Kosarev, and A. N. Papyrin, A Method of Cold Gas-Dynamic Deposition, *Doklady Akademii Nauk SSSR*, 1990, 35, pp. 1062–1065.
11. S. Yin, X. Wang, W. Li, and Y. Li, Numerical Study on the Effect of Substrate Size on the Supersonic Jet Flow and Temperature Distribution Within the Substrate in Cold Spraying, *J. Therm. Spray Technol.*, 2012, 21, pp. 628–635.
12. S. Yin, X. Suo, Z. Guo, H. Liao, and X. Wang, Deposition Features of Cold Sprayed Copper Particles on Preheated Substrate, *Surf. Coat. Technol.*, 2015, 268, pp. 252–256.

13. W. Li, H. Liao, C. Li, G. Li, C. Coddet, and X. Wang, On High Velocity Impact of Micro-Sized Metallic Particles in Cold Spraying, *App. Surf. Sci.*, 2006, 253, 2852–2862.

14. H. Y. Lee, S. H. Jung, S. Y. Lee, Y. H. You, and K. H. Ko, Correlation between Al_2O_3 Particles and Interface of Al–Al_2O_3 Coatings by Cold Spray, *Appl. Surf. Sci.*, 2005, 252, 1891.

15. A. Sturgeon, B. Dunn, S. Celotto, and W. O'Neill, Cold Sprayed Coatings for Polymer Composite Substrate, *ESA (SP)*, September 2006, 616, 1–5.

16. A. Faheem, A. Tyagi, S.M. Pandey, F. Hasan, and Q. Murtaza, A Sustainable Ecofriendly Additive Manufacturing Approach of Repairing and Coating on the Substrate: Cold Spray, *Austr. J. Mech. Eng.*, 2022, pp.1–18.

17. D. Giraud, F. Borit, V. Guipont, M. Jeandin, and J.M. Malhaire, Metallization of a Polymer Using Cold Spray: Application to Aluminum Coating of Polyamide 66, *International Thermal Spray Conference 2012*, Houston, TX, 2012, pp. 265–270.

18. A. Ganesan, J. Affi, M. Yamada, and M. Fukumoto, Bonding Behavior Studies of Cold Sprayed Copper Coating on the PVC Polymer Substrate, *Surf. Coat. Technol.*, 2012, 207, pp. 262–269.

19. M. Gardon, A. Latorre, M. Torrell, S. Dosta, J. Fernandez, and J.M. Guilemany, Cold Gas Spray Titanium Coatings onto a Biocompatible Polymer, *Mater. Lett.*, 2013, 106, pp. 97–99.

20. H. Ye and J. Wang, Preparation of Aluminum Coating on Lexan by Cold Spray, *Mater. Lett.*, 2014, 137, pp. 21–24.

21. M. Fukumoto, H. Terada, M. Mashiko, K. Sato, M. Yamada, and E. Yamaguchi, Deposition of Copper Fine Particle by Cold Spray Process, *Mater. Transactions*, 2009, 50(6), pp. 1482–1488.

22. B. Jodoin, P. Richer, G. Berube, L. Ajdelsztajn, A. Erdi-Betchi, and M. Yandouzi, Pulsed-Gas Dynamic Spraying: Process Analysis, Development and Selected Coating Examples, *Surf. Coat. Technol.*, 2007, 201(16–17), pp. 7544–7551.

Chapter 9

Anodization of implantable metal and alloy surfaces

Purpose and scope

Sikta Panda, Chandan Kumar Biswas,
and Subhankar Paul
National Institute of Technology, Rourkela, Rourkela, Odisha, India

CONTENTS

9.1 INTRODUCTION

Anodization is a form of wet-chemical method involving a chemical solution to yield an attenuated nanostructural layer in the large-scale production for various real-world applications such as bone tissue engineering, pharmaceuticals degradation, sensing, solar energy harvesting, antifouling, and various aesthetics and decorative purposes [1–5]. It can also be explained as a surface engineering process to tailor the natural surface oxide growth by the use of electrical voltage to a metal or alloy inside a suitable composition of electrolytes. The use of anodization dates back to the early 20th century [6]. In the initial phase of the development, the process was limited to value addition and aesthetics unlike recently, where the global demand for anodized aluminium has shot up to 4.65 million metric tons [7]. The process of anodization is a fundamental passivation method achieved by essentially connecting the positive terminal (anode) to the metal/alloy to be engineered and letting the current flow through the circuit. This allows the metal to form metal oxide by bonding with the free oxygen radicals evolved in the electrolyte. After a set time interval, the substrate is disconnected and

DOI: 10.1201/9781003319375-9

rinsed to get rid of excess electrolytes. The reactions taking place during the course of the process are as mentioned below:

Firstly, the metal gets oxidized as

$$M \rightarrow M^{a+} + ae^-$$ (9.1)

Then, the cathodic reduction takes place as

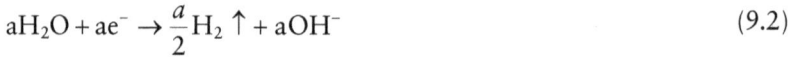

$$aH_2O + ae^- \rightarrow \frac{a}{2}H_2 \uparrow + aOH^-$$ (9.2)

Then, M^{a+} reacts with oxygen to form oxides as

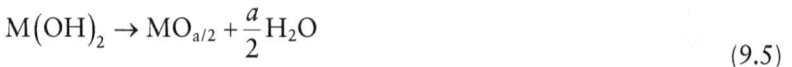

$$M + \frac{a}{2}H_2O \rightarrow MO_{a/2} + aH^+ + ae^-$$ (9.3)

$$M^{a+} + aH_2O \rightarrow M(OH)_2 + aH^+$$ (9.4)

$$M(OH)_2 \rightarrow MO_{a/2} + \frac{a}{2}H_2O$$ (9.5)

The schematic diagram for the process is shown in Figure 9.1(a).

Three possible situations in the case of anodization of metal/alloy may arise depending on electrolyte composition and process parameters such as the following: (i) the M^{a+} ions get solvitized in the electrolytic solution, thus the metal gets dissolved, (ii) the M^{a+} ions react with O^{2-} and a dense oxide film is formed, or (iii) a contest of dissolution and oxide formation occurs

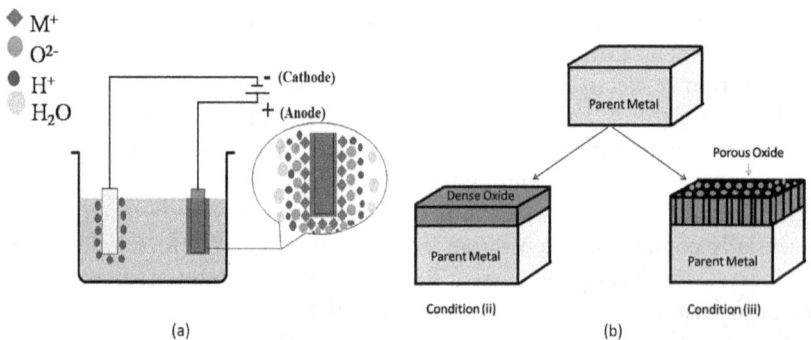

Figure 9.1 (a) Schematic diagram for the anodization process and (b) types of oxide layers formed on a parent material as a result of anodization.

and this contest gives rise to a porous oxide layer onto the parent metal/ alloy [6]. This process is shown in Figure 9.1(b). The above reactions take place at the juncture of the metal–oxide interface, forcing the ions to drift all the way through the oxide under the applied electric field. In the case of condition (ii), oxide formation can be described by

$$E = \Delta U / t_{layer} \tag{9.6}$$

where E denotes the strength of the electric field, ΔU is the applied voltage, and t_{layer} is oxide layer thickness.

With the increase in the oxide layer thickness, the field strength decreases to a threshold until ion migration becomes insignificant; this states that there is a practical maximum oxide thickness which can be achieved (around a few hundred nanometres). The current is expressed by the following equation:

$$i = i_o e^{(\beta E)} \tag{9.7}$$

where i represents the oxide formation current and i_o and β are material constants. On the other hand, in condition (iii) mentioned above, there exists an opposition between solvation and oxide formation leading to the formation of pores in the oxide. These pores can be initiated by adding an agent (e.g., fluoride) capable of etching through the oxide layer in the electrolytic solution.

The application of anodization for growing a thick and protective oxide on metallic implants is not very old and has been in practice for the past couple of decades [1]. The properties of these oxide layers can be governed by tuning the process parameters like current, voltage, electrolyte composition, agitation of electrolyte, stand-off distance, and temperature [9–11]. The most abundantly used metals/alloys for bone tissue engineering include titanium, tantalum, zirconium, niobium, magnesium, stainless steel, and cobalt–chromium alloys. These materials have an inherent propensity to form a thin natural oxide film when exposed to atmospheric oxygen. However, in recent years, numerous attempts have been made to enhance the thickness of the oxide layers with an aim to provide barrier action against corrosion and functionalizing them to enhance their performance *in vivo*.

We shall discuss the importance of the anodization process and its various applications in the field of biomedical engineering in this chapter. Also, some light will be shed on the applications beyond medicine. The effect of the process parameters on the final outcome will also be discussed in due course. Finally, the scope and future development in this field shall be focused on.

9.2 EFFECT OF PROCESS PARAMETERS

The applicability of an oxide layer depends on its major physiochemical properties, e.g., its optical properties, non-cytotoxicity, corrosion resistance, and strength. To achieve these properties in the desirable range, the oxide layer formed by anodization needs to be optimized for the hardness, thickness, porosity, and morphology of the nanostructures. These properties have been found to be greatly influenced by the input process parameters such as current density, electrolyte temperature, electrolyte concentration, stand-off distance, process time, and applied voltage. Following are some examples of how these parameters affect the material properties.

The distribution of micropores and nanopores resulting from anodization depends on the transmission of defects of the substrate to the surface of the oxide layer and the interaction of the oxide fibers, respectively. In addition to this, the shape of these pores depends on the pH and concentration of the electrolyte used [9]. One can get an entirely different set of material properties of the oxide layer than the substrate by the correct application of parameters. For example, Cu_2O resulting from anodization at a controlled temperature and electrolyte concentration is nonconductive and way much harder than copper [10]. Along with the physicochemical parameters, mechanical properties like fatigue strength, ultimate tensile strength, elongation, microhardness, and corrosion resistance of the material are also greatly influenced by anodizing bath parameters [11, 15, 20, 21]. The barrier effect of the oxide films against corrosion is dependent on the thickness and porosity of the film which is further found to be governed by the anodization parameters like ramp time and current density [12, 13, 16, 20]. Apart from this, the biological response of the oxide layer is also greatly affected by these parameters [18]. These dependencies of the productivity and quality of anodization on the input parameters compelled the researchers to explore the precise values of these parameters for the superlative output factors by using both experimental and predictive empirical modeling [9–21]. The effects of the above parameters are discussed lucidly in Table 9.1, and Figure 9.2 depicts them with the help of a fishbone diagram.

So, like any other surface engineering process, the optimization of the input factors is of significant importance for desired output in the anodization process. By tuning these factors, not only can we alter the physical properties of the oxide but also its functionality can be augmented for numerous applications. Many researchers have attempted to study the process and optimize the process parameters to draw meaningful conclusions in the past. Table 9.1 summarizes the key effects of the process parameters on the anodization of some relevant metals/alloys.

So, we can infer here that by controlling the parameters of the anodizing process not only physical properties but the functionalities depending on these properties can also be customized. In Section 9.3, we shall discuss the

Table 9.1 Effect of the input parameters on anodization

Authors	Metal/alloy	Input parameters	Output parameters	Inferences
Bara et al. [9]	EN AW-5251 Alloy	Current density, process time, electrolyte temperature	Porosity, pore density, contact angle	Increased current density decreases the number of pores. Hydrophilicity of the surface is influenced by the temperature of the electrolyte and time.
Mahmood et al. [10]	Copper	Electrolyte concentration, temperature, voltage	Microhardness	Hardness was increased after anodization where temperature was the most and voltage was the least influencing factors. Lower temperature and higher concentration yielded higher microhardness.
Kudari et al. [11]	2024-T-351 Alloy	Temperature, voltage, time	Fatigue strength, ultimate stress, elongation	Fatigue strength decreased after anodizing due to hydrogen embrittlement. Increase in time increases fatigue strength.
Rattanasatitkul et al. [12]	Aluminium	Current density, ramp time, anodizing time	Film thickness	A long ramp time will decrease burn defect. A longer anodizing time renders greater film thickness.
Chand et al. [13]	AA 7075	Electrolyte concentration, electrolyte temperature, voltage	Film thickness, porosity, average roughness	For maximum film thickness, a combination of lowest concentration and high temperature and voltage was recommended.
Kao et al. [14]	Aluminium	Voltage, electrolyte concentration, temperature	Volumetric expansion of oxide film	Voltage and temperature were the significant factors in volumetric expansion of the film.
Prisley et al. [15]	AA 6061	Voltage, electrolyte concentration, temperature	Thickness and microhardness	Temperature and voltage are the most significant factors affecting thickness and microhardness, respectively. Higher concentrations gave higher thickness and microhardness.

(Continued)

Table 9.1 (Continued) Effect of the input parameters on anodization

Authors	Metal/alloy	Input parameters	Output parameters	Inferences
Chowdhury et al. [16]	Aluminium	Voltage, electrolyte temperature, anodizing time	Pore dimensions and growth rate	Vertical growth rate of the pores showed exponential rise with voltage and linear increase with temperature. Pore diameter was proportional to voltage.
Diamanti et al. [17]	cp Ti	Electrolyte molarity, current density	Morphology and crystal phase	Increase in current density and decrease in electrolyte molarity supports anatase TiO_2 growth. Morphology is hugely affected by applied potential. Low current density gave lower oxidation.
Manjaiah et al. [18]	cp Ti (Grade 4)	Anodizing voltage	Surface composition, appearance, thickness, surface topography, cell viability	A variety of colors were obtained by varying voltage. Rutile phase gets converted to anatase by increasing the voltage. The increase in thickness was linear in the beginning but was exponential with higher voltages and same is true for surface roughness. Anodization significantly increased the cell proliferation.
Chai et al. [19]	AZ 31 alloy	Voltage, current density, electrolyte concentration, temperature	Film stability, surface morphology, corrosion protection	Stable anodic films produced with relatively lower potential. Better corrosion protection was obtained with increased current density. Higher temperature yielded weak porous film.
Mohan et al. [20]	cp Ti, Ti6Al4V, Ti6Al7Nb, Ti13Nb13Zr	Electrolyte temperature, anodization time	Pore diameter, wall thickness	Increased time yielded thicker TiO_2 layer. Higher pore diameter and elongated pores were formed at elevated bath temperatures.
Sowa et al. [21]	Tantalum	Voltage	Film thickness, corrosion protection	Meticulous tuning of voltage can control thickness, corrosion potential, and roughness.

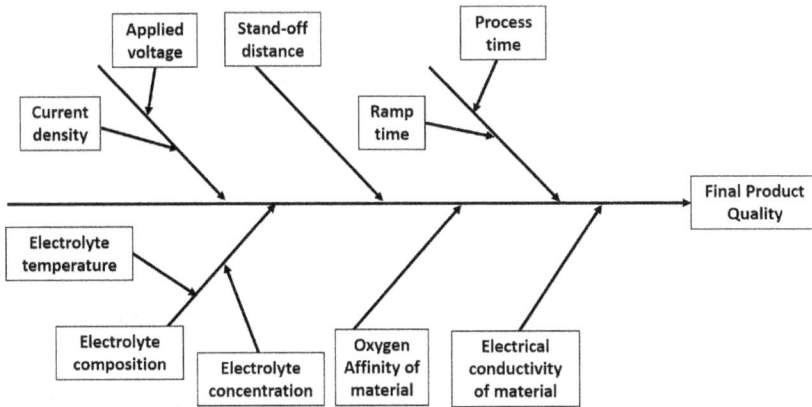

Figure 9.2 Fishbone diagram showing the main process parameters affecting the anodized product.

effect of anodizing on the bio-activity of some widely used metals/alloys in the field of bone tissue engineering.

9.3 ANODIZATION OF METALLIC IMPLANTS

For several years, metallic implants have been in extensive use in case of unrecoverable damage to the skeletal system caused mostly by accidents and age-related ailments. These metals also find applicability in maxillofacial implantation and other medical use like heart stents, screws, and non-implantable auxiliary medical devices. Metals are preferably used for long-term implants owing to their excellent mechanical traits. Some of the popular metals/alloys used in prosthetic devices are given in Table 9.2.

Table 9.2 Popular metallic bio-implants

Titanium and its alloys	Tantalum	Surgical grade SS	Co-Cr alloys	Zirconium-based alloys	Magnesium alloys
cp Ti	Porous tantalum	AISI SS 316L	CoCrMo	ZrNb-based alloys	AZ 31 alloy (MgAlZn alloys)
Ti6Al4V	Cold formed tantalum			ZrNbTi-based alloys	ZK 40 (MgZn-based alloys)
Ti6Al4V Extra Low Terrestrials (ELI)	Annealed tantalum	AISI SS 304L	CoNiCrMo	ZrMo-based alloys ZrSiNb-based alloys	MgCa-based alloys Mg-rare element-based alloys

It is of utmost importance for any material to have biocompatibility in order to find applicability as an implant material. Biocompatibility here is an extensive expression and is comprised of very widespread aspects related to the material. Here, we would consider the two main aspects of biocompatibility of an implantable metal/alloy; the first one is surface or superficial compatibility, which corresponds to the chemical and biological response of the material to the host tissue. This includes the ability of the surface to aid the attachment, growth, and proliferation of the host tissue on it without causing any toxic effect or morphological distortion of the cells. The second one is mechanical compatibility which deals with their structural behavior, including their shear strength, fatigue strength, corrosion resistance, and hardness.

Despite possessing comparable mechanical strength as human bone, metals also have a fair share of disadvantages too. They are bio-inert and also vulnerable to electrochemical and fretting corrosion. These disadvantages, if unattended may lead to localized leaching of harmful ions, bacteria colonization, and ultimately implant failure. Thus, a very meticulous alteration of the surface properties of implant materials by the application of surface engineering is chosen to avoid the discussed complications without compromising on their bulk properties. Some of these surface engineering strategies involve the application of biocompatible coating, changing surface morphologies by sand/grit blasting or laser ablation, machining, ion implantation and deposition, acid/alkali treatment, and anodizing (anodic oxide deposition). These implantable metals generally exhibit great affinity toward oxygen, forming an inherent oxide layer which has been found to be bioactive and anti-corrosive in nature [1, 22, 23].

Though metals are known to grow superficial oxides *suo motu*, it has been exposed to anodic oxidation either to thicken this layer or tailor their physical and chemical properties [1]. Generally, a three- or two-electrode electrochemical cell is used for carrying out the process in an acidic electrolytic media. Aqueous buffered solution or non-aqueous electrolytes with F^- ions can also be used as electrolytes apart from aqueous acidic solutions. As discussed earlier, the oxide formation takes place in two stages, i.e., first, the thick metal dioxide formation and second, the nanostructure formation with the balance in solvation and oxidation [8]. HF is the most preferred electrolyte for anodizing and it selectively etches away the surface giving rise to the formation of nanotubes. For proper growth of the nanostructures, the control of process parameters is very important as discussed earlier. The process of growth of oxides by anodization has been discussed extensively in the available articles so we shall discuss the biological relevance of the process applied to some popular metallic biomaterials here.

It has been established by former investigations that a rougher or patterned (micro/nano) supports rapid cell growth and proliferation [18, 20, 31, 35]. This can be attributed to the enhanced wettability (hydrophilicity) of the surface as a result of the irregular or patterned surface [21]. These surfaces are highly capable of adsorbing and retaining the proteins to help cell attachment [48]. The patterns on the surface of metallic implants can be generated by surface engineering techniques like abrasion, etching, laser ablation, and anodization. Not only the bioactivity but corrosion and wear resistance have been found to have improved by the application of anodic oxidation [24, 25, 32]. The as-grown oxide film not only serves as a barrier to corrosion and wear but it also serves as an intermediate layer for the application of secondary coatings and hence enhances its adhesion to the parent anodized material [21, 27–30]. Apart from being able to promote cell attachment and differentiation, anodized surfaces were found to have promising performance as drug delivery systems [37, 46]. It is also worth noting that the biocompatibility of the surfaces is greatly influenced by the morphology of the oxide films (pore diameter, shape, distribution, and thickness) [40–42] which is a function of the input parameters as discussed in Section 9.2. Table 9.3 elucidates some of the path-breaking findings regarding the biocompatibility of the anodized surfaces.

Table 9.3 Anodization of bio-implants

Author	Substrate	Oxide	Key findings
Manjaiah et al. [18]	cp Ti (Gr 4)	TiO_2	Cell viability enhanced significantly.
Mohan et al. [20]	Ti_6Al_7Nb	TiO_2	Growth of nanotubular structures on titanium alloys for better cell adhesion.
Sowa et al. [21]	Tantalum	HAP/ Ta_2O_5	Improved coating adhesion and lower contact angle were achieved
Li et al. [24]	TiNbTaZr	Nb_2O_5	Improved wear resistance.
Al-Mobarak et al. [25]	cp Ti	TiO_2	Lower TiO_2 nanotubes diameter ensures better corrosion protection.
El-Wassefy et al. [26]	cp Ti	TiO_2	Increased surface roughness, surface area with larger time of anodization.
Mosialek et al. [27], Ossowska et al. [28], Ossowska et al. [29], Zielinski et al. [30]	Ti13Nb13Zr	TiO_2	Better corrosion protection, Better adhesion of secondary coating (HAP) post anodization.
Park et al. [31]	cp Ti	TiO_2	Improved osteoblasts binding confirming better biocompatibility.
Kim et al. [32]	$Ti_{29}NbxZr$	Nb_2O_5, ZrO_2	Elevated corrosion resistance owing to the existence of Nb_2O_5, ZrO_2.

(Continued)

Table 9.3 (Continued) Anodization of bio-implants

Author	Substrate	Oxide	Key findings
Saji et al. [33]	$Ti_{35}Nb_5Ta_7Zr$		Heat treatment enhances barrier layer and nanotubes fusion.
Cheung et al. [34]	Ti_6Al_4V	TiO_2	Microstructures and morphology are greatly influenced by electrolyte composition; optical properties of anodized titanium can be applied for photo-bioluminescence purposes.
Huang et al. [35]	cp Ti	TiO_2	Increased cell adhesion (MG-63 cell line).
Fialho et al. [36]	Tantalum sheet	Ta_2O_5	Ta_2O_5 can be grown with the use of an HF-free electrolyte; a two-step anodization led to highly ordered Ta_2O_5 layer.
Kunrath et al. [37]	Grade II titanium plate	TiO_2	The oxides were found to be non-cytotoxic and highly promising drug delivery systems.
Mohan et al. [38]	AZ 31 alloy	MgO	Better cell attachment (MG63) and corrosion resistance were achieved.
Choi et al. [39]	Nb foils	Nb_2O_5	The aspect ratio of the pores was not satisfactory.
Bauer et al. [40]	Zirconium foils	ZrO_2	Mesenchymal stem cell adhesion and proliferation were found to depend on ZrO_2 nanotube diameter; optimum diameter was found to be 15 nm for best cellular activity.
Oh et al. [41], Bauer et al. [42]	Titanium sheet/foils	TiO_2	Cell adhesion and growth (without differentiation) and guided differentiation were strongly influenced by the size of nanotubes.
Lai et al. [43], Bauer et al. [44]	Titanium foils	TiO_2	Functionalization of the TiO_2 nanotubes with other bio-molecules (BMP2) significantly increases mesenchymal stem cell differentiation.
Wang et al. [45]	Tantalum foils	Ta_2O_5	Anticorrosion, biocompatibility, and osteoinduction can be induced to pure tantalum for implantation and implant coating application.
Yao et al. [46]	Titanium	TiO_2	Easy drug loading by precipitation method and a prolonged delivery can be achieved.

(Continued)

Table 9.3 (Continued) Anodization of bio-implants

Author	Substrate	Oxide	Key findings
Hatakeyama et al. [47]	TiNbSn	TiO_2	Improved wear resistance.
Uslu et al. [48]	Tantalum	Ta_2O_5	Protein adsorption enhanced significantly, improved bone cell proliferation (50% increase), and enhanced cellular spreading (23% increase).
Luz et al. [49]	$Ti_{35}Nb$	TiO_2	Improved tribological properties (lower wear rate).
Ohtsu et al. [50]	NiTi alloy	TiO_2	Enhanced growth of EA.hy926 endothelium cells.

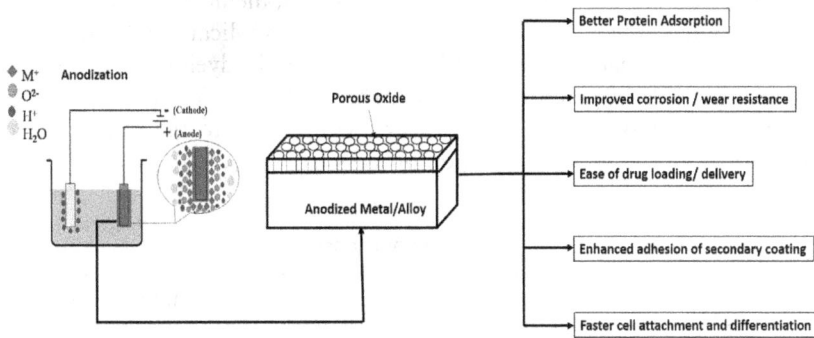

Figure 9.3 Effects of anodization on metallic implants.

From the above discussion, it is quite clear that not only the physical traits like surface roughness, hardness, and hydrophilicity of a surface can be altered by anodization but associated properties like corrosion resistance, wear resistance, and biocompatibility can also be augmented by growing nanostructured oxide layers and functionalizing them by anodization. These nanostructures, when optimized for their dimensions, may act as reservoirs of essential drugs and exhibit a very controlled release and also facilitate ease of loading. The oxide layers can be highly ordered if grown with a careful selection of parameters and act as an intermediate layer for the secondary bioactive coatings facilitating better adhesion. Figure 9.3 elucidates the effects of the anodization process on metallic implants.

9.4 OTHER APPLICATIONS OF ANODIZATION

In Section 9.3, we discussed the biological aspect of the nanostructures grown on metallic surfaces by the application of anodic oxidation. But it has been well established in the past that metal oxides have commendable photoluminescence and thermal conductivity along with other mechanical properties which makes them suitable for an assortment of applications other than the biomedical sector. This section will devotedly shed light on the application of anodization beyond tissue engineering.

Anodization can also be applied to form an oxide layer on a non-native substrate. Mor et al. [2] showed from their experiment that highly ordered TiO_2 nanotube arrays can be formed on a glass substrate via anodization and these were found to have excellent photo-conversion efficiency to be used in solar cells. Similarly, tungsten oxide layer grown on titanium foil can be employed for water splitting (photo-electrolysis) applications [5]. The right selection of input parameters can yield super-hydrophobic surfaces in contrast to the porous hydrophilic surfaces intended for biocompatibility as was discussed in Section 9.2 [3]. Along with the applications in solar cells due to excellent photo-conversion efficiency, the major applications of the anodization process include the fabrication of heavy-duty metallic surfaces for aerospace and automobile sectors [3, 54, 55], medical devices [51, 52], spectroscopic applications [56], Flame retarding applications [61], sensors [63, 67], supercapacitors [64, 70], non-electrolytic dyeing [71], and dye degradation [78–80] among others. Table 9.4 jots down some of the important findings of the last few years with regard to the multitudes of applications of the anodic oxidation process.

Table 9.4 Application of anodization beyond tissue engineering

Author	Substrate	Oxide	Applications
Mor et al. [2]	Glass substrate	TiO_2	Dye-sensitized solar cells (excellent photo-conversion efficiency)
Roy et al. [3]	SS T-304	TiO_2	Super-hydrophobic surface for heavy-duty applications
Shankar et al. [4]	Titanium foils	TiO_2	Water photo-electrolysis and dye-sensitized solar cells
Lai et al. [5]	Titanium foils	WO_3/TiO_2	Photo-electrochemical (PEC) water-splitting performances
Hang et al. [51], Hang et al. [52]	NiTi alloy (Nitinol)	NiTiO	Glucose sensing, photocatalytic activity, medical implants
Venturini et al. [53]	Titanium foils	Cobalt-doped TiO_2	Water splitting application, catalyst eco-friendly fuels
Kozhukharov et al. [54]	AA 2024-T3 aircraft alloy	CeO_2/Al_2O_3	Aerospace application
Yoganandan et al. [55]	AA 2024 aircraft alloy	Mn–Mo oxyanions	Aerospace application

(Continued)

Table 9.4 (Continued) Application of anodization beyond tissue engineering

Author	Substrate	Oxide	Applications
Baek et al. [56]	Gold thin films on Silicon wafer	$Au(OH)_2$	Surface-enhanced Raman spectroscopy applications
Lucas-Granados et al. [57]	Iron rod	Fe_2O_3	Photoelectrochemical applications
Zakir et al. [58]	Titanium	TiO_2	Photocatalytic application
Kozhukhova et al. [59]	Al (Al6082)	Al_2O_3	Catalyst for catalytic hydrogen combustion applications
Sakai et al. [60]	Gold electrode	Porous gold nanostructures	Electrodes in direct-electron-transfer-type bioelectrocatalysis
Silva et al. [61]	Aluminium	$Al(OH)_3$	Flame retarding applications
Yousif et al. [62]	Titanium foil	TiO_2	Dye-sensitized solar cell
Lashkov et al. [63]	Titanium wires	TiO_2	One-electrode gas sensor
Tantray et al. [64]	Zinc foil	ZnO	Electrodes used in supercapacitors
Domene et al. [65]	Tungsten rods	WO_3	Photoelectrocatalytic applications
Salehi et al. [66]	Aluminum	Al_2O_3	Solar panel cooling systems
Rosli et al. [67]	Titanium foil	TiO_2	Chromium sensing and removal
Lee et al. [68]	AA 3104 H18	Al_2O_3	Self-cleaning application in catalyst for hydrodeoxygenation (HDO) to produce the bio-fuels by controlling wettability through anodizing
Ponchio et al. [69]	Copper sheet	Cu_2O	Photoelectrocatalytic cell
Wang et al. [70]	Nickel sheet	NiO	Supercapacitor electrodes
Kongvarhodom et al. [71]	AA 6063	Al_2O_3	Non-electrolytic dyeing

It is quite clear from the above table that not only does anodization help in enhancing biocompatibility by enhancing surface area, surface roughness, and wettability of metallic bio-implants, but it also amends the thermo-physical, tribological, and photo-luminescence properties, making a surface capable of being used in diverse fields. Figure 9.4 shows the major areas of application of anodization.

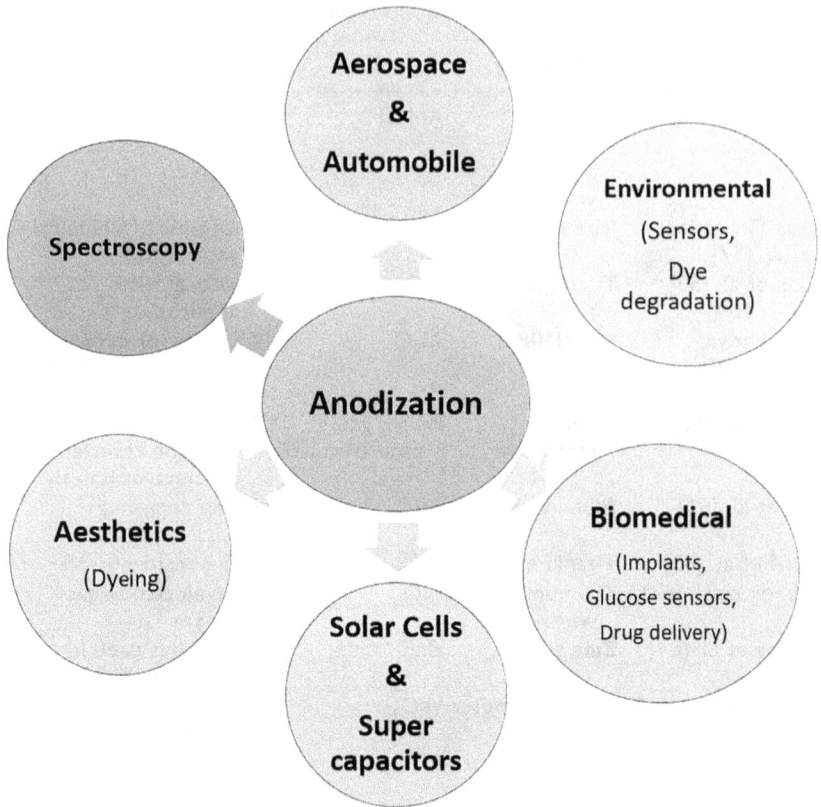

Figure 9.4 Major areas of application of anodization.

9.5 SUMMARY

Previous sections explain how the controlled application of anodization parameters enhances the surface properties of a material. By careful selection of the current source and combination of electrolytes, the surface of a conducting material can be functionalized in a highly precise manner with the development of nanostructured oxide films for multifaceted applications. Further, non-conducting surfaces like glass and silicon can also be subjected to anodization with the application of some pre-treatments like physical vapor deposition (PVD) and electron beam evaporation for being used in specialized applications like super-capacitors, electrodes, and other spectroscopic instruments [2, 56]. Along with the other parameters, the current source for anodization may also vary as DC, AC, or pulsed, depending upon the desired surface [75, 76].

Yasuda and Schmuki [72] showed that by applying repeated anodization to a surface, multilayer nanostructures of ZrO_2 can be formed. These staked layers can be produced by filling the voids between the existing nanostructures making the nanotubes' length and diameter adjustable. In another study, the duo also proved a similar tendency for titanate nanotubes [73], establishing the fact that self-organizing stacked nanotubes can be produced by repeated anodization steps. It is obvious from the available literatures that oxides can also be developed on non-native substrates by the anodization process [3, 54].

It has been established in the past that with the assistance of anodization, a layer of multi-element oxides can also be produced for specialized functional surfaces [5, 32, 55, 74]. Anodization not only creates a functional surface itself but also helps in enhancing the adhesion strength of secondary layer coatings (bioactive HAP and other dyes) by promoting an intermediate adhesive oxide layer [21, 30, 71, 77]. Anodization is a very effective technique toward cleaner energy generation by building solar cells and water-splitting applications [2, 4, 53, 60, 68]. This process can also contribute toward a cleaner environment through its applications in gas sensors [63], removal of toxic elements [67], and dye degradation [78–80].

9.6 CONCLUSIONS

It can be concluded from previous sections that the anodization process is a hassle-free surface engineering approach to modifying the surface of a substrate by depositing an oxide layer onto it. The morphology of the oxide layer can be tailored for specific applications by careful selection and optimization of process parameters. Further, this process also provides means for the generation of green energy from bio-fuel and solar energy harvesting. Its contribution toward a sustainable environment by application in dye degradation and adsorption of toxic elements cannot be overlooked. Anodization can also be a great tool for making an implant comparatively more biocompatible with the presence of an adherent oxide layer formed on its surface. It also helps in enhancing the adhesion of the secondary bioactive coating onto the metallic implant to enhance cell adhesion and proliferation. Not only the biological performance of an implant but its physical traits like wear and corrosion resistance and bioluminescence can be enhanced with anodization. So, in conclusion, we can say that anodization is a multipurpose surface engineering process having myriads of applications and scope of improvements. Thus, research must be continued to optimize and explore its applicability in various fields.

REFERENCES

1. Minagar S, Berndt CC, Wang J, Ivanova E, Wen C. A review of the application of anodization for the fabrication of nanotubes on metal implant surfaces. *Acta Biomaterialia*. 2012 Aug 1 8(8):2875–88.

2. Mor GK, Shankar K, Paulose M, Varghese OK, Grimes CA, Use of highly-ordered TiO$_2$ nanotube arrays in dye-sensitized solar cells. *Nano Letters.* 2006;6:215–218.

3. Roy P, Kisslinger R, Farsinezhad S, Mahdi N, Bhatnagar A, Hosseini A, Bu L, Hua W, Wiltshire BD, Eisenhawer A, Kar P, Shankar K, All-solution processed, scalable superhydrophobic coatings on stainless steel surfaces based on functionalized discrete titania nanotubes. *Chemical Engineering Journal.* 2018;351:482–489.

4. Shankar K, Mor GK, Prakasam HE, Yoriya S, Paulose M, Varghese OK, Grimes CA, Highly-ordered TiO$_2$ nanotube arrays up to 220 μm in length: Use in water photoelectrolysis and dye-sensitized solar cells. *Nanotechnology.* 2007;18:065707.

5. Lai CW, Sreekantan S. Preparation of hybrid WO$_3$–TiO$_2$ nanotube photoelectrodes using anodization and wet impregnation: Improved water-splitting hydrogen generation performance. *International Journal of Hydrogen Energy.* 2013 Feb 19;38(5):2156–66.

6. Runge JM. *A Brief History of Anodizing Aluminum, in the Metallurgy of Anodizing Aluminum: Connecting Science to Practice* (Springer International Publishing, Cham, 2018), 65–148.

7. *Technavio, Aluminum Extrusion Market by Product, End-user, and Geography – Forecast and Analysis 2021–2025* https://www.technavio.com/report/aluminum-extrusion-market-industry-analysis (Apr 2021).

8. Fojt J, Moravec H, Joska L. Nanostructuring of titanium for medical applications. In *Nanocon 2010, 2nd International Conference* 2010 (pp. 209–13). Olomouc, Czech Republic, EU: Tanger Ltd.

9. Bara M, Niedźwiedź M, Skoneczny W. Influence of anodizing parameters on surface morphology and surface-free energy of Al$_2$O$_3$ layers produced on EN AW-5251 alloy. *Materials.* 2019 Jan;12(5):695.

10. Mahmood MH, Al Hazza MH, Haider FI. Study the influence of the anodizing process parameters on the anodized copper hardness. In *MATEC Web of Conferences* 2017 (Vol. 130, p. 08003). EDP Sciences.

11. Kudari SK and Sharanaprabhu CM. The effect of anodizing process parameters on the fatigue life of 2024-T-351-aluminium alloy. *Fatigue of Aircraft Structures* 2017;2017(9):109–115. https://doi.org/10.1515/fas-2017-0009

12. Rattanasatitkul A, Prombanpong S, Tuengsook P. An effect of process parameters to anodic thickness in hard anodizing process. In *Materials Science Forum* 2016 (Vol. 872, pp. 168–172). Trans Tech Publications Ltd.

13. Chand S, Jalali K, Oommen G, Chakravarthy P, Murugesan N, Arun PN, Manickavasagam A. Effect of process parameters on anodization of AA7075. In *Materials Science Forum* 2015 (Vol. 830, pp. 643–646). Trans Tech Publications Ltd.

14. Kao TT, Liu TK, Tsai YW. Optimization of anodizing process parameters for the volume expansion of anodic aluminum oxide film by Taguchi method. In *11th IEEE International Conference on Control & Automation (ICCA)* 2014 Jun 18 (pp. 590–595). IEEE.

15. Prisley VM, Shajan K, Sunny KG. Study of process parameters on the thickness and micro hardness during the sulphuric acid anodization of AA6061. *International Journal of Engineering and Innovative Technology.* 2013;3(6):110–4.

16. Chowdhury P, Raghuvaran K, Krishnan M, Barshilia HC, Rajam KS. Effect of process parameters on growth rate and diameter of nano-porous alumina templates. *Bulletin of Materials Science.* 2011 Jun 1;34(3):423–7.

17. Diamanti MV, Pedeferri MP. Effect of anodic oxidation parameters on the titanium oxides formation. *Corrosion Science.* 2007 Feb 1;49(2):939–48.

18. Manjaiah M, Laubscher RF. Effect of anodizing on surface integrity of Grade 4 titanium for biomedical applications. *Surface and Coatings Technology.* 2017 Jan 25;310:263–72.

19. Chai L, Yu X, Yang Z, Wang Y, Okido M. Anodizing of magnesium alloy AZ31 in alkaline solutions with silicate under continuous sparking. *Corrosion Science.* 2008 Dec 1;50(12):3274–9.

20. Mohan L, Dennis C, Padmapriya N, Anandan C, Rajendran N. Effect of electrolyte temperature and anodization time on formation of TiO_2 nanotubes for biomedical applications. *Materials Today Communications.* 2020 Jun 1;23:101103.

21. Sowa M, Woszczak M, Kazek-Kęsik A, Dercz G, Korotin DM, Zhidkov IS, Kurmaev EZ, Cholakh SO, Basiaga M, Simka W. Influence of process parameters on plasma electrolytic surface treatment of tantalum for biomedical applications. *Applied Surface Science.* 2017 Jun 15;407:52–63.

22. Liu X, Chu P, Ding C. Surface modification of titanium, titanium alloys, and related materials for biomedical applications. *Materials Science and Engineering R,* 2004;47:49–121

23. Khudhair D, Bhatti A, Li Y, Hamedani HA, Garmestani H, Hodgson P, Nahavandi S. Anodization parameters influencing the morphology and electrical properties of TiO_2 nanotubes for living cell interfacing and investigations. *Materials Science and Engineering: C.* 2016 Feb 1;59:1125–42.

24. Li SJ, Yang R, Li S, Hao YL, Cui YY, Niinomi M, Guo ZX. Wear characteristics of Ti–Nb–Ta–Zr and Ti–6Al–4V alloys for biomedical applications. *Wear.* 2004;257:869–76.

25. Al-Mobarak NA, Al-Swayih AA. Development of titanium surgery implants for improving osseointegration through formation of a titanium nanotube layer. *International Journal of Electrochemical Science* 2014 Jan 1;9:32–45.

26. El-Wassefy NA, Hammouda IM, Habib AN, El-Awady GY, Marzook HA. Assessment of anodized titanium implants bioactivity. *Clinical Oral Implants Research.* 2014 Feb;25(2):e1–9.

27. Mosiałek M, Nawrat G, Szyk-Warszyńska L, Żak J, Maciej A, Radwański K, Winiarski A, Szade J, Nowak P, Simka W. Anodic oxidation of the Ti–13Nb–13Zr alloy. *Journal of Solid State Electrochemistry.* 2014 Nov 1;18(11):3073–80.

28. Ossowska A, Sobieszczyk S, Supernak M, Zielinski A. Morphology and properties of nanotubular oxide layer on the 'Ti–13Zr–13Nb' alloy. *Surface and Coatings Technology.* 2014 Nov 15;258:1239–48.

29. Ossowska A, Zieliński A, Supernak M. Formation of high corrosion resistant nanotubular layers on titanium alloy Ti13Nb13Zr. In *Solid State Phenomena* 2012 (Vol. 183, pp. 137–142). Trans Tech Publications Ltd.

30. Zieliński A, Antoniuk P, Krzysztofowicz K. Nanotubular oxide layers and hydroxyapatite coatings on 'Ti–13Zr–13Nb' alloy. *Surface Engineering.* 2014 Sep 1;30(9):643–9

31. Park JW, Kim YJ, Jang JH, Kwon TG, Bae YC, Suh JY. Effects of phosphoric acid treatment of titanium surfaces on surface properties, osteoblast response and removal of torque forces. *Acta Biomaterialia.* 2010 Apr 1;6(4):1661–70.

32. Kim JU, Kim BH, Lee K, Choe HC, Ko YM. Corrosion behavior of nanotubular oxide on the Ti–29Nb–xZr alloy. *Journal of Nanoscience and Nanotechnology.* 2011 Feb 1;11(2):1636–9.

33. Saji VS, Choe HC. The effect of post-heat treatment on the ion dissolution behavior of nanotubular titanium alloys. *Metals and Materials International.* 2011 Apr;17(2):275–8.

34. Cheung KH, Pabbruwe MB, Chen WF, Koshy P, Sorrell CC. Thermodynamic and microstructural analyses of photocatalytic TiO_2 from the anodization of biomedical-grade Ti6Al4V in phosphoric acid or sulfuric acid. *Ceramics International.* 2021 Jan 15;47(2):1609–24.

35. Huang BH, Lu YJ, Lan WC, Ruslin M, Lin HY, Ou KL, Saito T, Tsai HY, Lee CH, Cho YC, Yang TS. Surface properties and biocompatibility of anodized titanium with a potential pretreatment for biomedical applications. *Metals.* 2021 Jul;11(7):1090.

36. Fialho L, Alves CA, Marques LS, Carvalho S. Development of stacked porous tantalum oxide layers by anodization. *Applied Surface Science.* 2020 May 1;511:145542.

37. Kunrath MF, Penha N, Ng JC. Anodization as a promising surface treatment for drug delivery implants and a non-cytotoxic process for surface alteration: A pilot study. *Journal of Osseointegration.* 2020 Jan 7;12(1).

38. Mohan L, Kar S, Nandhini B, Kumar SS, Nagai M, Santra TS. Formation of nanostructures on magnesium alloy by anodization for potential biomedical applications. *Materials Today Communications.* 2020 Dec 1;25:101403.

39. Choi J, Lim JH, Lee SC, Chang JH, Kim KJ, Cho MA. Porous niobium oxide films prepared by anodization in HF/H3PO4. *Electrochimica Acta.* 2006 Jul 28;51(25):5502–7.

40. Bauer S, Park J, Faltenbacher J, Berger S, von der Mark K, Schmuki P. Size selective behavior of mesenchymal stem cells on ZrO_2 and TiO_2 nanotube arrays. *Integrative Biology.* 2009 Aug 25;1(8–9):525–32.

41. Oh S, Brammer KS, Li YJ, Teng D, Engler AJ, Chien S, Jin S. Stem cell fate dictated solely by altered nanotube dimension. *Proceedings of the National Academy of Sciences.* 2009 Feb 17;106(7):2130–5.

42. Bauer S, Park J, Mark K, Schmuki P. Improved attachment of mesenchymal stem cells on super-hydrophobic TiO_2 nanotubes. *Acta Biomaterialia.* 2008;4(5):1576–1582.

43. Lai M, Cai K, Zhao L, Chen X, Hou Y, Yang Z. Surface functionalization of TiO_2 nanotubes with bone morphogenetic protein 2 and its synergistic effect on the differentiation of mesenchymal stem cells. *Biomacromlolecules.* 2011;12(4):1097–1105.

44. Bauer S, Park J, Pittrof A, Song Y, von der Mark K, Schmuki P. Covalent functionalization of TiO_2 nanotube arrays with EGF and BMP-2 for modified behavior towards mesenchymal stem cells. *Integrative Biology.* 2011;3(9):927–936.

45. Wang N, Li H, Wang J, Chen S, Ma Y, Zhang Z. Study on the anticorrosion, biocompatibility, and osteoinductivity of tantalum decorated with tantalum oxide nanotube array films. *ACS Applied Materials & Interfaces.* 2012;4(9):4516–4523.

46. Yao C, Webster TJ. Prolonged antibiotic delivery from anodized nanotubular titanium using a co-precipitation drug loading method. *Journal of Biomedical Materials Research. Part B, Applied Biomaterials.* 2009;91(2):587–95.
47. Hatakeyama M, Masahashi N, Michiyama Y, Inoue H, Hanada S. Mechanical properties of anodized TiNbSn alloy for biomedical applications. *Materials Science and Engineering A.* 2021 Aug 11:141898.
48. Uslu E, Mimiroglu D, Ercan B. Nanofeature size and morphology of tantalum oxide surfaces control osteoblast functions. *ACS Applied Bio Materials.* 2021 Jan 4;4(1):780–94.
49. Luz AR, De Souza GB, Lepienski CM, Siqueira CJ, Kuromoto NK. Tribological properties of nanotubes grown on Ti-35Nb alloy by anodization. *Thin Solid Films.* 2018 Aug 30;660:529–37.
50. Ohtsu N, Yamasaki K, Taniho H, Konaka Y, Tate K. Pulsed anodization of NiTi alloy to form a biofunctional Ni-free oxide layer for corrosion protection and hydrophilicity. *Surface and Coatings Technology.* 2021 Apr 25;412: 127039.
51. Hang R, Liu Y, Zhao L, Gao A, Bai L, Huang X, Zhang X, Tang B, Chu PK. Fabrication of Ni-Ti-O nanotube arrays by anodization of NiTi alloy and their potential applications. *Scientific Reports.* 2014 Dec 18;4(1):1–9.
52. Hang R, Liu Y, Gao A, Bai L, Huang X, Zhang X, Lin N, Tang B, Chu PK. Highly ordered Ni–Ti–O nanotubes for non-enzymatic glucose detection. *Materials Science and Engineering: C.* 2015 Jun 1;51:37–42.
53. Venturini J, Bonatto F, Guaglianoni WC, Lemes T, Arcaro S, Alves AK, Bergmann CP. Cobalt-doped titanium oxide nanotubes grown via one-step anodization for water splitting applications. *Applied Surface Science.* 2019 Jan 15;464:351–9.
54. Kozhukharov S, Griginov C. Enhancement of the cerium oxide primer layers deposited on AA2024-T3 aircraft alloy by preliminary anodization. *Journal of Electrochemical Science and Engineering.* 2018 May 15;8(2):113–27.
55. Yoganandan G, Balaraju JN, Low CH, Qi G, Chen Z. Electrochemical and long term corrosion behavior of Mn and Mo oxyanions sealed anodic oxide surface developed on aerospace aluminum alloy (AA2024). *Surface and Coatings Technology.* 2016 Feb 25;288:115–25.
56. Baek KM, Kim J, Kim S, Cho SH, Jang MS, Oh J, Jung YS. Engraving high-density nanogaps in gold thin films via sequential anodization and reduction for surface-enhanced raman spectroscopy applications. *Chemistry of Materials.* 2018 Aug 6;30(17):6183–91.
57. Lucas-Granados B, Sánchez-Tovar R, Fernández-Domene RM, Garcia-Anton J. Iron oxide nanostructures for photoelectrochemical applications: Effect of applied potential during Fe anodization. *Journal of Industrial and Engineering Chemistry.* 2019 Feb 25;70:234–42.
58. Zakir O, Idouhli R, Elyaagoubi M, Khadiri M, Aityoub A, Koumya Y, Rafqah S, Abouelfida A, Outzourhit A. Fabrication of TiO2 nanotube by electrochemical anodization: Toward photocatalytic application. *Journal of Nanomaterials.* 2020 Dec 29;2020.
59. Kozhukhova AE, Du Preez SP, Shuro I, Bessarabov DG. Development of a low purity aluminum alloy (Al6082) anodization process and its application as a platinum-based catalyst in catalytic hydrogen combustion. *Surface and Coatings Technology.* 2020 Dec 25;404:126483.

60. Sakai K, Kitazumi Y, Shirai O, Kano K. Nanostructured porous electrodes by the anodization of gold for application as scaffolds in direct-electron-transfer-type bioelectrocatalysis. *Analytical Sciences*. 2018:18P302.

61. Silva FK, Santa RA, Fiori MA, de Aquino TF, Soares C, Martins MA, Padoin N, Riella HG. Synthesis of aluminum hydroxide nanoparticles from the residue of aluminum anodization for application in polymer materials as antiflame agents. *Journal of Materials Research and Technology*. 2020 Jul 1;9(4):8937–52.

62. Yousif QA, Haran NH. Fabrication of TiO2 nanotubes via three-electrodes anodization technique under sound waves impact and use in dye-sensitized solar cell. *Egyptian Journal of Chemistry*. 2021 Jan 1;64(1):125–32.

63. Lashkov AV, Fedorov FS, Vasilkov MY, Kochetkov AV, Belyaev IV, Plugin IA, Varezhnikov AS, Filipenko AN, Romanov SA, Nasibulin AG, Korotcenkov G. The Ti wire functionalized with inherent TiO_2 nanotubes by anodization as one-electrode gas sensor: A proof-of-concept study. *Sensors and Actuators B: Chemical*. 2020 Mar 1;306:127615.

64. Tantray AM, Mir JF, Mir MA, Rather J, Shah MA. Random oriented ZnO nanorods fabricated through anodization of zinc in KHCO3 electrolyte. *ECS Journal of Solid State Science and Technology*. 2021 Aug 5;10(8):081003.

65. Fernández-Domene RM, Roselló-Márquez G, Sánchez-Tovar R, Cifre-Herrando M, García-Antón J. Synthesis of WO_3 nanorods through anodization in the presence of citric acid: Formation mechanism, properties and photoelectrocatalytic performance. *Surface and Coatings Technology*. 2021 Sep 25;422:127489.

66. Salehi R, Jahanbakhshi A, Golzarian MR, Khojastehpour M. Evaluation of solar panel cooling systems using anodized heat sink equipped with thermoelectric module through the parameters of temperature, power and efficiency. *Energy Conversion and Management: X*. 2021 Aug 19:100102.

67. Rosli SA, Alias N, Bashirom N, Ismail S, Tan WK, Kawamura G, Matsuda A, Lockman Z. Hexavalent chromium removal via photoreduction by sunlight on titanium–dioxide nanotubes formed by anodization with a fluorinated glycerol–water electrolyte. *Catalysts*. 2021 Mar;11(3):376.

68. Lee J, Jung SY, Kumbhar VS, Uhm S, Kim HJ, Lee K. Formation of aluminum oxide nanostructures via anodization of Al3104 alloy and their wettability behavior for self-cleaning application. *Catalysis Today*. 2021 Jan 1;359:50–6.

69. Ponchio C, Srevarit W. Photoelectrocatalytic improvement of copper oxide thin film fabricated using anodization strategy application in nitrite degradation and promoting oxygen evolution. *Chemical Papers*. 2021 Mar;75(3):1123–32.

70. Wang S, Liu H, Hu J, Jiang L, Liu W, Wang S, Zhang S, Yin J, Lu J. In situ synthesis of NiO@ Ni micro/nanostructures as supercapacitor electrodes based on femtosecond laser adjusted electrochemical anodization. *Applied Surface Science*. 2021 Mar 1;541:148216.

71. Kongvarhodom C, Khumsa-Ang K, Siripornmongkolchai B, Jearasupat S, Turner CW. Anodic aluminum oxide film fabricated with galvanostatic anodization for non-electrolytic dyeing. *Materials Letters*. 2020 Feb 15;261:126992.

72. Yasuda K, Schmuki P. Electrochemical formation of self-organized zirconiumtitanate nanotube multilayers. *Electrochemistry Communications* 2007;9:615–9.

73. Yasuda K, Schmuki P. Formation of self-organized zirconium titanatenanotube layers by alloy anodization. *Advanced Materials* 2007;19:1757–60.

74. Ghicov A, Aldabergenova S, Tsuchyia H, Schmuki P. TiO2–Nb2O5 nanotubes with electrochemically tunable morphologies. *Angewandte Chemie International Edition.* 2006;45(42):6993–6. Schiavi PG, Altimari P, Rubino A, Pagnanelli F. Electrodeposition of cobalt nanowires into alumina templates generated by one-step anodization. *Electrochimica Acta.* 2018 Jan 1;259:711–22.

75. Kape JM. Comparison of AC and DC sulphuric acid based anodizing processes. *Transactions of the IMF.* 1988 Jan 1;66(1):41–6.

76. Raj V, Rajaram MP, Balasubramanian G, Vincent S, Kanagaraj D. Pulse anodizing—An overview. *Transactions of the IMF.* 2003 Jan 1;81(4):114–21.

77. Blackwood DJ, Seah KH. Influence of anodization on the adhesion of calcium phosphate coatings on titanium substrates. *Journal of Biomedical Materials Research Part A: An Official Journal of the Society for Biomaterials, the Japanese Society for Biomaterials, and the Australian Society for Biomaterials and the Korean Society for Biomaterials.* 2010 Jun 15;93(4):1551–6.

78. Kar A, Smith YR, Subramanian V. Improved photocatalytic degradation of textile dye using titanium dioxide nanotubes formed over titanium wires. *Environmental Science & Technology.* 2009 May 1;43(9):3260–5.

79. Hao L, Liu T, Yan J, Hu Y, Jiang F, Zhao Q, Lu Y, Ping X. Anodized BiOI coatings and their photocatalytic activity of organic dye degradation. *Surfaces and Interfaces.* 2020 Sep 1;20:100562.

80. Sohn YS, Smith YR, Misra M, Subramanian VR. Electrochemically assisted photocatalytic degradation of methyl orange using anodized titanium dioxide nanotubes. *Applied Catalysis B: Environmental.* 2008 Dec 1;84(3–4):372–8.

Chapter 10

Effect of surface treatment of cenospheres on the mechanical properties of cenosphere/recycled-PET composites

B. Krishna Prabhu

Canara Engineering College, Bantwal, Karnataka, India

S. M. Kulkarni

National Institute of Technology Karnataka, Surathkal, Karnataka, India

Ajith G. Joshi

Canara Engineering College, Bantwal, Karnataka, India

CONTENTS

10.1 INTRODUCTION

Polymer matrix composites comprise polymers as their matrix materials with several types of reinforcement materials. Though, artificial as well as synthetic polymers are employed for various engineering applications. The packaging and food industries application primarily utilizes thermoplastic polymers such as polyethylene terephthalate (PET), high-density polyethylene (HDPE), polyvinyl chloride (PVC), low-density polyethylene (LDPE), polypropylene (PP), and polysulphide (PS) due to their unique characteristics such as hydrophobic and biological inertness. Besides, the recycling

DOI: 10.1201/9781003319375-10

property of thermoplastic has made it popular in recent decades. It is one of the major concerns in the sustainable engineering of engineering materials. Thereby, contributing towards energy saving and greenhouse gas emission control. PET is highly recyclable, which belongs to the aromatic–aliphatic polyester family. PET exhibits better mechanical and tribological properties in addition to its hard and rigid nature. Perhaps its high melting points and glass transition temperature (Tg) aid to retain its good mechanical properties at elevated temperatures. Also, it offers resistance to a variety of solvents and chemicals [1]. In the past few years, PET has been widely employed in engineering applications such as gears, bearings, housings, impellers, pulleys, switch parts, bumper extensions, and as a better alternative to glass for beverage containers.

Despite the fact of recycling of thermoplastics being practicable, the process of recycling deteriorates its mechanical properties and poor aesthetic appearance at each cycle [2]. The degradation of mechanical properties owes to mechanical processing routes employed for its reduction [3]. Besides, recycled PET (r-PET) has shown better post-processing property retention characteristic with a marginal increase of brittleness [4–5]. Thus, r-PET is gaining importance and demonstrating as a promising alternative material for engineering applications.

Several researchers have attempted to produce r-PET with fillers and fibre-reinforced composites to reduce brittleness with retained mechanical properties [6]. The review of research reports presented by [7] has illustrated that the mechanical properties of PET can be enhanced with the incorporation of glass fibres in either fibre or filler form. Novello et al. [8] have shown that short fibre reinforcement into the r-PET/PA66 blend enhances flexural, tensile, and impact behaviour of composites. The incorporation of fillers has revealed promising results in enhancing mechanical properties of r-PET composites. Calcium carbonate added r-PET/PP composites have depicted increased thermal stability, stiffness, and impact strength but with decreased yield strength [9]. Thumsorn et al. [10] have illustrated that talc incorporation into r-PET improves mechanical properties and an increase in the percentage of talc from 5% to 10% has drastically increased the flexural and tensile properties of talc/r-PET composites. The addition of wollastonite to r-PET /PP blend has exhibited [11] increased Young's modulus and thermal characteristics at the cost of elongation and impact strength while the addition of wollastonite to r-PET and polybutylene terephthalate composites showed increased Young's modulus and impact strength [12]. Further, recent literatures have revealed that nanofiller incorporated r-PET composites yield better mechanical properties than those blended with other thermoplasts and neat r-PET. Chowreddy et al. [13] have shown that carbon nanotubes (CNTs) incorporated r-PET composites exhibited enhanced mechanical properties compared to neat r-PET. The purified multi-walled CNTs (MWCNTs) were found to increase the tensile strength and hardness of MWCNTs/r-PET -thermoplastic polyurethane composites [14]. The r-PET composites with graphene, nano-clay, and zirconia (ZnO) nanoparticles fillers have also shown similar results with improved mechanical properties.

Fly ash is another non-conventional filler material, which is widely used with thermoplastics to improve their mechanical, thermal and tribological characteristics. Fly ash (FA) is one of the waste-by-products and is abundantly available. It is gaining interest due to its suitability as filler material for polymer- and alloy-based composite systems [15]. The study of Cazan et al. [16] has shown that FA addition to rubber/HDPE/r-PET blend has improved its tensile, impact and compression strength. Zaichenko and Nefedov [17] have modified fly ash with sulphuric acid to improve adhesion characteristics with r-PET in the composite system. It has resulted in increased compressive strength of the composite material.

Fly ash cenospheres (FACs) are one of the major constituents of fly ash, which is a hollow microsphere [18]. They possess low-density constituents of FA, which is predominantly inorganic residue obtained from the flue gases of furnaces at pulverized coal power plants during the combustion of coal for energy production. Only 30% of FA is used by the cement industry, while a large proportion often goes to landfill and ash ponds. Therefore, there are chances of environmental risk due to possible long-term adverse effects of transition metal oxides dependent on processing conditions [19–20]. However, it has found its applications in recent years due to environment-linked issues associated with subsequent disposal of FAC [21]. Besides, nearly uniform resin distribution with FAC fillers can be achieved due to their spherical shape. Also, ashes are essentially a mixture of solid, hollow, and composite particles demonstrating nearly isotropic properties; hence, developing newer and utilitarian systems using them should be an interesting and challenging task [22]. Sampathkumar et al. [23] revealed that cenosphere addition to thermoplastic polymers helps to reduce its density and improve the hardness of composites. While compressive strength and impact strength have decreased, a gradual increase in wear resistance of composites was reported in the study.

Kushnoore et al. [24] concluded that FAC-based composites are suitable for lightweight engineering applications such as aerospace and marine due to superior wear resistance and anti-corrosion properties associated with low density. The utilization of FAC as filler material will help to reduce greenhouse emissions and contribution to sustainable engineering. Kasar et al. [15] presented the state-of-art literature on the mechanical and tribological behaviour of fly ash reinforced composites. The authors have stated that surface treatment of fly ash particles is necessary to improve wettability and adhesion characteristics. Perhaps, it helps to enhance mechanical properties in a remarkable manner compared to untreated fly ash incorporated polymer composites. Further, FACs would float on the matrix during processing and forms functional gradient microstructure on solidification. Hence, it will lead to better tribological characteristics. Thus, surface modification of fly ash and cenospheres play a vital role in improving mechanical and tribological behaviour through enhanced adhesion between matrix and fillers. Silane coupling agents have demonstrated promising results in improving surface characteristics. Also, interfacial bonding property improved when silane-treated fillers were incorporated into composites. Silane treatment of reinforcements help to achieve better bonding and adhesive characteristics. While in the case of untreated reinforcement, dirt or greasy/oily layers envelop the fillers. Thereby, the effectiveness of the medium or matrix to wet reinforcements/fillers reduces [25]. Their presence also affects the properties such as the mechanical behaviour. The mechanical properties of polymer-FAC composite are inferior, owing to poor interfacial interactions between the hydrophilic FAC surface and hydrophobic polymer [26]. However, surface-treated FAC is found to improve the interfacial interactions [27]. Joseph et al. [28] have found that both amino silane and vinyl silane coupling agents treated FA caused an increase in the tensile strength and flexural strength of FA/PET composites due to their improved interfacial interaction. They observed that the elongation value increases drastically with the addition of FA treated with coupling agents associated with low agglomeration. The impact resistance of low-impact FA/PET composites could be improved by silane coupling agents. Suresha et al. [29] concluded that surface-treated cenospheres filled polymer composites exhibited better tribological characteristics compared to untreated FAC/r-PET.

The discussed literatures have reported on the effect of several types of reinforcements on the mechanical behaviour of thermoplastics composites. The r-PET matrix is widely used as thermoplastic matrix material compared to other types. Fly ash has been widely studied as a reinforcement material for thermoplastic polymer composite systems. But, there is large scope to investigate FAC surface treatment effect on the mechanical properties of FAC/r-PET composites. Thus, in the current study FAC are treated initially with silane with the aim to improve their wettability characteristics. The treated FAC fillers were incorporated with varying percentages in the r-PET matrix. The mechanical properties such as flexural strength, toughness, fracture strain, and elastic modulus of the prepared specimen were studied. FTIR analysis of silane-treated FAC was carried out to ensure the presence of coated layer. SEM analysis of the fractured surface was studied to understand the failure mechanism of composites.

10.2 MATERIALS AND METHODS

10.2.1 Materials

r-PET was employed as the matrix material. PET beverage container bottles were retrieved from various local recycling agencies. The collected bottles were then cleaned to remove any traces of beverages and other contaminants with water and acetone. Later, they were subjected to mechanical

Table 10.1 Compositional details of fly ash cenosphere particles

Components	SiO_2	Al_2O_3	Fe_2O_3	Cao	Na_2O	K_2O	MgO
Wt%	56–60	25–35	1–6	0.2–0.6	0.5–2	1–2	0.5–2

pulverization to obtain r-PET granules which make them suitable for the compression moulding process. FACs were used as reinforcing particulates in the current research. As discussed in the Introduction section, they are low-density hollow constituents of fly ash with approximately unit aspect ratio. Therefore, it was estimated to depict nearly isotropic characteristics. The major advantage of FACs is their low cost associated with good mechanical properties. For the current work, fly ash was collected from Raichur Thermal Power Plant Corporation Ltd., Raichur, India. It is an ASTM class 'C' fly ash (ASTM C 618) with a bulk density of about 900 kg/m³ and a true particle density of 2250 kg/m³ similar to the fly ash reported elsewhere [30]. It was found to consist of a mixture of solid and hollow spheres of different sizes. FACs were extracted from fly ash by water floating method. Table 10.1 illustrates the composition of cenospheres by weight% [31]. The average density of FAC, obtained, was found to be 391 kg/m³ with an average size of about 57 microns.

10.2.2 Surface treatment of cenospheres

Further, to improve surface characteristics (3-aminopropyl) trimethoxy silane (3APTMS) surface treatment was carried out to achieve silane compound coating on a partial amount of obtained cenospheres [28]. The adsorption technique was adopted to yield silane coating on the surface of cenospheres. Initially, FACs were cleaned with acetone and dried at 60°C in a hot-air circulating oven for around 1 hour. The 3APTM silane was mixed at different percentages by varied volumes of isopropyl alcohol in the beaker along with FACs. Continuous stirring was carried out by a magnetic stirrer for around 4–5 hours, subsequently; the mixture was kept in a hot circulating oven at 65–70°C for about 14–16 hours. The mixture was heated until all the solvent evaporates, leaving behind the silane-coated FAC. The coated filler was preserved in a desiccator to prevent any attack of moisture and atmospheric oxygen. Based on reported literatures, it was anticipated that silane coating results in enhancing wettability and adhesion characteristics with the employed matrix material. The primary chemical group of interest is the radical amine group, NH_2, which aids the cenospheres for better bonding with the r-PET matrix, thereby enhancing the mechanical properties of the resultant composite material. The chemical structure of silane coupling agent 3APTMS is illustrated in Figure 10.1. It can be observed that the NH_2 group helps in better binding of cenospheres with the r-PET/r-LDPE blend.

$$H_3CO-\underset{\underset{OCH_3}{|}}{\overset{\overset{OCH_3}{|}}{Si}}\diagup\diagup NH_2$$

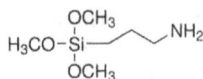

Figure 10.1 Chemical structure of silane coupling agent (3-aminopropyl) trimethoxy silane.

Table 10.2 Properties of 3APTM silane coupling agent

Properties	Values
Physical state	Liquid
Density	1.025 g/cm³ at 25°C
Boiling point (°C)	91–92

The physical properties of the utilized silane coupling agent are summarized in Table 10.2.

10.2.3 Specimen preparation

The specimens were fabricated through a compression moulding route. A programme compression moulding machine (supplied by MACHINEFABRIK, Belagavi, India) was employed in the study for the fabrication of specimens. The predetermined quantity of r-PET granules and FACs was blended. The blend was continuously electrically stirred to achieve uniform dispersion of particles in the matrix. Later, the mixture was used as a charge for the compression moulding process. A temperature in the range of 140–160°C and a pressure of about 10 N/mm² was employed for a predetermined period of about 1 hour with preheated moulds for processing. The fabricated specimen sample, compression moulding machine, and moulds used are shown in Figure 10.2. On processing, solid composite parts were unloaded from the moulding machine. Subsequently, secondary machining of moulded composites was performed to yield the desired size for mechanical characterization. Thus, varied weight fractions (0%, 5%, 10%, 15%) of FAC and surface-treated FAC (T-FAC) incorporated r-PET composite specimens were fabricated.

10.2.4 Mechanical characterization

The mechanical properties evaluated in the present study were flexural strength, toughness, fracture strain, and elastic modulus. Thus, to assess mechanical properties considered, a widely used three-point bending test was employed in accordance with ASTM D790. The rectangular cross-section specimen bearing dimensions of 80 mm × 14 mm × 4 mm and deflected at a predetermined specific strain rate. The crosshead speed used for the

Figure 10.2 (a) Compression moulding machine, (b) mould used for specimen preparation, and (c) sample fabricated specimen.

test is about 2 mm/min on a universal testing machine with a maximum deflection of 5%. Each test was repeated five times and average results were considered for analysis. The fractured specimen surfaces were observed with the help of a scanning electron microscope (SEM) to understand the fracture behaviour of the studied composites.

10.3 RESULTS AND DISCUSSIONS

The recycled PET matrix was incorporated with untreated and silane-treated FACs. The weight fraction of incorporated fillers (treated and untreated FAC) was varied from 0% to 15% in the steps of 5%. The mechanical characterization of prepared composites was studied as their flexural properties. The flexural properties studied were flexural strength, fracture strain, toughness, and elastic modulus. Experimental results obtained were analysed and compared for studied specimens as illustrated in Figure 10.3.

10.3.1 Flexural properties of r-PET/FAC composites

In Figure 10.3, we can observe the flexural properties of the r-PET/FAC composite (continuous curve). Figure 10.3(a) depicts the effect of FAC on the flexural strength of the composite. It can be observed from the graph that the flexural strength of the composite drops with the increase in the content of the FAC. A decrease of 32% in strength was observed when 15% of FAC is added to the matrix. Figure 10.3(b) indicates the effect of FAC on the fracture strain of the composite. The decrease of up to 55% in the

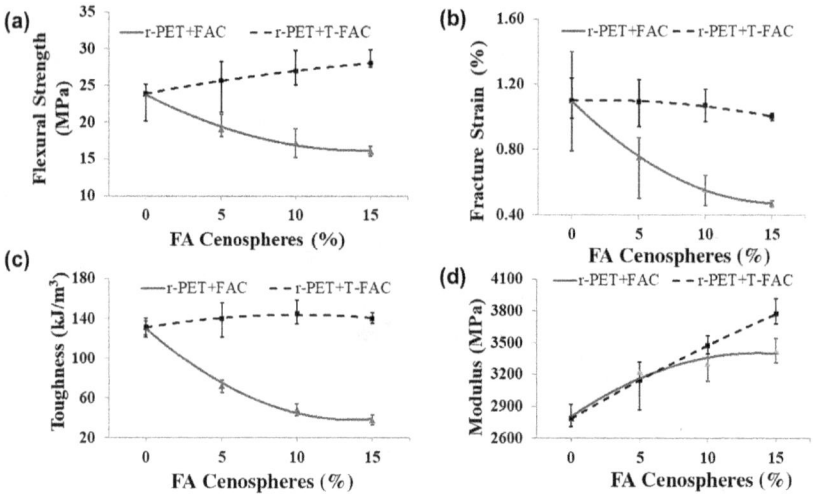

Figure 10.3 Flexural properties of studied composites illustrating (a) flexural strength, (b) fracture strain, (c) toughness, and (d) elastic modulus.

property was observed as a contribution from FAC in the composite. The composite seems to become more and more brittle owing to the addition of FAC. The steep drop in the toughness of the composite can be seen in Figure 10.3(c). The composite loses 69% of its toughness when 15% of FAC was added to r-PET. In contrast to these observations on strength, strain, and toughness, an improvement of 21% in the modulus was observed in the r-PET/FAC composite (Figure 10.3(d)).

It can be noted from Figure 10.3(a–c) that FAC content in the r-PET composite shows a steep downward trend. The reason could be poor binding between the matrix and the filler. Processing of r-PET/FAC composite was done entirely on a recycling basis without any addition of compatibilizer or surface treatments to FAC. The matrix is an organic aliphatic-aromatic polymer, and the reinforcing filler is an inorganic alumino-silicate. Such a combination causes poor wetting between the constituents resulting in inferior binding and thus, poor load sharing. Under stress, differential deformations of matrix and filler lead to the formation of new surfaces at the interfaces through which the microcracks propagate and grow, leading to the failure of the component. Joseph et al. [28] reported pulling filler out from the matrix, resulting in the formation of holes in a polymer-filled composite, when matrix and reinforcement are not bound properly, under a tensile load.

As indicated in Figure 10.4, the SEM micrograph of the fracture surface reveals the dents formed due to the de-bonding of FAC from the matrix. It is also evident from the SEM image that the de-bonding of FAC from the r-PET matrix has occurred. Thus, for enhanced wettability FACs were surface treated before incorporating into the matrix. Treatment of FAC was

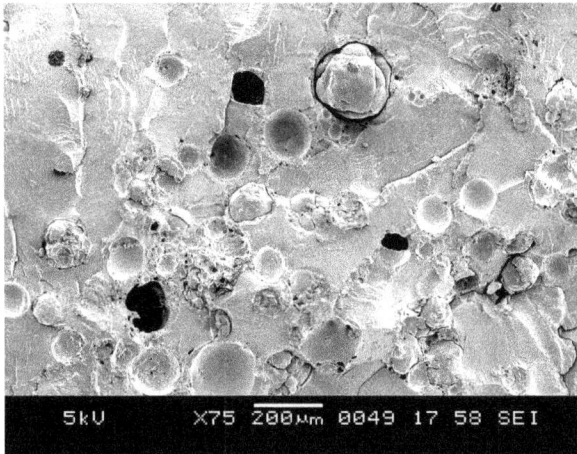

Figure 10.4 SEM pictures showing debonding of FAC from the r-PET matrix.

predicted to improve the interface between the matrix and the reinforcement. FAC was chemically treated with 3APTMS, and the coated FAC are tested for coating through FTIR spectroscopy. Figure 10.5 is a spectrogram for FAC treated with 10% 3APTMS. The spectrograms reveal amine peaks in the wavelength range of 3300 cm^{-1} to 3500 cm^{-1} and C-H stretch in the wavelength range from 2900 cm^{-1} to 3000 cm^{-1}, which are characteristic of 3APTMS. The presence of these chemical groups is helpful in the binding of FAC to r-PET.

10.3.2 Flexural properties of r-PET/T-FAC composites

Figure 10.3(a) also reveals the details of the flexural properties of r-PET/T-FAC composites (dotted curves). Flexural strength, strain toughness, and modulus of r-PET/T-FAC composite are found to be superior to its counterpart, r-PET/FAC composite. In Figure 10.3(a), the effect of T-FAC on the flexural strength of r-PET can be observed. It can be observed from Figure 10.3(a) that the addition of 15% of T-FAC imprwoves the strength of r-PET by 17%. Figure 10.3(b) illustrates the details of the fracture strain of the composite. It can be observed from the figure that a marginal decrease of 9% in strain is evident owing to the addition of T-FAC to r-PET. In Figure 10.3(c), a marginal increase of 8% in the toughness can be attributed to the addition of T-FAC. An increase of 34% in the modulus results from reinforcement with T-FAC, as illustrated in Figure 10.3(d).

The analysis of data pinpoints that T-FAC reinforced r-PET depicts appreciable improvement in flexural strength, fracture strain, and modulus. Fracture toughness, however, drops marginally at lower concentrations of T-FAC. The improvement in the properties could be the result of the binding

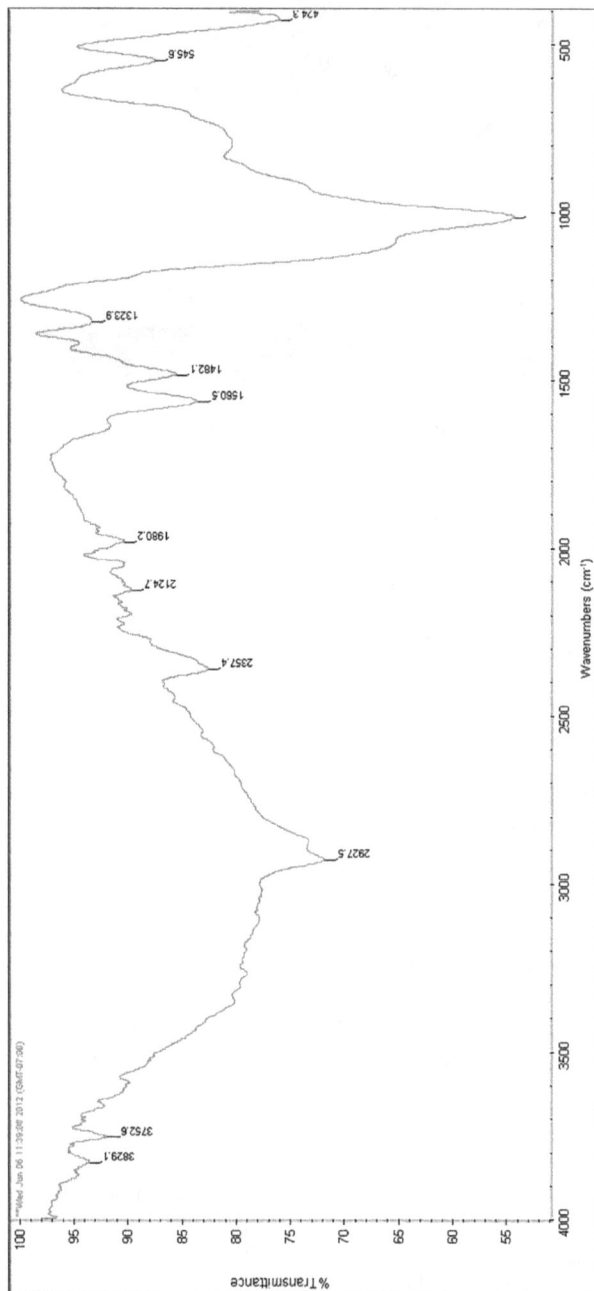

Figure 10.5 FTIR spectrogram showing the presence of silane coating on FAC treated with 10% 3APTMS.

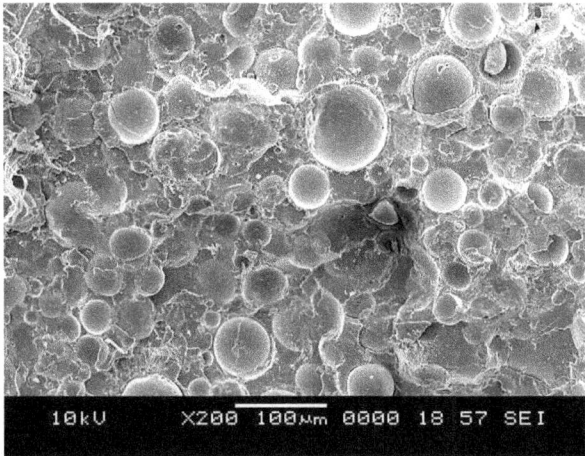

Figure 10.6 SEM micrograph of fracture surface showing binding between T-FAC and r-PET matrix.

of T-FAC to the matrix as illustrated in Figure 10.6 for 10% 3APTMS-treated FAC composite. It also shows the improvement in binding between FAC and matrix when compared to the untreated FAC composite.

The results presented in this section indicate that the objectives to improve strength by properly binding the FAC to the r-PET matrix can be met by treating it with 3APTMS. The evidence for improved binding was noticed in the SEM micrograph. The loss in the strength of the composite due to matrix modifications can be recovered through surface treatment of the FAC.

10.4 CONCLUSIONS

A systematic study was carried out with r-PET reinforced with cenospheres which are treated with (3-aminopropyl) trimethoxy silane. The obtained results have led to the following conclusions. Reinforcing r-PET with 3APTMS (10% by wt.) treated FAC improved flexural strength of r-PET/T-FAC composite. An increase of 34% strength at 5% T-FAC, 57% increase at 10%, and 120% improvement in strength at 15% of T-FAC was observed, in comparison to untreated FAC. Such an improvement can be attributed to surface treatments given to FAC. An improvement in the fracture strain to the extent of 36% at 5% and 50% each at 10% and 15% of T-FAC was observed. The increase in these properties leads to an improvement in the toughness of the composite. Toughness improves by 95% at 5% of T-FAC, 200% at 10%, and an increase of 271% in toughness was observed when r-PET is reinforced with 15% T-FAC. Treating FAC also improves modulus marginally; 20%, 1%, and 10% increase at 5%, 10%, and 15% T-FAC, respectively, were recorded. The results of flexural properties of r-PET/T-FAC composite indicates an improvement in the properties owing to the treatments given to FAC. The SEM micrographs illustrated good bonding between the matrix and the reinforcement in a brittle matrix. These results are comparative with respect to r-PET/FAC composites.

REFERENCES

1. Seidel A. (ed), Fillers. In: *Encyclopedia of Polymer Science and Technology* 4th edition. John Wiley & Sons, Inc., Vol. 10 (2012), p. 2. https://doi.org/10.1002/0471440264.
2. da Costa H.M., Ramos VD, de Oliveira MG, Degradation of polypropylene (PP) during multiple extrusions: Thermal analysis, mechanical properties and analysis of variance. *Polymer Testing* 26 (2007) 676–684. https://doi.org/10.1016/j.polymertesting.2007.04.003
3. Pawlak A., Pluta M., Morawiec J., Galeski A., Pracella M., Characterization of scrap polyethylene terephthalate. *European Polymer Journal* 36 (2000) 1875–84.
4. Torres N., Robin J.J., Boutevin B., Study of thermal and mechanical properties of virgin and recycled poly (ethylene terephthalate) before and after injection molding. *European Polymer Journal* 36(10) (2000) 2075–2080.
5. Gaymans R.J., Toughening of semicrystalline thermoplastics. In: Paul D.R., Bucknall C.B., Eds. *Polymer Blends: Performance*. New York: John Wiley & Sons (2000), pp. 178–219.
6. Avila A.F., Marcos V.D., A mechanical analysis on recycled PET/HDPE composites. *Polymer Degradation and Stability* 80 (2003) 373–382.
7. Scelsi L., Hodzic A., Soutis C., Hayes S. A., Rajendran S., AlMa'adeed M. A., Kahraman R., A review on composite materials based on recycled thermoplastics and glass fibres. *Plastics, Rubber and Composites* 40(1) (2011) 1–10
8. Novello M.V., Carreira L.G., Canto L.B., Post-consumer polyethylene terephthalate and polyamide 66 blends and corresponding short glass fiber reinforced composites. *Materials Research* 17 (2014) 1285–94. https://doi.org/10.1590/1516-1439.281914
9. Thumsorn S., Yamada K., Leong Y.W., Hamada H., Development of blending sequence on CaCO3 reinforced recycled PET/recycled PP blend. In: *18th International Conference on Composite Materials*; 2011.
10. Thumsorn S., Negoro T., Thodsaratpreeyakul W., Inoya H., Okoshi M., Hamada H., Effect of ammonium polyphosphate and fillers on flame retardant and mechanical properties of recycled PET injection molded. *Polymers for Advanced Technologies* 28 (2017) 979–85. https://doi.org/10.1002/pat.3730
11. Chaiwutthinan P., Suwannachot S., Larpkasemsuk A., Luang K., Thani P., Recycled poly (ethylene terephthalate)/polypropylene/wollastonite composites using PP-g-MA as compatibilizer: mechanical, thermal and morphological properties. *Journal of Metals, Materials and Minerals* 28 (2018) 115–23. https://doi.org/10.14456/jmmm.2018.34

12. Chaiwutthinan P., Riyaphan T., Simpraditpan A., Larpkasemsuk A., Influence of poly (butylene terephthalate) and wollastonite on properties of recycled poly(ethylene terephthalate) preforms. *Journal of Metals, Materials and Minerals* 30 (2020) 124–32. https://doi.org/10.14456/jmmm.2020.15

13. Chowreddy R.R., Nord-Varhaug K., Rapp F., Recycled polyethylene terephthalate/ carbon nanotube composites with improved processability and performance. *Journal of Material Science* 53 (2018) 7017–29. https://doi.org/10.1007/s10853-018-2014-0

14. Fang C., Yang R., Zhang Z., Zhou X., Lei W., Cheng Y., et al. Effect of multi-walled carbon nanotubes on the physical properties and crystallisation of recycled PET/ TPU composites. *RSC Advances* 8 (2018) 8920–8. https://doi.org/10.1039/c7ra13634j

15. Kasar A.K., Gupta N., Rohatgi P. K., Menezes L.P., A brief review of fly ash as reinforcement for composites with improved mechanical and tribological properties, *JOM* 72 (2020) 2340–2351. https://doi.org/10.1007/s11837-020-04170-z

16. Cazan C., Cosnita M., Isac L., The influence of temperature on the performance of rubber-PET-HDPE waste - based composites with different inorganic fillers. *Journal of Cleaner Production* 208 (2019) 1030–40. https://doi.org/10.1016/j.jclepro.2018.10.045

17. Zaichenko N., Nefedov V., Composite material based on the polyethylene terephthalate polymer and modified fly ash filler. In: *MATEC Web of Conferences* 2018;245: 03007. https://doi.org/10.1051/matecconf/201824503007

18. Mohapatra, R., Rao R., Some aspects of characterisation, utilisation and environmental effects of fly ash. *Journal of Chemical Technology and Biotechnology* 76(1) (2001) 9–26.

19. Alkan C.M. Arslan M. Cici M. Kaya M. Aksoy M., A study on the production of a new material from fly ash and polyethylene. *Resources Conservation and Recycling* 13 (1995) 147–154.

20. Ward C.R., French D., Determination of glass content and estimation of glass composition in fly ash using quantitative X-ray diffractometry. *Fuel* 85 (2006) 2268–2277.

21. Kulkarni S.M., Effects of surface treatments and size of fly ash particles on the compressive properties of epoxy based particulate composites. *Journal of Materials Science*, 37(20) (2002) 4321–4326.

22. Kulkarni, S.M., Sharathchandra, S., Sunil, D., On the use of an instrumented set-up to characterize the impact behavior of an epoxy system containing varying fly ash content. *Polymer Testing* 21(7) (2002) 763–771.

23. Sampathkumaran P., Kishore S.S., Pattanashetti V.V., Kumar V.V., Kumar, N.H.B., Fly ash cenospheres as reinforcement in different polymer composites - a comparative study of physical and mechanical properties. *Indian Journal of Engineering & Material Sciences* 22 (2015) 354–362.

24. Kushnoore S., Kamitkar N., Atgur V., Uppin M.S., Satishkumar M., A review on utilization of light weight fly ash cenosphere as filler in both polymer and alloy-based. *Journal of Mechanical Engineering Research*, 03(02) (2020) 17–23.

25. Farinha J. P. S., Winnik, M. A. and Hahn K. G., Characterization of oil droplets under a polymer film by laser scanning confocal fluorescence microscopy. *Langmuir* 16(7) (2000) 3391–3400.

26. Guhanathan S., Sarojadevi M., Murugesan, V., Effect of coupling agents on the mechanical properties of fly ash/ polyester particulate composites. *Journal of Applied Polymer Science* 82 (7) (2001) 1755–60.
27. Thongsang S., Smbatsompap N., Effect of NaOH and Si69 treatments on the properties of fly ash/natural rubber composites. *Polymer Composites* 27(1) (2006) 30–40.
28. Joseph S., Bambola V.A., Sherhtukade V.V., Mahanwar P.A., Effect of fly ash content, particle size of flyash, and type of silane coupling agents on the properties of recycled poly(ethylene terephthalate)/flyash composites. *Journal of Applied Polymer Science* 119 (2010) 201–208.
29. Suresha B., Chandramohan G., Jayaraju T., Influence of cenosphere filler additions on the three-body abrasive wear behavior of glass fiber-reinforced epoxy composites. *Polymer Composites* 29 (3) (2008) 307–312.
30. Narasimha Rao A.V., Characteristics of Fly Ash and Its Application- A Brief Review. In: *Proc. of national seminar on flyash characterization and its geotechnical applications*; 1999. IISc, Bangalore, India.
31. Chand N., Sharma P., Fahim M., Correlation of mechanical and tribological properties of organosilane modified cenosphere filled high density polyethylene. *Materials Science and Engineering: A* 527 (2010) 5873–5878. https://doi.org/10.1016/j.msea.2010.06.02

Chapter 11

Effects of performance parameters, surface failure and mitigation techniques on steam turbine blades

Satyajeet Kumar and Shailesh Mani Pandey
National Institute of Technology Patna, Patna, Bihar, India

CONTENTS

11.1 INTRODUCTION

In today's scenario, developing country like India is facing a huge energy crisis because of the ever-increasing demand for energy. Due to modernisation and overall economic growth of the country, the consumption of energy has almost outpaced its ability to supply energy. As a result of this, the country is witnessing a rapid increase in the price of fossil fuels and instability in the market. Industrial sectors and other reliability problems related to human comfort are also hampered to a large extent. Steam turbines create economic development and environmental compatibility due to the usage of renewable energies [1–3]. Engineers from the powerplant sectors have been meticulously working especially in the field of turbine blade design and development of new categories of material in order to enhance the performance,

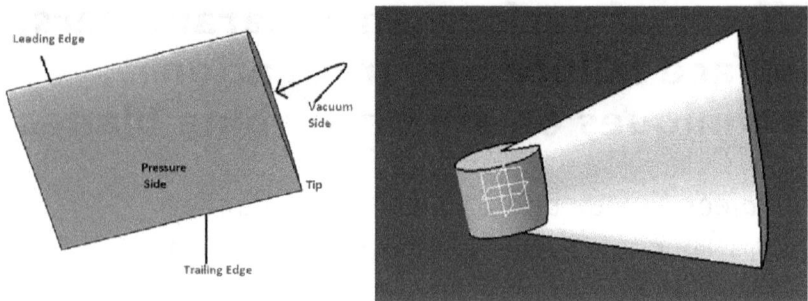

Figure 11.1 Steam turbine blade profile, classifying the location of both the leading and trailing edges, and low- and high-pressure striking zones.

thermodynamic efficiency and reliability. The thermal power plant is one of the major sources of electric power generation across the globe.

Steam turbines are turbomachines which utilize steam obtained from a boiler at very high pressure and temperature. The pressure energy of the steam is converted into kinetic energy by the rotodynamic action of the turbine's moving blades and the nozzle. The heat energy possessed by the steam is transformed into kinetic energy and then into mechanical and electrical energy through several stages of fixed and moving blades. Steam under supercritical pressure and temperature is isentropically expanded through different zones, i.e., high-pressure (HP) zone, medium- or intermediate-pressure (IP) zone and low-pressure (LP) zone before entering the condenser. However, the major portion of the energy is extracted with the help of low-pressure (LP) turbine blades than using other intermediate (IP) and high-pressure (HP) turbine blades. Hence, the low-pressure zone of the blade section is more prone to catastrophic failures without warning. Turbine blades are considered to be the critical components of a steam power plant as these are exposed to supercritical pressure, temperature and other environmental factors. These factors may directly or indirectly reduce the service life of the blades. Both the leading and trailing edge of the steam turbine blades exhibits variation in steam pressure, kinetic energy, mechanical loads and stresses, resulting in a decrease in the performance of the steam turbine and its functionalities. Figure 11.1 shows the typical blade profile, classifying the location of both the leading and trailing edges, and low- and high-pressure striking zones.

11.2 FACTORS AFFECTING THE PERFORMANCE OF A STEAM TURBINE

The factors which affect the overall performance and efficiency of a steam turbine are shown in Figure 11.2. Brief details of each of the factors are discussed below.

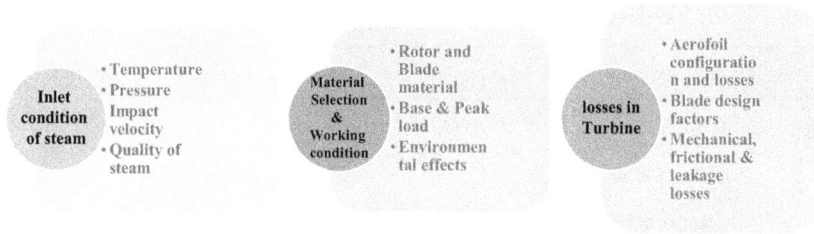

Figure 11.2 Factors affecting the performance of a steam turbine.

11.2.1 Inlet condition of steam

The performance of a steam turbine largely depends upon the quality of steam at which it strikes the runner attached to the rot. The service lives of various steam power plant equipment such as control valves, heat exchangers, compressors, guide vanes, etc., are reduced by the poor quality of steam. Both impact pressure and temperature affect the inlet condition of steam as these are categorised as one of the major affecting performance parameters [4]. In order to retain the thermodynamic and design efficiency, the steam inlet pressure should be maintained. Enthalpy or the total heat content of the system is a function of inlet pressure as well as temperature. A decrease in inlet pressure will reduce the turbine efficiency and on the other hand, the specific steam consumption (SSC) substantially increases. An increase in inlet temperature also increases the amount of heat extracted from the turbine. Turbine blades have to withstand the effect of two-phase fluid. Steam with moisture content adversely affects the striking zone of turbine blades leading to scale deposition, crack initiation and other damage mechanisms such as corrosion, pitting and wear. The quality of steam can be improved either by using a superheater or reheating the steam before entering the condenser. Centrifugal force induces stresses in the turbine blades and tends to rotate the steam at almost 1800 RPM while the fluid and other forces are responsible for fracture, yielding and creep deformation.

11.2.2 Material selection and working environment

An improper selection of material depending on the working environment is another important factor in the performance of a steam turbine. The suitable selection of special characteristic materials and new technologies permits engineers to achieve the desired performance of the steam power plant with low chances of component failure. Steam turbine blades are subjected to severe working conditions and operating environments. The harsh condition of steam turbine blades has the tendency to alter the surface integrity, especially in the low-pressure zone of the blade section. The disturbance created due to the said condition would result in the steam path inaccuracy

Table 11.1 Materials for steam turbine components

S. No.	Components	Material	Ref.
1	Rotor	Low alloy steel, CrMoV steel 10%CrMoVNbN 1%CrMoV	[6]
2	Turbine blade	Martensitic stainless steel, **nickel-based superalloys**, Nimonic 80A, titanium alloys, especially Ti-6Al-4V	[7–9]
3	Casing	low-alloy CrMo steels (e.g., the 1-2CrMo steel, cast 9CrMoVNb alloys	—

and eventually lead to a decrease in the performance of the steam turbine [5]. Various components of the steam turbine like rotor, moving and stationary blades, casing and others are made of specially alloyed materials. Table 11.1 shows a brief detail of materials for turbine components.

11.2.3 Losses in turbine

The turbine blades are configured and designed in such a way that they have a significant effect on energy transfer and power generation. This configuration is commonly known as "aerofoil". Steam flows aerodynamically from the leading edge to the trailing edge at a wider range of angles. An aerodynamic force is created when steam passes through the aerofoils and hence delivers a typical steam flow pattern to work with enhanced efficiency. The change in the angle of attack or the angle of inclination is to increase the lift-to-drag ratio, which ultimately affects the efficiency of the aerofoil attached and also contributes to the potential of the turbine. When the blade geometry (aerofoil) is placed at a suitable angle of attack and impact velocity, an increase in the performance and efficiency of the turbine can be easily visualised. Aerofoil blade configuration and various associated losses in the steam path of the steam turbine need to be modified and minimized for efficiency enhancement. Surface roughness, turbine blade geometry, variation in Mach number, steam flow turbulence, aerofoil camber angle, etc., are some of the important parameters which affect steam turbine blade and aerodynamic losses [10–12]. Some other aerodynamic losses may include wake and potential interaction of the turbine blades. The following techniques are adopted by power plant engineers to eliminate or reduce aerodynamic losses (Figure 11.3). The gliding action of the steam jet on the blade creates a frictional loss in the steam turbine. This loss is termed blade friction loss and occurs predominantly on the surface of fixed and moving blades. To minimise the effect of blade friction losses, relative velocity should be reduced. There is friction between the shaft, bearing (journal and thrust

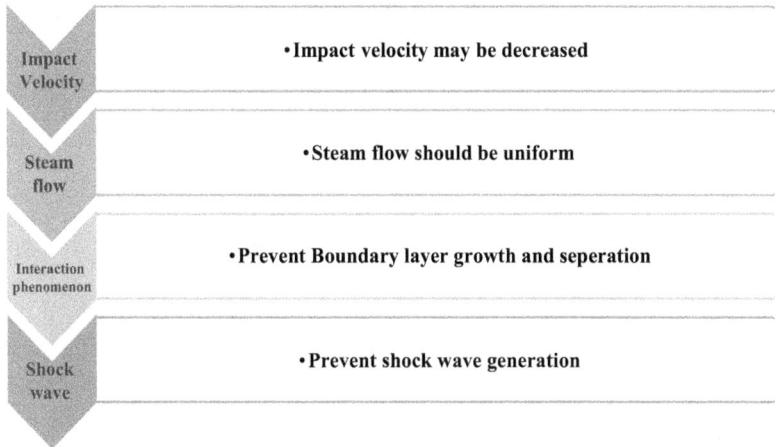

Impact Velocity	• Impact velocity may be decreased
Steam flow	• Steam flow should be uniform
Interaction phenomenon	• Prevent Boundary layer growth and seperation
Shock wave	• Prevent shock wave generation

Figure 11.3 Methods to reduce aerodynamic losses.

bearing), regulating valves and other moving parts of the turbine, resulting in mechanical losses. The selection and use of proper lubrication will reduce mechanical frictional losses. One can observe losses due to leakage is different in impulse and reaction turbine. In the case of an impulse turbine, components like nozzles, shafts, bearings and stationary blades experience leakage of steam. The reaction turbine inhibits leakage losses at the tip of the blade during each stage of blading. Out of total turbine loss, leakage loss accounts for nearly about 1–2%.

11.3 BLADE FAILURE MECHANISMS, CAUSES AND MITIGATION TECHNIQUES

Steam turbine blades and rotors are the essential components of a thermal power plant. Blades are bolted together with the rotor that exerts maximum amount of centrifugal and bending stresses arising due to the combined effect of mechanical loading and rotational speed. These stresses and adverse operating environmental conditions induce a series of blade failure mechanisms. The most common types of blade failure mechanisms (Figure 11.4) include erosion and wear, corrosion, vibration, fatigue, pitting, fretting, creep deformation, stress corrosion cracking (SCC) and many more [13] The performance, reliability and service life of steam turbine blades are very much affected by these damage mechanisms. This section deals with the various mechanisms of blade failure, root causes and its mitigation techniques.

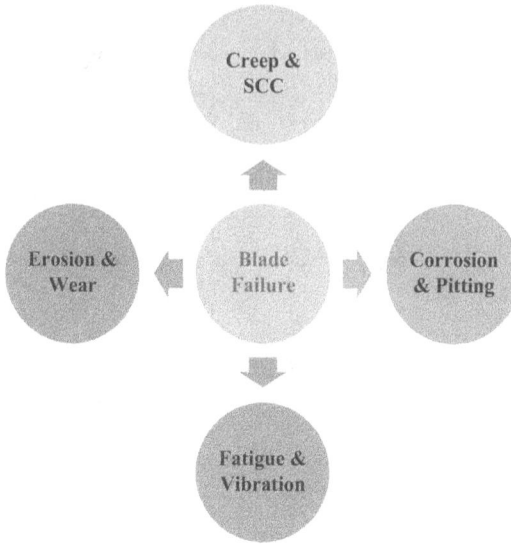

Figure 11.4 Root causes of failure of steam turbine blades.

11.3.1 Failure mechanisms and root causes

11.3.1.1 Corrosion and pitting

Corrosion in industries, power plants, structures, ship-building, etc., has been a matter of great concern in the past. Corrosion in steam turbine blades refers to the degradation in the quality and quantity of surface material as well, as its critical properties. The top surface of the blade is attacked by electrochemical and metal reactions (mainly oxidation) due to the existence of several impurities present in the steam. The phenomenon is accelerated mainly due to the high temperature of steam and exposure of metals to air, moisture, acid and salts. Oxides, silicates, chlorides and sulphates are the group of compounds that contributes to the development of the undesirable phenomenon. Mass imbalance of turbine blades due to corrosion is one of the serious problems that leads to vibration in the structure. Furthermore, corrosion accelerates the fatigue, fretting fatigue and wear, which may give rise to the deformation or the fracture of metals [14–20].

Pitting is another random and erratic blade failure mechanism. Hence, it is very difficult to sense the effect on the surface of turbine blades, rotors and disks. Since the chloride, sulphides, SiO_2 and salts present in wet steam are more compared to those in dry steam, they strike the blade section during the isentropic expansion of steam and form small grooves called pits or corrosion pits. The pit progressively grows and can take various shapes. Several corrosion pits are developed on turbine blades and become severe when these pits are converged into the crack opening, leading to the failure

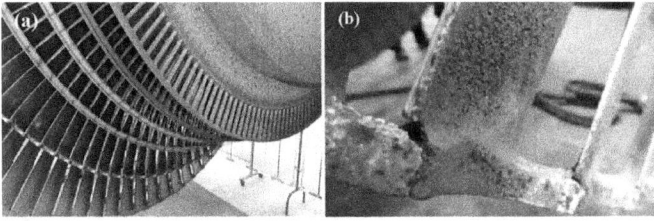

Figure 11.5 (a) Evidence of severe corrosion [21] and (b) failure due to surface-corroded layers on the steam turbine components [21].

Figure 11.6 (a) Development of corrosion pits [21] and (b) salt and oxide deposition on turbine blades [22].

of turbine blades. Structural failure of the turbine blades can be observed when the corrosion pit pierces the metal surface and damages the entire turbine. Stress concentration becomes active around the corrosion pits. It has been confirmed from literature that a corrosion pit or groove takes nearly about 3–11 years of time span for initiation of crack to failure. So, there is ample time for the engineers to inspect the failure and modify the necessary surface conditions. From the fracture surface of the blade, it is concluded that the crack is initiated from the corrosion pits at the leading edge of the blade vane and propagated inside by fatigue due to the vibration of the blade. Phenomena of corrosion, pit formation and salt deposition are shown in Figures 11.5(a,b) and 11.6(a,b).

11.3.1.2 Fatigue, creep and vibration

Both fatigue and creep are turbine-related material problems. Fatigue is a cyclic loading phenomenon that occurs mainly due to the presence of alternating stresses. Together, corrosion and pitting action on the surface of steam turbine blades give rise to the corrosion fatigue phenomenon. Steam turbine blades experience alternating or cyclic stresses due to the combined effect of centrifugal, mechanical and steam load as well as variation in pressure and temperature. Stresses produced by centrifugal forces are much more vulnerable as compared to steam and mechanical loads (bending and twisting) [23]. These stresses are induced during the start-up and shut-down stages of a thermal power plant. Also, the rotor experiences acceleration and deceleration during this stage and the blades are subjected to transient resonant vibrations. Variation in pressure distribution from the leading to trailing edge, flow recirculation, backflow, and negative impact angle are a few factors which develop vibratory stresses further causing fatigue fracture of the blades [8, 24].

The microcracks or grooves generated during the process take time to propagate and depend upon the intensity and frequency of stress cycles, despite the failure being brittle in nature. Thermal stresses are induced by the cyclic variation in temperature during the alternate stages of low-pressure turbine blades. Salts like sodium chloride (NaCl) and sodium hydroxide (NaOH) are the most deadly and deleterious impurities present on the layers of blades that reduce fatigue strength. So, it is required to restrict the effect of cyclic stresses under such regimes to avoid fatigue failure and enhance the overall performance of a powerplant [7, 13, 25]. In order to predict the service life of steam turbine blades, the most common type of test performed in the laboratory includes the determination of crack growth as well as the plot of the S-N curve (intensity of stress vs. number of cycles).

Creep deformation in a thermal power plant is another failure mechanism that mostly affects the rotor of the turbine. Creep failure is not so much important for low-pressure turbine blades as most of the thermal energy is taken away by the rotor itself. The deformation in rotor material increases with time under constant loading conditions. Time, temperature and loading condition are the important factors that affect the creep life of various components. Especially the components that are subjected to long-term high thermal stress are more prone to catastrophic creep failure. Hence, the life estimation and creep behaviours of these components are of great importance. A creep rupture test is generally done to determine the service life of the turbine rotor. The creep strengths of the blades are enhanced by the incorporation of molybdenum as one of the alloying elements [26–29].

11.3.1.3 Erosion and wear

Erosion and wear are the major problems that create irregular, rough and patchy surfaces of the steam turbine blade. These surfaces strongly affect the direction of steam flow and ultimately reduce the thermal efficiency of the plant [30–32]. Erosion by solid particles and water droplets are the two most common erosion phenomenon that takes place on the surface of steam turbine blades. Both these erosion mechanisms are predominant in the leading and trailing edge of the blade section [33–35]. The major concern towards the design and development of turbine blades lies in the fact of finding which erosion mechanisms would induce catastrophic failures. As per the practical experience of an engineer, erosion and wear created by water droplet is the most prominent and serious failure aspects of steam turbine blades. Due to this, the performance and service life of the steam turbine are deteriorated. Huge pressure is generated when the steam strikes the blade with high impact velocity. This pressure is termed "water hammer pressure" [36, 37]. The relative velocity of steam also affects the erosion behaviour caused by water droplets. Though erosion on the surface of a steam turbine blade by solid particles is of little significance as compared to water droplets, it cannot be neglected at any cost. Scale deposition, salts and oxides present in the steam are the root causes of development of solid particles. These particles are accelerated to high velocities and create small pits or grooves on the surface of blades [38, 39].

11.3.2 Mitigation techniques

In order to enhance the performance of the steam turbine, the occurrences of failure mechanism need to be investigated and proper vindication techniques should be imposed. Recent advancements and development of special characteristic material and surface modification techniques have shown a path to achieving desired efficiency and increased power output. Till now, we have seen that the maximum damage is being done on the rotor and turbine blades since these are subjected to high pressure and temperature as well as harsh environmental conditions. Precautionary measures must be taken to protect these essential components from corrosion, pitting, fatigue, erosion and wear. Coating steam turbine blades or surface modification is the rational way to increase surface properties. Microstructural and compositional modification (Figure 11.7) in the steam turbine blades can be done to increase the surface-related properties in order to resist the attack of harsh environmental conditions. Microstructural modification techniques

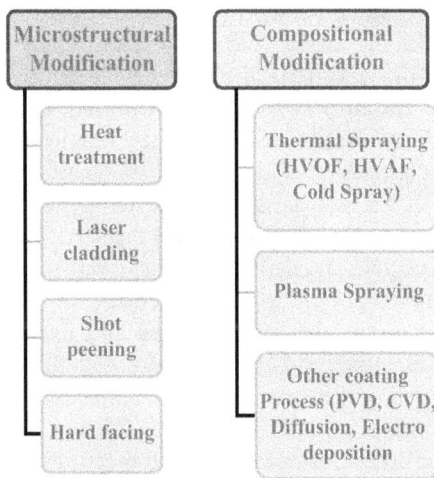

Figure 11.7 Techniques to improve surface behaviour of steam turbine blades.

may include heat treatment, laser cladding, hard facing, shot peening, etc. Various surface coating techniques like thermal spraying (high-velocity oxy fuel (HVOF), high-velocity air fuel (HVAF), cold spray), plasma spraying, physical vapour deposition (PVD) and chemical vapour deposition (CVD), diffusion and electrodeposition improve compositional and surface behaviours of the steam turbine blades [40, 41].

Corrosion and pitting are the main source of other damage mechanisms like fatigue, erosion, wear, SCC, etc.; hence, prevention of this prime source of failure is of utmost importance to minimize major losses. Chromium is the most common corrosion-resistant material. The surface of the turbine blades is coated with sophisticated coatings. Nickel-based superalloys are the most preferred modern turbine blade material. It contains a group of alloying elements like boron, cobalt, chromium, tantalum, rhenium and tungsten. Besides the superior mechanical and chemical properties, these superalloys are good corrosion and creep resistant. Owing to the use of Ni-based superalloys, oxidation of metals and sulphide deposits on the surface of turbine blades are largely eliminated. Electroplating, galvanization, anodization and corrosion protective coatings are some of the common methods to prevent corrosion of steam turbine blades.

Hard coating materials like nitride and carbide are widely used to reduce erosion and wear on the surfaces of steam turbine blades. The coating processes employed to coat the substrate include PVD, magnetron sputtering and cathode arc PVD. Phenomena like water droplet erosion can be reduced by the application of thermal spraying methods mainly due to HVOF and HVAF. These coating methods provide hard, compact and ductile deposits that prevent corrosion, erosion, abrasion and surface restoration of blade particles. Tungsten and chromium carbide stellites (alloys of chromium and cobalt) are the first choice of materials for the thermal spraying of turbine blades [42, 43]. Since the coating is ductile and hence produces very high bond strength which in turn requires high mechanical and steam force to peel off or erode the blade surface. Other mechanical properties like impact strength, fatigue and stress resistance are also significantly increased. Coating processes and their associated characterization techniques are mentioned in Chapter 3.

Microstructural changes may appear on the surface of turbine blades since these are subjected to excessively high temperatures. Heat treatment processes are generally employed to refine the structure and improve the mechanical properties like strength, hardness, ductility corrosion and creep resistance. Different classes of materials require different heat treatment processes depending upon the requirements and application. Restorative heat treatment process like hot isostatic pressing (HIP) is generally used to heal the microcracks or voids present on the blade surface.

Solid particles containing deleterious impurities get dissolved and create the original grain structure. Blade hardening mechanisms like precipitation hardening, flame hardening, laser hardening, induction hardening and various others are used to strengthen the surface properties and also protect from erosion-related problems. Surface treatment process such as shot peening, hammer peening and laser shock peening (LSP) is used to increase resistance to fatigue, wear or corrosion. These processes cause severe plastic deformations on the sub-layers of the blade surface, leading to the development of dislocation mechanisms, further refinement of grain microstructure and significant hardening as well [44].

Shot peening, a cold working process has become an attractive tool to increase the gap between failure and success to a greater extent. The fatigue life of metallic components is generally improved by this process. Other failures of turbine blades related to SCC and intergranular cracks are also reduced by shot peening. The introduction of compressive residual stress on the substrate surface is the major highlight of this process. The presence of this residual compressive stress strongly opposes crack propagation and enhances fatigue life. As compared to conventional shot peening, the LSP process has emerged to be an effective surface enhancement process. This process offers a better surface finish of substrate surface without process-generated defects [45–48].

11.4 CONCLUSION

Failure of steam turbine blades significantly affects the performance of thermal power plants. The performance-dependent parameters like inlet condition of steam, material selection, working environment and various losses are mentioned in the current literature. Damage mechanisms such as corrosion, fatigue, erosion and wear are the root causes of blade failure. The presence of corrosion pits and intergranular cracks led to the activation of corrosion fatigue. The study revealed that crack initiation and propagation due to corrosion pits and fatigue phenomena are the major sources of blade failure. Coating steam turbine blades or surface modification is the rational way to increase surface properties. Preventive techniques like thermal barrier coating, heat treatment, laser cladding, hard facing, shot peening, etc., are some of the methods to improve the compositional and microstructural behaviour of steam turbine blades. HVOF and HVAF coatings provide hard, compact and ductile deposits that prevent corrosion, erosion, abrasion and surface restoration of blade particles. Tungsten and chromium carbide stellites are the first choice of materials for the thermal spraying of turbine blades. Blade design and proper material selection are equally important to avoid severe damage. Special characteristic materials like martensitic stainless steel, **nickel-based superalloys**, Nimonic 80A and titanium alloys (Ti-6Al-4V) are highly recommended for turbine components, especially for blade material. Electroplating, galvanization, anodization and corrosion protective coatings are some of the common methods to prevent corrosion of steam turbine blades. Hard coating materials like nitride and carbide are widely used to reduce erosion and wear on the surfaces of steam turbine blades.

REFERENCES

1. Dolatabadi, A.M., Lakzian, E., Heydari, M. and Khan, A., 2022. A modified model of the suction technique of wetness reducing in wet steam flow considering power-saving. *Energy*, *238*, p. 121685.
2. Sharifi, N., Boroomand, M. and Kouhikamali, R., 2012. Wet steam flow energy analysis within thermo-compressors. *Energy*, *47*(1), pp. 609–619.
3. Yang, Y., Karvounis, N., Walther, J.H., Ding, H. and Wen, C., 2021. Effect of area ratio of the primary nozzle on steam ejector performance considering nonequilibrium condensations. *Energy*, *237*, p. 121483.
4. Lucacci, G., 2017. Steels and alloys for turbine blades in ultra-supercritical power plants. In *Materials for Ultra-Supercritical and Advanced Ultra-Supercritical Power Plants* (pp. 175–196). Woodhead Publishing.
5. Di Gianfrancesco, A. ed., 2016. *Materials for Ultra-Supercritical and Advanced Ultra-Supercritical Power Plants*. Woodhead Publishing.
6. Vanstone, R.W. and Osgerby, S., 2011. Steam turbine materials selection, life management and performance improvement. In *Power Plant Life Management and Performance Improvement* (pp. 518–534). Woodhead Publishing.
7. Rivaz, A., Anijdan, S.M., Moazami-Goudarzi, M., Ghohroudi, A.N. and Jafarian, H.R., 2022. Damage causes and failure analysis of a steam turbine blade made of martensitic stainless steel after 72, 000 h of working. *Engineering Failure Analysis*, *131*, p. 105801.
8. Azevedo, C.R.F. and Sinátora, A., 2009. Erosion-fatigue of steam turbine blades. *Engineering Failure Analysis*, *16*(7), pp. 2290–2303.
9. Zhu, M., 2019, August. Design and analysis of steam turbine blades. In *Journal of Physics: Conference Series* (Vol. *1300*, No. 1, p. 012056). IOP Publishing.
10. Tanuma, T., 2017. Design and analysis for aerodynamic efficiency enhancement of steam turbines. In *Advances in Steam Turbines for Modern Power Plants* (pp. 109–126). Woodhead Publishing.
11. Manda, A.Y., Chada, J.S.R., Surapaneni, S.P. and Geeri, S., 2020. Flow behaviour on aerofoils using CFD. *Journal of Mechanical Engineering, Automation and Control Systems*, *1*(1), pp. 26–36.
12. Ligrani, P., 2012. Aerodynamic losses in turbines with and without film cooling, as influenced by mainstream turbulence, surface roughness, airfoil shape, and mach number. *International Journal of Rotating Machinery*, *2012*.
13. Cano, S., Rodríguez, J.A., Rodríguez, J.M., García, J.C., Sierra, F.Z., Casolco, S.R. and Herrera, M., 2019. Detection of damage in steam turbine blades caused by low cycle and strain cycling fatigue. *Engineering Failure Analysis*, *97*, pp. 579–588.
14. Katinić, M., Kozak, D., Gelo, I. and Damjanović, D., 2019. Corrosion fatigue failure of steam turbine moving blades: A case study. *Engineering Failure Analysis*, *106*, p. 104136.
15. Luo, S. and Wu, S., 2016. Fatigue failure analysis of rotor compressor blades concerning the effect of rotating stall and surge. *Engineering Failure Analysis*, *68*, pp. 1–9.

16. Kushwaha, A.D., Soni, A. and Garewal, L., 2014. Critical review paper of steam turbine blades corrosion and its solutions. *International Journal of Scientific Research and Engineering Trends*, 3(4).

17. Jonas, O. and Machemer, L., 2008. Steam Turbine Corrosion and Deposits– Problems and Solutions. In *Proceedings of the 37th Turbomachinery Symposium*. Texas A&M University, Turbomachinery Laboratories.

18. Rivaz, A., Mousavi Anijdan, S.H. and Moazami-Goudarzi, M., 2020. Failure analysis and damage causes of a steam turbine blade of 410 martensitic stainless steel after 165,000 h of working. *Engineering Failure Analysis*, 113, p. 104557.

19. Katinić, M., Kozak, D., Gelo, I. and Damjanović, D., 2019. Corrosion fatigue failure of steam turbine moving blades: A case study. *Engineering Failure Analysis*, 106, pp. 104–136.

20. Bedaiwi, B.O. and Abd, A.K., 2017. Enhancement of corrosion resistance in steam turbines blades using nanoparticles coatings. *Al-Nahrain Journal for Engineering Sciences*, 20(5), pp. 1172–1181.

21. Wei, Y., Li, Y., Lai, J., Zhao, Q., Yang, L., Lin, Q., Wang, X., Pan, Z. and Lin, Z., 2020. Analysis on corrosion fatigue cracking mechanism of 17-4PH blade of low-pressure rotor of steam turbine. *Engineering Failure Analysis*, 118, p. 104925.

22. Grossel, S.S., 2007. *Luiz Otavio Amaral Affonso, Machinery Failure Analysis Handbook*, Gulf Publishing Company, Houston, TX, 2006 (308pp. $155).

23. Bhamu, R.K., Shukla, A., Sharma, S.C. and Harsha, S.P., 2021. Low-cycle fatigue life prediction of LP steam turbine blade for various blade–rotor fixity conditions. *Journal of Failure Analysis and Prevention*, 21(6), pp. 2256–2277.

24. Mazur, Z., Garcia-Illescas, R., Aguirre-Romano, J. and Perez-Rodriguez, N., 2008. Steam turbine blade failure analysis. *Engineering Failure Analysis*, 15(1–2), pp. 129–141.

25. Vyas, N.S., Gupta, K. and Rao, J.S., 1987. Transient response of turbine blade. In *Proceedings: 7th World Congress IFToMM*, Sevilla, Spain (p. 697).

26. Tabrizian, H., 2015. Reverse engineering and production of steam turbine blades. *Magazine Technical Engineering Products*, 50, pp. 47–53.

27. Ghalambaz, M., Abdollahi, M., Eslami, A. and Bahrami, A., 2017. A case study on failure of AISI 347H stabilized stainless steel pipe in a petrochemical plant. *Case studies in Engineering Failure Analysis*, 9, pp. 52–62.

28. Bahrami, A., Anijdan, S.M., Taheri, P. and Mehr, M.Y., 2018. Failure of AISI 304H stainless steel elbows in a heat exchanger. *Engineering Failure Analysis*, 90, pp. 397–403.

29. Bahrami, A. and Taheri, P., 2019. Creep failure of reformer tubes in a petrochemical plant. *Metals*, 9(10), p. 1026.

30. Mann, B.S., 1999. Solid-particle erosion and protective layers for steam turbine blading. *Wear*, 224(1), pp. 8–12.

31. Campos-Amezcua, A., Gallegos-Muñoz, A., Romero, C.A., Mazur-Czerwiec, Z. and Campos-Amezcua, R., 2007. Numerical investigation of the solid particle erosion rate in a steam turbine nozzle. *Applied Thermal Engineering*, 27(14–15), pp. 2394–2403.

32. Cao, L., Tu, C., Hu, P. and Liu, S., 2019. Influence of solid particle erosion (SPE) on safety and economy of steam turbines. *Applied Thermal Engineering*, 150, pp. 552–563.

33. Quintanar-Gago, D.A., Nelson, P.F., Diaz-Sanchez, A. and Boldrick, M.S., 2021. Assessment of steam turbine blade failure and damage mechanisms using a Bayesian network. *Reliability Engineering & System Safety*, 207, p. 107329.

34. Elhadi Ibrahim, M. and Medraj, M., 2020. Water droplet erosion of wind turbine blades: Mechanics, testing, modelling and future perspectives. *Materials*, *13*(1), p. 157.
35. Herring, R., Dyer, K., Martin, F. and Ward, C., 2019. The increasing importance of leading-edge erosion and a review of existing protection solutions. *Renewable and Sustainable Energy Reviews*, *115*, p. 109382.
36. Ahmad, M., Schatz, M. and Casey, M.V., 2013. Experimental investigation of droplet size influence on low pressure steam turbine blade erosion. *Wear*, *303*(1–2), pp. 83–86.
37. Ibrahim, M.E. and Medraj, M., 2022. Prediction and experimental evaluation of the threshold velocity in water droplet erosion. *Materials & Design*, *213*, p. 110312.
38. Rajendra, K.D. and Arakerimath, R., 2019, February. Analysis of Steam Turbine Blade Failure Causes. In *International Conference on Reliability, Risk Maintenance and Engineering Management* (pp. 46–52). Springer, Singapore.
39. Behera, A., 2022. Superalloys. In *Advanced Materials* (pp. 225–261). Springer, Cham.
40. Pradeep, D.G., Venkatesh, C.V. and Nithin, H.S., 2022. Review on tribological and mechanical behavior in HVOF thermal-sprayed composite coatings. *Journal of Bio-and Tribo-Corrosion*, *8*(1), pp. 1–19.
41. Bastidas, F.G.S. and Bergmann, C.P., 2022. Jet slurry erosion of CERMET Nanocoatings obtained by HVOF. In *Technological Applications of Nanomaterials* (pp. 1–33). Springer, Cham.
42. Bolelli, G., Berger, L., Bonetti, M. and Lusvarghi, L., 2014. Comparative study of the dry sliding wear behaviour of HVOF-sprayed WC-(W, Cr)2C-Ni and WC-CoCr hardmetal coatings. *Wear*, *309*, pp. 96–111.
43. Mahdipoor, M.S., Tarasi, F., Moreau, C., Dolatabadi, A. and Medraj, M., 2015. HVOF sprayed coatings of nano-agglomerated tungsten-carbide/cobalt powders for water droplet erosion application. *Wear*, *330*, pp. 338–347.
44. Mordyuk, B.N., Milman, Y.V., Iefimov, M.O., Prokopenko, G.I., Silberschmidt, V.V., Danylenko, M.I. and Kotko, A.V., 2008. Characterization of ultrasonically peened and laser-shock peened surface layers of AISI 321 stainless steel. *Surface & Coatings Technology*, *202*, pp. 4875–4883.
45. Sundar, R., Pant, B.K., Kumar, H., Ganesh, P., Nagpure, D.C., Haedoo, P., Kaul, R., Ranganathan, K., Bindra, K.S., Oak, S.M. and Kukreja, L.M., 2014. Laser shock peening of steam turbine blade for enhanced service life. *Pramana*, *82*(2), pp. 347–351.
46. James, M.N., Newby, M., Hattingh, D.G. and Steuwer, A., 2010. Shot-peening of steam turbine blades: Residual stresses and their modification by fatigue cycling. *Procedia Engineering*, *2*(1), pp. 441–451.
47. Ortolano, R.J., Kleppe, R.L., Chetwynd, R.W. and Engineer, T., 2005. Shot peening in steam turbines. *The Shot Peener Magazine*, *19*.
48. Prabhugaunkar, G.V., Rawat, M.S. and Prasad, C.R., 1999. Role of Shot Peening on life extension of 12% Cr turbine blading martensitic steel subjected to SCC and Corrosion Fatigue. In *The 7th International Conference on Shot Peening: Proceedings*, Warsaw, Poland (pp. 177–183).

Chapter 12

Performance analysis of tidal turbine blades for different composite materials

R. Deepak Suresh Kumar, Justin Joseph,
K. B. Vigneshwara, K. T. Yugendheran, and N. Adithya
Chennai Institute of Technology, Chennai, Tamil Nadu, India

CONTENTS

12.1 INTRODUCTION

The tidal turbine is one of the best ways of renewable power generation. A tidal stream is a quick streaming waterway made by tides. A turbine is a machine that takes energy from a progression of liquid in which the fluid can be liquid (water) or air (wind) [1]. Since water is notably denser than air, tidal energy is more prominent than wind energy. Unlike wind,

DOI: 10.1201/9781003319375-12

tides are predictable and stable. While we use the flow of water to run the turbine, it might raise some ruckus in the propeller, and the perennial interaction between the water and the propeller causes layers of greenery on the propellers, which creates extra load on the propeller. And due to this supplementary load, the productivity of electrical energy reduces [2]. The significant obstacle to the advancement of tidal ranches is their expense of upkeep. To decrease the development of greenery on the propeller, we need to select the material accordingly. As they have less resistance, tidal turbines ought to be conveyed at destinations with solid conditions (high flows, choppiness, waves, and tempests). Besides, the tidal turbine components will be exposed to different marine animosities, for example, biofouling, disintegration, and consumption. Therefore, selecting the aptest material for tidal turbine blades in such a critical environment is vital to reduce the risks of error, minimize the costly maintenance load, and prolong their service periods (by more than 25 years) [3]. FRP has many potentials such as light weight, weakness obstruction, high elasticity, hostility to erosion, and warm protection. Owing to these properties, FRP is the most adequate material against corrosion and erosion of propeller blades. In this chapter, we will examine the consumption of greenery on the propeller cutting edges, the harm brought about by them, and the investigation of quality will be shown. We will discuss the FRP (Fiber-Reinforced Polymer) composites on the propeller, and then the improvements brought by the FRP composites [4]. After the investigation of both, we will analyze the FRP composites on the sharp edges of the tidal turbine using the "Ansys" analysis software. The FRP composites include Carbon Fiber-Reinforced Polymer (CFRP), Graphite Fiber-Reinforced Polymer (GFRP), and Aramid Fiber-Reinforced Polymer (AFRP). The recorded composite materials are examined, and the visits underneath will clarify the contrast between each composite and between structural steel, and we will settle for the best material for the propeller in the turbine.

12.2 MATERIALS

The blades play a major role in the efficiency of the tidal turbine. Therefore, the material using which the blades are manufactured can be amended in order to improve its performance drastically.

Here, we compare four materials, namely structural steel, CFRP, GFRP, and AFRP to determine the best alternative to structural steel.

12.2.1 Structural steel

It is also one of the other broadly utilized building materials in the construction sector, as well as the most learned and best acknowledged. Structural steel is 100% recyclable and perhaps one of the many highly reused materials around the globe [5]. Steel can be various such as carbon steel, heat-treated carbon steel, high-strength low-composite steel, and heat-treated constructional combination steel. Structural steel below the sustained effect of operational aspects such as temperature, cyclic loads, pressure, radiation, and the surrounding can precede embrittlement as an outcome of thermal aging and corrosion damage as well as fatigue.

12.2.2 CFRP

CFRP is a tremendously solid, profoundly conductive, and light-weight FRP. There are no limitations to its application. The common applications are in the field of Automotive, Aerospace, Robotics, Civil Engineering, Sports and Leisure, and Wind and Tidal Energy for Turbines [6]. CFRP is one of many composites using which strengthening of concrete structures as well as repairing can be done. Environmental effects such as humidity and temperature have intense effects on the FRP composites.

12.2.3 GFRP

Graphite fibers have the advantage of high surface area. Because of a high surface area, it can generate a maximum power of 68.4 W/m cubes and it has low ohmic resistance which improves the efficiency of the turbine [7]. Graphite fiber-reinforced metal has been developed with a combination of properties like stiffness, strength, thermal conductivity, and electrical conductivity. This improves the underlying attributes.

12.2.4 AFRP

Aramid fibers are most commonly known as Kevlar fibers. Aramid fibers are more expensive compared to glass fibers. Impact-resistant structural products such as body armor are usually manufactured with these materials [8]. The excellence of AFRP is its advanced impact properties, significantly lower fiber elongation, high firmness, higher tensile strength, high rigidity, high modulus, low gauge, and thickness, as well as good high-temperature properties. Long Aramid fibers bonded with resins are anticipated to be applied as concrete reinforcement displacing steel.

12.3 METHODOLOGY

With the help of the CAD software, the tidal turbine blade is drawn and is used as a modeling part in different analyses in Ansys software, where the turbine blade is researched for different tests like stress, strain, equivalent stress (von Mises criteria), and fatigue [9]. The modal analysis is also done for the turbine blade. The behavior of the turbine blade is analyzed for structural steel, CFRP, GFRP, and AFRP after observing the reaction of the blade to the given input parameters in the Ansys software [10]. A comparative study is made between these materials about how they would react to the same environment when tested under the same properties and conclusions are drawn.

12.3.1 Von Mises stress

Von Mises pressure is a measure used to decide whether a given material will yield or crack. The von Mises yield basis expresses that if the von Mises pressure of a material under load is equivalent to or more prominent than the yield strength of the same material under uniaxial loading and under straightforward strain, then the material will yield. Table 12.1 shows the von Mises stress for various materials.

Table 12.1 von Mises stress for various materials

S. No	Material	Analysis result	von Mises stress
1	Structural steel		1.0793e+10
2	GPFP		1.0848e+10
3	CRPF		1.0819e+10
4	ARFP		1.0837e+10

12.3.2 Total deformation

The behavior of the structural steel and CFRP under a cantilever load of 5000 N is given below. The maximum and minimum effects of the force can be seen in Table 12.2.

Table 12.2 Total deformation for various materials

S. No	Material	Analysis result	Total deformation
1	Structural steel		0.020517
2	GPFP		0.020571
3	CRPF		0.056366
4	ARFP		0.066066

12.3.3 Normal stress

A typical pressure is a pressure that happens when a part is stacked by a hub power. The estimation of the ordinary power for any kaleidoscopic segment is basically the power partitioned by the cross-sectional region [11]. An ordinary pressure will happen when a part is under strain or pressure (Table 12.3).

Table 12.3 Normal stress for various materials

S. No	Material	Analysis result	Normal stress
1	Structural steel		1.000e+00
2	GPFP		1.000e+00
3	CRPF		1.000e+00
4	ARFP		1.000e+00

12.3.4 Modal analysis

A modular investigation is the investigation of the unique properties of frameworks in the frequency domain (Table 12.4).

Table 12.4 Modal frequency for various materials

S. No	Material	Analysis result	Modal frequency
1	Structural steel		16.634
2	GPFP		31.832
3	CRPF		33.303
4	ARFP		38.889

12.3.5 Safety factor

In designing, the safety factor or Factor of Safety (FOS) communicates how much more grounded a framework is than it should be for an expected burden [12]. Security factors are frequently determined by utilizing point-by-point investigation since extensive testing is unfeasible on numerous tasks, like scaffolds and structures; however, the construction's capacity to convey a heap should be resolved to a sensible precision (Table 12.5).

Table 12.5 Safety factors for various materials

S. No	Material	Analysis result	Factor of safety
1	Structural steel		15
2	GPFP		15
3	CRPF		15
4	ARFP		15

12.4 FATIGUE SENSITIVITY

Ansys Workbench is utilized to examine the pressure and the exhaustion strength of the injector body, getting the consequences of the most extreme worth of stress and the base worth of weakness strength if the fuel stacks and dumps quickly [13]. It was additionally used to track down pressure fixation. The graph in Figure 12.1 shows the fatigue behavior of the turbine blade when different loads are applied to it.

12.5 NORMAL STRESS VERSUS ELASTIC EQUIVALENT STRAIN

The stress–strain diagram provides valuable information about how much force a material can withstand before permanent deformation or failure occurs. Engineering stress and strain data are commonly used because it is easier to generate the data and the tensile properties are adequate for engineering calculations.

A normal stress is a stress that occurs when a member is loaded by an axial force. The value of the normal force for any prismatic section is simply the force divided by the cross-sectional area. A normal stress will occur when a member is placed in tension or compression.

The equivalent plastic strain is the total strain energy of this plastic deformation value on a material. Figure 12.2 shows the relationship between normal stress and elastic equivalent strain.

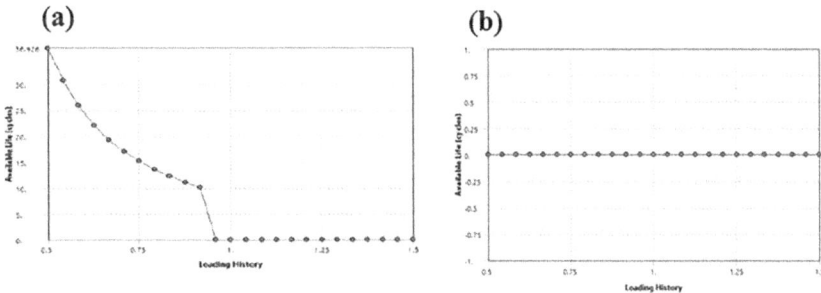

Figure 12.1 (a) When 2000 N load is applied and (b) when 5000 N force is applied.

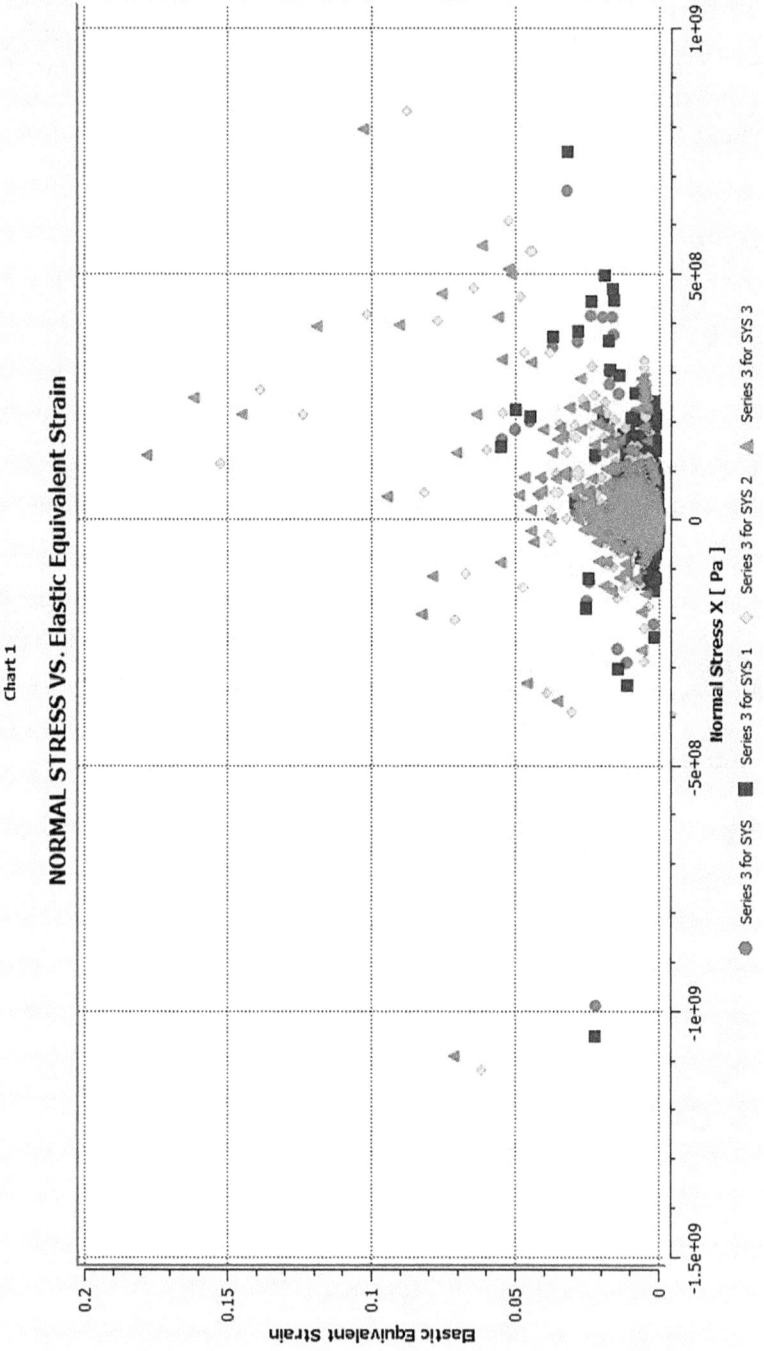

Figure 12.2 Relationship between normal stress and elastic equivalent strain.

12.6 RESULTS AND DISCUSSIONS

12.6.1 von Mises stress

Von Mises pressure is a measure used to decide whether a given material will yield or crack. The von Mises yield basis expresses that if the von Mises pressure of a material under load is equivalent to or more prominent than the yield strength of the same material under uniaxial loading and under straightforward strain, then the material will yield. The von Mises pressure is equivalent for the four materials and it's underneath the yielding point which shows that any material can be replaced. In any case, when the structural distortion is viewed, then GFRP and AFRP are more disfigured than CFRP, which demonstrates that it is the most ideal alternative. A graph chart (Table 12.1) is given between the principal stress and the percentage of volume.

12.6.2 Total deformation

The total deformation of the structural steel and CFRP, GFRP, and AFRP composites are practically the same for the stress applied. At the point when the yield of the four investigations is thought about, clearly, CFRP is the nearest to the estimation of the structural steel which shows that it will be the correct substitute. For example, the strength and hardness given by the steel can be accomplished by supplanting it with the CFRP than GFRP and AFRP, hence, decreasing the weight and cost in this manner and improving the effectiveness of the sharp edge.

12.6.3 Normal stress

Mechanical strain is a mathematical proportion of deformity addressing the general relocation between particles in a material body. Strain is brought about by outside requirements or burdens. Warm strains are strains created at the point when a material is warmed or cooled. They can be the most important aspect from an engineer's point of view that they are not thought of as materials that fit impeccably at one temperature and can crack or drop out when their ecological temperature changes. A visual representation is given in Figure 12.2.

12.6.4 Modal analysis

The modular examination empowers the plan to predict the full vibration or to vibrate at a particular recurrence and gives designs a thought of how the plan will react to various unique load types. From our analysis, it is clear that the CFRP materialized blade will deform less at a particular frequency when compared to structural steel blades, GFRP, and AFRP.

12.6.5 Safety factor

It is noticed that for a similar power applied to the four materials, the CFRP fabricated technique affects its functionability and use in view of its advanced properties and adaptable nature. In spite of the fact that the GFRP and AFRP look like CFRP, when inspected intently, the impact applied because of power is significantly less on the underlying surface of the CFRP. The FOS is essentially used to ensure the essential arranging doesn't lead to any alarming disillusionment or presence of deformation or blemish.

12.7 CONCLUSION

The water that goes through the turbine creates some uproar and produce layers of greenery on the propeller which diminishes the productivity of the turbine. Subsequent to testing various materials, CFRP can be supplanted rather than steel in the turbine when contrasted with various materials like GFRP and AFRP. The motivation behind why CFRP stands out from the rest is that it has a low deformity rate and it is adaptable in nature. Since the water that goes through the turbine creates some uproar and produce layers of greenery on the propeller which diminishes the productivity of the turbine, a substitute material is required. FRP is a substitute material for steel. It has properties like high flexibility, resistance to disintegration, light weight, shortcoming hindrance, etc. The improvement of utilizing FRP composites on the sharp edge of the propeller is shown and demonstrated via "Ansys" analysis software.

REFERENCES

1. Z. C. Sun, Y. F. Mao, M. H. Fan, "Performance optimization and investigation of flow phenomena on tidal turbine blade airfoil considering cavitation and roughness", *Applied Ocean Research*, 106 (2021): 102463.
2. I. S. Abbood, S. A. Odaa, K. F. Hasan, M. A. Jasin, "Properties evaluation of fiber reinforced polymers and their constituent materials used in structures – A review", *Materials Today: Proceedings*, 43 (2021): 1003–1008.
3. M. G. Borg, Q. Xiao, S. Allsop, A. Incecik, C. Peyrard, "A numerical performance analysis of a ducted, high-solidity tidal turbine", *Renewable Energy*, 159 (2020): 663–682.
4. H. M. Su, T. Y. Kam, "Reliability analysis of composite wind turbine blades considering material degradation of blades", *Composite Structures*, 234 (2020): 111663.
5. J. S. Walker, R.B. Green, E.A. Gillies, C. Phillips, "The effect of a barnacle-shaped excrescence on the hydrodynamic performance of a tidal turbine blade section", *Ocean Engineering*, 217 (2020): 107849.

6. E. Hassan, L. Zekos, "Erosion mapping of through-thickness toughened powder epoxy gradient glass-fiber-reinforced polymer (GFRP) plates for tidal turbine blades", *Lubricants*, 9, no. 3 (2021): 22.

7. C. Floreani, C. Robert, P. Alam, "Mixed-mode interlaminar fracture toughness of glass and carbon fibre powder epoxy composites—For design of wind and tidal turbine blades", *Materials* 14, no. 9 (2021): 2103.

8. W. Finnegen, E. Fagan, T. Alanagan, "Operational fatigue loading on tidal turbine blades using computational fluid dynamics", *Renewable Energy*, 152 (2020): 430–440.

9. H. Gonabadi, A. Oila, A. Yadav, S. Bull, "Structural performance of composities tidal turbine blades", *Composite Structures*, 278 (2021): 114679.

10. M. Guisser, "Fatigue analysis of a tidal turbine blade in composite material", 2021.

11. M. G. Borg, Q. Xiao, S. Allsop, "A numerical structural analysis of ducted, high-solidity, fibre-composite tidal turbine rotor configurations in real flow conditions", *Ocean Engineering*, 233 (2021): 109087.

12. C. Robert, T. Pecur, J. M. Maguire, "A novel powder-epoxy towpregging line for wind and tidal turbine blades", *Composites Part B: Engineering*, 203 (2020): 108443.

13. I. R. Chowdhury, N. P. O'Dowd, "Experimental study of hygrothermal ageing effects on failure modes of noncrimp basalt fibre-reinforced epoxy composite", *Composite Structures* 275 (2021): 114415.

Chapter 13

Surface texturing for reducing sliding friction and wear under dry and lubricated conditions

Peddakondigalla Venkateswara Babu
Vasireddy Venkatadri Institute of Technology, Guntur,
Andhra Pradesh, India

Syed Ismail and Vasavi Boggarapu
National Institute of Technology Warangal, Warangal, Telangana, India

CONTENTS

13.1 INTRODUCTION

Lubrication improvement is one of the crucial factors to improve the tribological behavior of mechanical components, which are in relative motion. The main function of a lubricant is to reduce friction and wear of the interacting surfaces by avoiding metal-to-metal contact. In addition, it can also reduce corrosion, temperature, contamination, and vibrations between the sliding surfaces [1]. The friction and wear of engine components can significantly affect engine performance. Proper lubrication enhances the working life as well as the performance of the engine [2].

In addition to proper lubrication, surface modification techniques also play a crucial role in enhancing the tribological behavior of sliding surfaces. In recent years, most researchers have focused on controlling friction and wear through the modification of surface topography. Well-controlled micro-surface features such as surface textures can be emerging in attaining

DOI: 10.1201/9781003319375-13

anti-frictional and wear behavior of the components [3]. In addition, significant breakthroughs have been achieved through surface texturing of many tribological surfaces such as mechanical seals [4–7], journal bearings/ thrust bearings [8–12], and piston ring-liner contact in internal combustion engines [13–17]. Henry et al. [18] investigated the impact of square-shaped dimples on the frictional performance of parallel thrust pad bearing. Their results showed that textured bearings reduced the friction by up to 30% at low loading conditions. Etsion et al. [19] investigated the laser-textured thrust bearing for better frictional performance. Spherical-shaped dimples were tested under hydrodynamic lubrication conditions for various load and speed conditions. Their results showed a 50% reduction in friction coefficient when compared with untextured bearings.

In this research, the combined effects of lubrication improvement and surface texture on the tribological behavior of interacting sliding surfaces had been discussed. Different lubricants were tested for better lubrication performance. Afterward, the circular-shaped protrusions were fabricated on the pin samples, and the sliding tests were performed to investigate their tribological performance under different contact conditions.

13.2 EXPERIMENTAL METHODS

In the following sections, the experimental methodologies for selecting the appropriate lubricant and the sliding tests of interacting surfaces are discussed.

13.2.1 Four-ball tester

Initially, the experiments were conducted in order to select the appropriate lubricant for lubricating the sliding contact. Three lubricants, namely SAE10W-30, SAE15W-40, and SAE20W-50, were selected which are commercially available and also synthetic. These lubricants have undergone the experiments on a four-ball tester for better lubrication conditions at the conjunction. The four-ball tester consists of four steel balls (AISI E-52100) out of which three balls remain stationary, while the fourth ball rotates against these three balls (see Figure 13.1). The detailed procedure can be found in [1]. The tests were conducted according to ASTM D4172. The test conditions are given in Table 13.1.

Each test was conducted for 1 hr, and new steel balls were used for each test. After each test, the frictional torque can be measured with the help of a torque sensor for the corresponding applied load. Afterward, the friction coefficient was calculated by Equation (13.1) [1].

$$\text{Friction coefficient} = \frac{T\sqrt{6}}{3W_1 r} \tag{13.1}$$

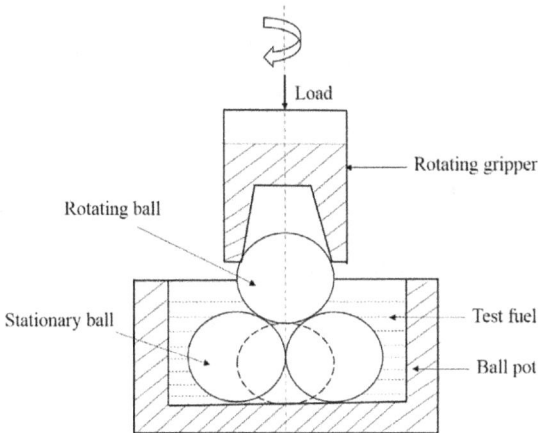

Figure 13.1 Four-ball tester assembly.

Table 13.1 Testing conditions of four-ball tester (ASTM D4172)

Parameter	Condition
Applied load (kg)	15
Speed (rpm)	600, 900, 1200, 1500
Test duration (hr)	1
Temperature (°C)	30
Test lubricants	SAE10W-30, SAE15W-40, SAE 20W-50

where T is the frictional torque in Kg-mm, W_1 is the applied load in Kg, and r (3.67 mm) is the center distance between the contacting lower ball surfaces to the axis of rotation.

After calculating the friction coefficient, the wear scar diameter of the worn steel balls was measured by using SCARVIEW software and the average wear scar diameter is utilized to characterize the anti-wear properties of tested lubricants.

13.2.2 Pin-on-disc setup

The pin-on-disc experiments were designed to assess the tribological performance of textured and untextured surfaces. The textured surface (pin) is considered stationary while, the untextured surface (disc) is considered moving, as shown in Figure 13.2. The material used for the pin was AISI 1020 steel, having a cross section of ø10 mm × 28 mm. The material used for the disc was EN31 steel, having a cross section of ø165 mm × 8 mm.

Circular-shaped textures were considered in the present study. A chemical etching process has been used to develop circular-shaped protrusions on the

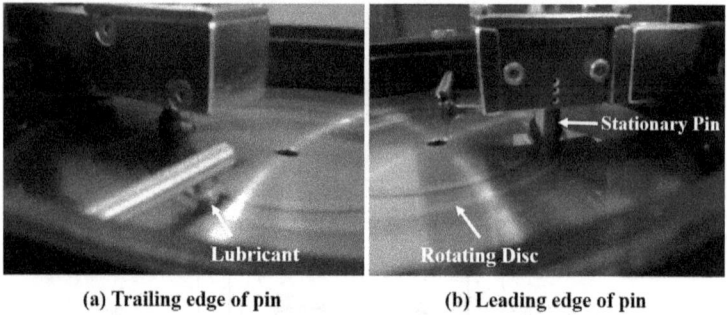

(a) Trailing edge of pin   (b) Leading edge of pin

Figure 13.2 Sliding condition between the stationary pin and rotating disc.

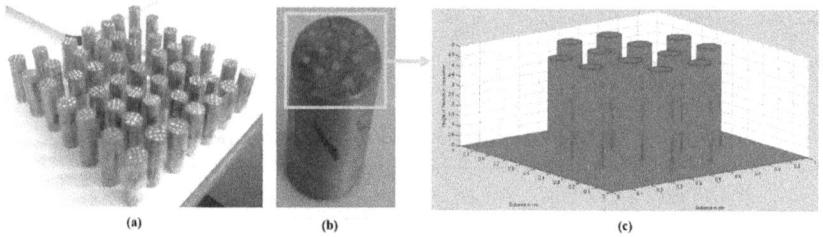

(a)   (b)   (c)

Figure 13.3 (a) Masked pin samples, (b) circular-shaped textures after etching process, and (c) enlarged view of surface textures.

pin surface. In this process, a mask of desired texture shape and distribution was used to make surface texture. The surface which was exposed to a chemical gets etched, while the remaining portion, which was under the mask, would not be etched. The detailed procedure can be found in [20–22]. The fabricated circular-shaped protrusions are provided in Figure 13.3.

After the fabrication of surface textures on pin samples, sliding tests were conducted to characterize the tribological behavior of interacting sliding contact. The frictional coefficient and wear rate were measured after each test for all pin surfaces. Table 13.2 shows the testing parameters used in the sliding experiment.

Table 13.2 Testing parameters

Parameter	Value
Applied load (N)	5
Sliding velocity (m/s)	0.26, 0.52, 0.78
Texture diameter (mm)	1
Texture height (µm)	5, 10, 20
Track radius (mm)	25

13.3 RESULTS AND DISCUSSION

The lubrication performance of different lubricants and the tribological performance of interacting sliding surfaces under different scenarios have been discussed in the following sections.

13.3.1 Selection of lubricant

Before sliding tests, the experiments were performed on the four-ball testers to study the lubrication behavior of different lubricants. The frictional and wear performances of lubricants were tested according to ASTM D472 and also by varying the rotational speed. Figure 13.4 shows the frictional coefficient and wear scar diameter of tested lubricants at different speeds.

It was observed that the frictional coefficient and wear scar diameter were increased with the rotating speed for all the lubricants (see Figure 13.4(a and b)). This may be due to the fact that for higher speeds, the lubricant moves away from the contact zone and allows direct contact of the surfaces, which increased the friction coefficient and wear scar diameter. However, lubricant SAE15W-40 exhibited superior frictional and wear performance among other considered lubricants. Based on the results, SAE15W-40 has been chosen to use between interacting sliding surfaces in further analysis of surface modification.

13.3.2 Sliding friction and wear

Sliding tests were performed to investigate the surface texture performance of sliding contact. The frictional coefficient and wear rate were measured after each test. The tests were conducted under dry as well as lubricated conditions. The selected lubricant, i.e., SAE15W-40, was supplied continuously at a rate of 15 s/drop.

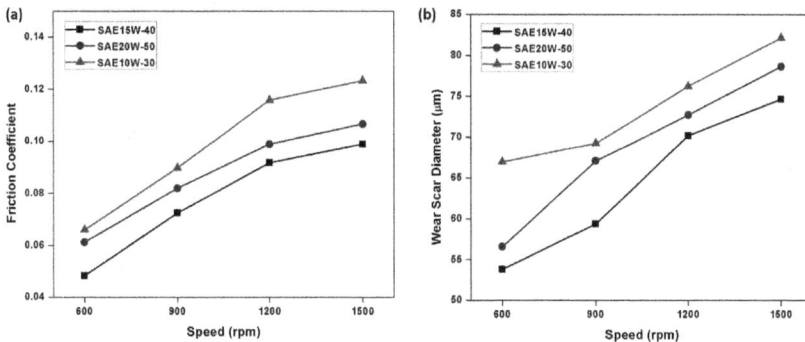

Figure 13.4 Tribological performance of various lubricants at different speeds.

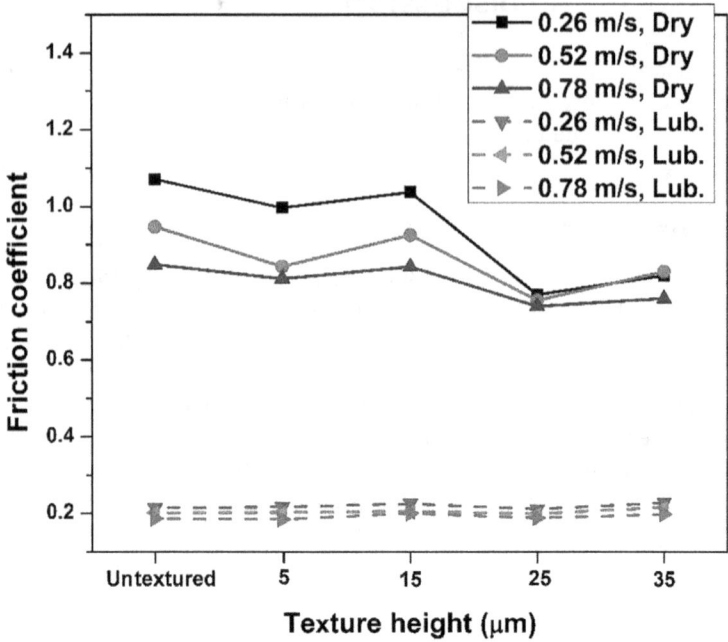

Figure 13.5 Frictional performance of pin samples at different contact conditions.

The frictional performance of textured surfaces is shown in Figure 13.5 under different scenarios. The friction coefficient under lubricated condition was shown to be lower compared to that under dry operating condition for all tested samples. The circular-shaped protrusions reduced the friction coefficient more than that by untextured samples in both dry as well as lubricated conditions. Furthermore, the effect of texture was more prominent in dry conditions in comparison to lubricated conditions. Initially, there was an increase in the coefficient of friction, which subsequently underwent a drastic drop with an increase in the height of the textures. A texture height of 25 μm exhibited better anti-frictional performance among all the tested samples. It was also observed that friction reduction was more at higher sliding velocities compared to lower velocities.

The wear rate of textured surfaces is shown in Figure 13.6 at different contact conditions. It was observed that the lower wear rate was found in the case of lubricated condition as compared to that of dry operating condition, which was ascribed to the lower friction force experienced by the pin samples during the lubrication condition. It was also observed that textured samples experienced lower wear rates than untextured samples at all tested velocities. This may be due to the ability of the textures that captures the wear debris more, which further reduced the wear rate. Furthermore, at a texture height of 25 μm, the pin sample was found the

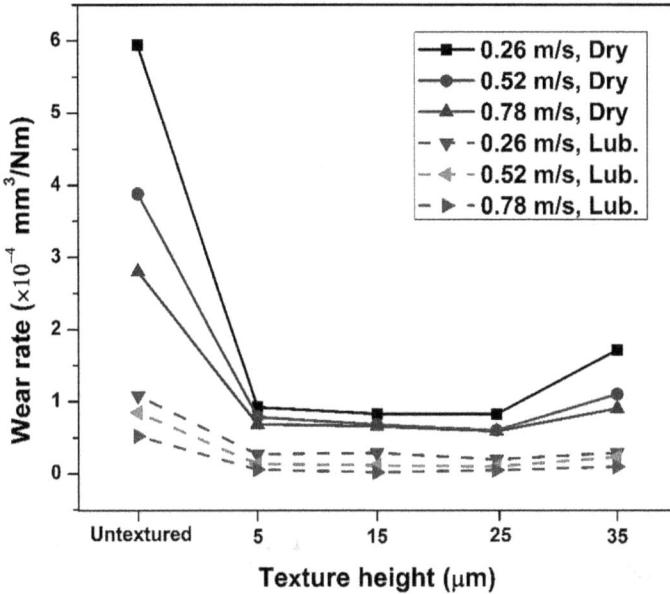

Figure 13.6 Wear rate of pin samples at different contact conditions.

lowest wear rate for the specified operating conditions taken into account. In addition, higher sliding velocities exhibited better anti-wear performance than lower velocities.

13.4 CONCLUSIONS

The present study investigated the combined effects of lubrication and surface texture on the frictional and wear behavior of sliding contact under different scenarios. From the present study, the following conclusions can be drawn:

- The lubrication behavior of different tested lubricants revealed that SAE15W-40 had excellent anti-friction and wear behavior.
- Sliding friction and wear tests revealed that textured pin samples experienced lower friction and wear than untextured pin samples under dry as well as lubricated conditions.
- The texture height was crucial in controlling friction and wear. A texture height of 25 μm exhibited better tribological improvement among all texture heights.
- Sliding velocity had shown a substantial effect on the fricwtional and wear behavior of sliding contact. In this context, at a higher sliding velocity such as 0.78 m/s, the friction and wear rate were found to be lowest.

REFERENCES

1. Yadav, G., Tiwari, S. and Jain, M.L., 2018. Tribological analysis of extreme pressure and anti-wear properties of engine lubricating oil using four ball tester. *Materials Today: Proceedings*, 5(1), pp. 248–253.
2. Farhanah, A.N. and Bahak, M.Z., 2015. Engine oil wear resistance. *Jurnal Tribologi*, 4, pp. 10–20.
3. Wang, J., Zhou, J., Zhu, S.S. and Zhang, J.S., 2017. Friction properties of groove texture on Cr12MoV surface. *Journal of Central South University*, 24(2), pp. 303–310.
4. Etsion, I. and Halperin, G., 2002. A laser surface textured hydrostatic mechanical seal. *Tribology Transactions*, 45(3), pp. 430–434.
5. Yagi, K., Takedomi, W., Tanaka, H. and Sugimura, J., 2008. Improvement of lubrication performance by micro pit surfaces. *Tribology Online*, 3(5), pp. 285–288.
6. Wan, Y. and Xiong, D.S., 2008. The effect of laser surface texturing on frictional performance of face seal. *Journal of Materials Processing Technology*, 197(1–3), pp. 96–100.
7. Yu, X.Q., He, S. and Cai, R.L., 2002. Frictional characteristics of mechanical seals with a laser-textured seal face. *Journal of Materials Processing Technology*, 129(1–3), pp. 463–466.
8. Brizmer, V., Kligerman, Y. and Etsion, I., 2003. A laser surface textured parallel thrust bearing. *Tribology Transactions*, 46(3), pp. 397–403.
9. Fowell, M.T., Medina, S., Olver, A.V., Spikes, H.A. and Pegg, I.G., 2012. Parametric study of texturing in convergent bearings. *Tribology International*, 52, pp. 7–16.
10. Lu, X. and Khonsari, M.M., 2007. An experimental investigation of dimple effect on the stribeck curve of journal bearings. *Tribology Letters*, 27(2), p. 169.
11. Tala-Ighil, N., Maspeyrot, P., Fillon, M. and Bounif, A., 2007. Effects of surface texture on journal-bearing characteristics under steady-state operating conditions. *Proceedings of the Institution of Mechanical Engineers, Part J: Journal of Engineering Tribology*, 221(6), pp. 623–633.
12. Wang, X., Kato, K., Adachi, K. and Aizawa, K., 2003. Loads carrying capacity map for the surface texture design of SiC thrust bearing sliding in water. *Tribology International*, 36(3), pp. 189–197.
13. Akbarzadeh, A. and Khonsari, M.M., 2018. Effect of untampered plasma coating and surface texturing on friction and running-in behavior of piston rings. *Coatings*, 8(3), p. 110.
14. Grabon, W., Pawlus, P., Wos, S., Koszela, W. and Wieczorowski, M., 2018. Evolutions of cylinder liner surface texture and tribological performance of piston ring-liner assembly. *Tribology International*, 127, pp. 545–556.
15. Morris, N., Rahmani, R., Rahnejat, H., King, P.D. and Howell-Smith, S., 2016. A numerical model to study the role of surface textures at top dead center reversal in the piston ring to cylinder liner contact. *Journal of Tribology*, 138(2), p. 021703.
16. Grabon, W., Pawlus, P., Wos, S., Koszela, W. and Wieczorowski, M., 2017. Effects of honed cylinder liner surface texture on tribological properties of piston ring-liner assembly in short time tests. *Tribology International*, 113, pp. 137–148.

17. Babu, P.V., Syed, I. and Ben, B.S., 2020. Optimization of Texture Geometry for Enhanced Tribological Performance in Piston Ring-Cylinder Liner Contact under Pure Hydrodynamic and Mixed Lubrication. In BBVL Deepak, DRK Parhi, Pankaj C. Jena (Eds.), *Innovative Product Design and Intelligent Manufacturing Systems* (pp. 799–808). Springer, Singapore.

18. Henry, Y., Bouyer, J. and Fillon, M., 2015. An experimental analysis of the hydrodynamic contribution of textured thrust bearings during steady-state operation: A comparison with the untextured parallel surface configuration. *Proceedings of the Institution of Mechanical Engineers, Part J: Journal of Engineering Tribology*, 229(4), pp. 362–375.

19. Etsion, I., Halperin, G., Brizmer, V. and Kligerman, Y., 2004. Experimental investigation of laser surface textured parallel thrust bearings. *Tribology Letters*, 17(2), pp. 295–300.

20. Babu, P.V., Ismail, S. and Ben, B.S., 2020. Experimental and numerical studies of positive texture effect on friction reduction of sliding contact under mixed lubrication. *Proceedings of the Institution of Mechanical Engineers, Part J: Journal of Engineering Tribology*. 10.1177/1350650120930911.

21. Syed, I. and Beera, S.B., 2019. Influence of positive texturing on friction and wear properties of piston ring-cylinder liner tribo pair under lubricated conditions. *Industrial Lubrication and Tribology*, 71(4), pp. 515–524.

22. Venkateswara Babu, P., Syed, I. and Beera, S.B., 2019. Modification of surface topography and analysis of its impact on friction and wear reduction of sliding contact. *International Journal of Engineering and Advanced Technology*, 9(1S3), pp. 23–26. 10.35940/ijeat.A1006.1291S319.

Chapter 14

Industry 5.0 for sustainable manufacturing

New product, services, organizational and social information

Shubhangi Chourasia
Global Institute of Technology and Management, Gurugram,
Haryana, India
Delhi Technological University, Delhi, India

Shailesh Mani Pandey
National Institute of Technology Patna, Patna, Bihar, India

Kalpana Gupta
Delhi Technological University, Delhi, India

Qasim Murtaza
Delhi Technological University, Delhi, India

R. S. Walia
Punjab Engineering College, Chandigarh, Punjab, India

CONTENTS

DOI: 10.1201/9781003319375-14

14.1 INTRODUCTION

In the present scenario, competition among the manufacturer and service-providing organizations has raised to obtain a feasible, sustainable manufacturing, and service provider system for customers. We are leading toward the adoption of Industry 5.0 with the enabling of digital technologies it is necessary to understand the requirement of these new technologies. Industry 5.0 required new arrangements of work environment, new markets, new rules, and new government policies where humans and robots can work together toward the prosperity of society. The primary purpose of crafting the concept of Industry 5.0 and Society 5.0 is to develop a digital manufacturing system that would be enabled by a novel digital society where collaborative efforts of social positioning and technical innovation can be inculpated. Industry 4.0 has given birth to Industry 5.0 through a global model of sustainable development of industries and society [24–27] somewhere, Industry 5.0 and Society 5.0 work like a propeller to push our industries and society toward a new concept model of sustainability for attaining better prosperous life. Figure 14.1 shows the various visions behind Industry 4.0 and Industry 5.0. Nowadays, manufacturing industries continuously face global challenges, competition, and constrained availability of resources on earth, increasing customized demand of the customers

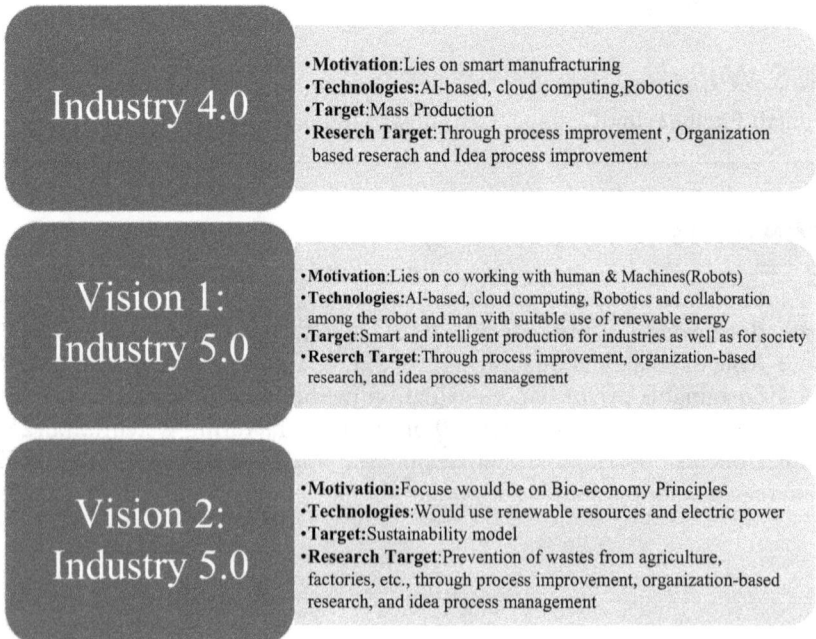

Industry 4.0
- **Motivation:** Lies on smart manufracturing
- **Technologies:** AI-based, cloud computing, Robotics
- **Target:** Mass Production
- **Reserch Target:** Through process improvement, Organization based reserach and Idea process improvement

Vision 1: Industry 5.0
- **Motivation:** Lies on co working with human & Machines(Robots)
- **Technologies:** AI-based, cloud computing, Robotics and collaboration among the robot and man with suitable use of renewable energy
- **Target:** Smart and intelligent production for industries as well as for society
- **Reserch Target:** Through process improvement, organization-based research, and idea process management

Vision 2: Industry 5.0
- **Motivation:** Focuse would be on Bio-economy Principles
- **Technologies:** Would use renewable resources and electric power
- **Target:** Sustainability model
- **Research Target:** Prevention of wastes from agriculture, factories, etc., through process improvement, organization-based research, and idea process management

Figure 14.1 Various visions behind Industry 4.0 and Industry 5.0.

and market demand. In this time, the role of sustainable manufacturing comes into the light to develop a sustainable manufacturing system for a product from product pre-manufacturing to post use where the concept of sustainable manufacturing uses sustainable materials, production methods, a sustainable supply chain, a sustainable multiple supplier's systems, and various distribution units to achieve the defined target without affecting the environment. The concept of sustainability is the stance on three strong pillars, i.e., environment, economic, and social. It involves the essential elements of manufacturing like the development of products, manufacturing processes, product services, value creation, and organizational, social, and economic growth. Although to improve the sustainable manufacturing system, all these elements must be illustrated individually and critically examined to acquire the advantages at various levels of sustainability such as environmental, economic, and social [5]. Many organizations have different definitions of sustainable manufacturing. As per the United States Department of Commerce's definition, sustainable manufacturing is non-polluting manufacturing with limited use of natural resources, conserving energy, safe for society, communities, and most economically [6].

In contrast, NCFAM, i.e., National Council for Advanced Manufacturing, explained products with sustainability and sustainable products [7]. Sustainable manufacturing is to provide full-fledged benefits to all the stakeholders so that negative influence in all the areas can be reduced and to endorse sustainable manufacturing. The aim of building sustainable Industry 5.0 cannot be achieved solely by focusing on the individual development of products, manufacturing processes, product services, value creation, and organizational, social, and economic growth; all these agendas are required to take into consideration simultaneously. Industry 5.0 works on the concept of humans, robots, and machines co-working culture where they work together to enhance system efficiency without eliminating humans from the system. Various digital technologies would be employed, such as cloud servers, artificial intelligence (AI), data mining, the Internet of Things (IoT), computational techniques, 5G and advanced robotics, quantum computing, and augmented reality [8]. Industry 5.0 focuses on a sustainable human-centric approach for all digital technologies, believing in enhancing the skill and re-skill of the workers by applying unique skills plans and a digital-based education plan strategy. Industries' dream of sustainable Industry 5.0 could be accomplished when contemporary resource-efficient manufacturing industries transition to circular sustainable manufacturing. Strict initiatives are required to achieve fewer defects production, zero wastage, eco-friendly products, constitutions innovative research policies, and economical and social benefits [9].

Industry 5.0 engrosses sustainable decision-making all over the value chain and sustainable manufacturing, which is based on the principle of TBL (triple bottom line), where it influences the production system from product pre-manufacturing to product use to entire manufacturing processes to dispatch. The sustainable revolution in Industry 5.0 is one of the most vital initiatives that could assist us in creating a desirable atmosphere for the alive things for the upcoming generation [10]. The concept of six R (6 R), i.e., Redesign, Reuse, Reduce, Recover, Recycle, and Remanufacture, is introduced [11] in sustainable green manufacturing, where redesign explains how to recognize the better approach to redesign and make the things better for future. Remanufacture emphasizes reusing products and components for refurbishment and believes in reusing them as much as possible. Recycling illustrates how industries can convert old materials into new materials or products by using several technologies. Recovery covers the assortment of products at the last stage and uses them for the following product lifecycle [12]. Implementation of all these 6R could help recover cradle-to-cradle products for the closed-loop cycle for attaining more sustainable manufacturing [13]. Figure 14.2. Shows the closed 6 R-based sustainable manufacturing, where R1 is Reduce, R2 is Reuse, R3 is Recycle, R4 is Recover, R5 is Redesign, and R6 is Remanufacture.

Sustainability is not a narrow area for research and development; it's a broad area that focuses on defining the problems, reporting, and observing the ecological balance, economic, and various aspects of the society and

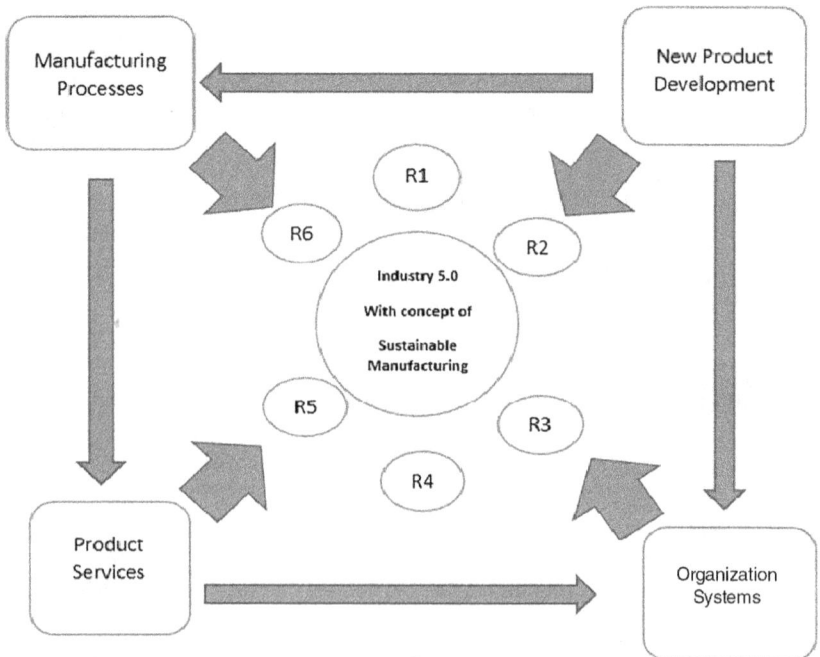

Figure 14.2 Role model for sustainable manufacturing by using 6R principles.

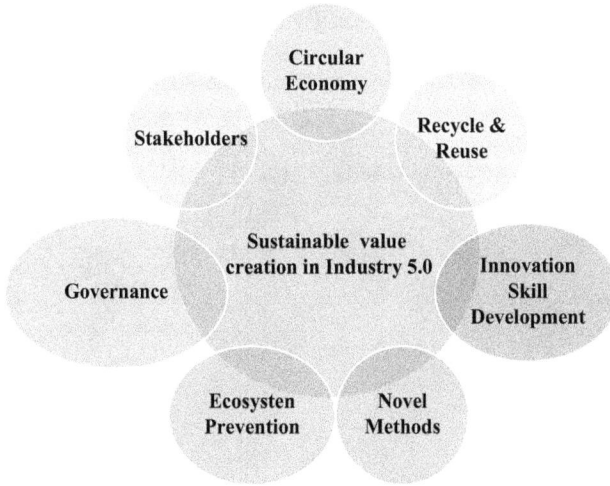

Figure 14.3 Sustainable value creation in Industry 5.0.

organizations [14, 15]. That could be accomplished by embracing the several measuring standards and different innovative models or frameworks of sustainability such as the Global Reporting Initiative (GRI), product sustainability index [16], sustainability accounting, projects on carbon footprints [15], sustainable circular economy, lean and green manufacturing, and closed-loop system. At present, researchers are giving much emphasis to the sustainable value conception based on the Circular Economy due to the requirement of a closed-loop flow system for materials to improve the utilization of resources more economically [17]. The role of circular economy in Industry 5.0 is to reduce resource wastage by properly utilizing available resources on the earth, concentrating on the economy of business models for the nebulous area of applications such as fashion, hotel, manufacturing, and medical. In contrast, sustainable manufacturing is concerned merely with manufacturing. Both concepts are employed interchangeably in various literature due to their interchangeability behavior. That's why it is easy to say that the circular economy is the main set and sustainability is its subset. It is used as a circularity precept for manufacturing. Figure 14.3. shows the linkage between sustainable value creation in Industry 5.0 in the collaboration of various parameters.

14.2 BACKGROUND OF INDUSTRY 5.0

Industry 5.0 has its origins in the idea of Industry 4.0, which Germany devised in 2011, where Germany has projected its future project as adopting a high-tech strategy in decision-making, science, engineering, and business

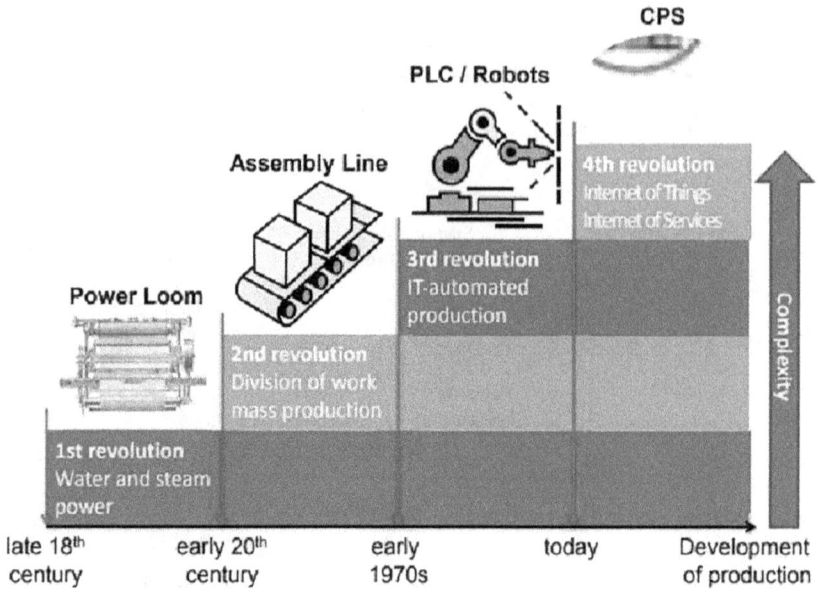

Figure 14.4 The journey from Industry 1.0 to Industry 4.0 [33].

[18]. For the last many years, the focus of Industry 4.0 has been on digitalization and the technologies driven by artificial intelligence for enhancing the efficiency and flexibility of production systems but less on the core principles of good social fairness and sustainability. Figure 14.4. shows the journey of Industry 1.0 to Industry 4.0. The sole concept of Industry 5.0 is to deliver a distinct emphasis and acmes on the innovative research that provides backing to the industry in long-lasting service to humanity within planetary boundaries. Industry 5.0 believes in enhancing the system's productivity without eliminating human beings from the system to trailing the prosperity in a sustainable manner [19]. Some researchers have stated that this revolution is fit into the all-over framework of several progressive new proposals for the "new and novel capitalism," rectifying its excesses which can continue to flourish solitary when it would be fair, wide-ranging, and sustainable [20]. Many complementary visions describe Industry 5.0 as more social-oriented, people-oriented, and measurable by using improved digital technologies [21].

All these are planned to take by using advanced 5G technology, robots, and AI, which are paired with the control and created by the human brain [22]. Another view of Industry 5.0 is to accentuate its budding economic, social, and ecologic impacts. It aims to keep sustainability as a more organized and efficient way to prevent pollution and industrial waste and is dedicated to industrial upcycling [23]. The wider axiom of Industry 5.0 is to break all barriers between the natural world and the cybernetic world where it constitutes the three critical basic parameters: human-centric, sustainable, and flexible [19]. Industry 5.0 will keep fundamental needs and welfares at the heart of manufacturing processes in a human-centric approach. It does not support the influence of new technologies on the workers under the fundamental rights, autonomy, right to privacy, and human respect. This revolution will answer us what this technology can do for us rather than what we can do with this technology. The manufacturing industries within terrestrial boundaries require to be sustainable, which means Industry 5.0 would have to work on the model of circularity where reuse, recycling of natural resources, remanufacturing, reducing wastage, and low carbon footprint are on high priority without hampering the biosphere. The concept of sustainability works on reducing energy consumption, low emission of greenhouse gasses, and reducing the depletion of earth's natural resources so that present generations' needs can be fulfilled without compromising the lack of future peers group. Digital technologies such as AI, Big Data, cloud computing, the IoT, and additive manufacturing will play a vital role in enhancing the efficiency of resources and reducing waste. Industry 5.0 focuses on resiliency, putting pressure on developing a high degree of stoutness in manufacturing and industrial production. Preparation of robust setup/infrastructure besides the disruptions during the crisis, geopolitical shifts, and natural calamities such as Covid-19 and tsunami acmes the globalized production. This plan can only be perceived by employing Industry 5.0 by developing flexible business strategies, a sustainable supply chain, adaptable production capacity, and flexible business processes.

14.3 ROLE OF SMART MANUFACTURING

Smart manufacturing is devised through many definitions by numerous people, industries, and countries. According to DoE @ NIST (Department of energy and National Institute of Standards and Technology), smart manufacturing works as a data-intensive system which keeps all the shop floor (intensive) information to help the advanced, intelligent, efficient, and responsive operating operations [34]. According to Choi [35] et al., smart manufacturing is the tool that is enabled with ICT (information and communication technology) and ADA (advanced data analytics); by using them, manufacturing operations can be improved at all the steps in the supply chain network [37]. A smart manufacturing system is furnished with several technologies such as the IoT, Big Data, automation, CPS (cyber-physical system), and cloud computing applications [37, 38] to understand the vision behind the linked supply chain and data-driven information. The thought process after smart manufacturing, Industry 4.0, and Industry 5.0 is not to eliminate/replace humans from the system but to enhance their intellectual skill, efficiency, and capabilities by using smart technologies designed for a definite area. Some agenda after the smart manufacturing areas trails: Plant resources optimization using a technology-driven approach, sustainable production, and agile supply chain. Smart manufacturing is enabled with several technologies such as AI, machine learning, block chain, predictive analysis, and edge computing. Artificial intelligence produced intelligent results from the provided data that would be based on the input information that helps manufacturers to decide the company's welfare. Block chain technology is digital technology used to distribute data, public ledger, and decentralized purposes across the global network. It profits in traceability and disintermediation to industries to store/record their data in a more efficient way [39]. Edge computing is the technology that converts massive machine data into confined data and improves the manufacturer's decision-making system. Predictive analytics is a tool used to analyze the massive amount of data from all data sources and provide the right forecasting decision to manufacturers for future perspective.

14.4 SUSTAINABLE PERFORMANCE EVALUATION METHODS

The most reliable method to evaluate and verify manufacturing industries' sustainability at every level is by using best-suited quantitative and qualitative metrics. Many studies are conducted appropriately to evaluate industries' performance toward sustainability and have also identified various factors relevant to evaluating the various parameters for attaining sustainability [24]. Haapala et al. have studied several aspects for assessing the sustainable performance of manufacturing industries [25]. Analytical Hierarchy Process (AHP), Graph Theory (GT), Life Cycle Assessment (LCA), Analytical Network Process (ANP), and others are the quantitative techniques that have been used to evaluate the sustainable performance of the system, product sustainability, sustainable index, etc. [26, 27]. Abundant studies have been done centered on developing tabulate form metrics, geographical indications (GIs), ranking methods, and measuring the performance of many manufacturing processes such as machining, riding, and welding by sustainable frameworks [25, 28, 29]. McKinney & Company has recognized the bunch of technologies and suggested that using these technologies in any manufacturing system can hold advantages across the globe at different manufacturing levels [30]. The various digital technologies would be employed in Industry 5.0, such as cloud servers, artificial intelligence, data mining, the IoT, computational techniques, 5G and advanced robotics, quantum computing, and augmented reality [8]. Figure 14.5 shows the transition framework of sustainable Industry 5.0.

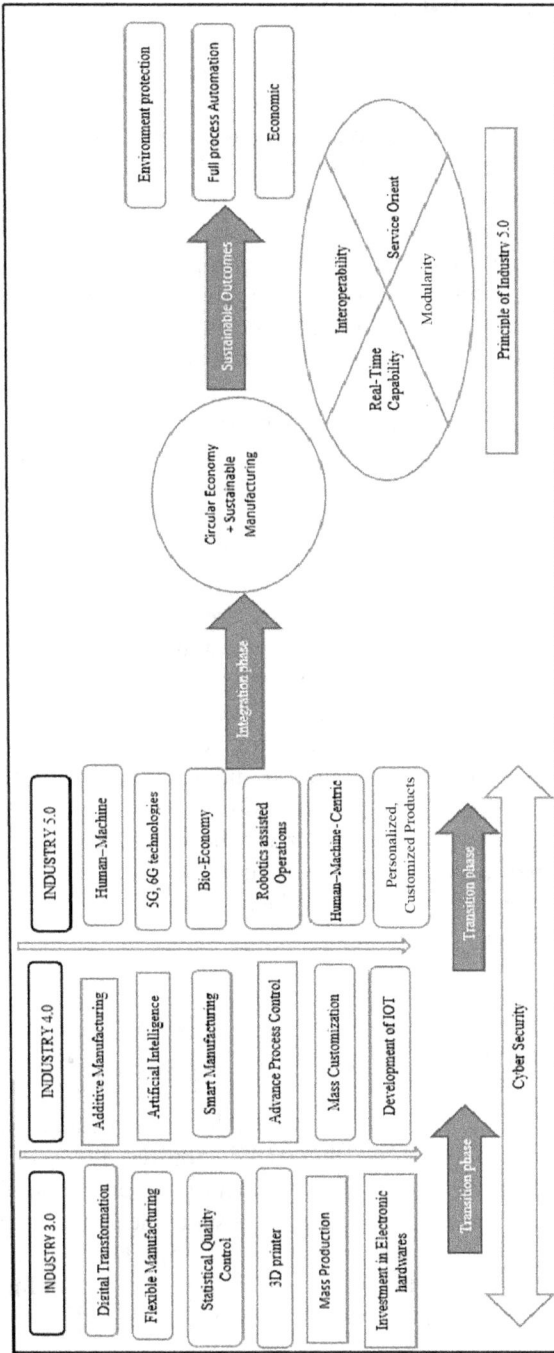

Figure 14.5 Sustainable Industry 5.0 framework.

14.5 FUTURE IMPACT OF INDUSTRY 5.0 ON THE MANUFACTURING SYSTEM

The prior industrial revolutions have demonstrated how the wide-ranging manufacturing system, planning, and strategies were changed to attain higher productivity and 100 percent efficiency. As per many industrialists, manufacturers, and conducted conferences and conventions held with the target of Industry 5.0, it is also believed that it is too early for entering the new revolution [31]. While on the other hand, for adopting the new industrial revolution with proper standardization and implementation of digital and novel technologies, industries should have required their infrastructure that would be enabled with all facilities and developments. During the time of Industry 5.0 adoption, enterprises will face lots of challenges in the area of HMI (Human–Machine Interaction) because this revolution would bring man and machine very close to each other in the daily routine life of every human and brings colossal job opportunities in the area of HMI and analysis of human computational factors. Industry 5.0 will make human work easy by deleting repetitive, dull, risky, and dirty tasks of the workers when possible. The AI-enabled robot's system would have entered into the manufacturing system, processes, supply chain, and production workshop to an uninterrupted level [32]. These future dreams would be only possible when high-quality, exceptional robots would be made using advanced materials that are strong enough, lightweight, highly enabled with optimized use of batteries, capable of handling massive data, and economical. This revolution would increase manufacturing efficiency and operational efficiency, reduce human accidents, and expand sustainability by keeping the economy, technology, and society as its first priority. Additionally, Industry 5.0 would give the solution of customized robots in the sense of hardware and software systems, which would construct a new sustainable ecosystem worldwide. This auxiliary would boost the cash flow, circular economy, and global economy throughout the world.

14.6 CONCLUSIONS

This chapter engaged a concept of sustainable green manufacturing with a future revolution to study the Impact of Industry 5.0 on sustainable manufacturing for more sustainable production of products. The study subsidizes the growing understanding of sustainable manufacturing, 6R principles, and circular economy through Industry 5.0. Finally, the transition framework of sustainable Industry 5.0 and the future impact of Industry 5.0 in a manufacturing system have been proposed.

REFERENCES

1. Önday, Ö. (2020). Society 5.0-its historical logic and its structural development. *Journal of Scientific Reports*, 2(1), pp. 32–42.
2. Salgues, B. (2018). *Society 5.0: Industry of the Future, Technologies, Methods, and Tools*. John Wiley & Sons.
3. Potočan, V., Mulej, M., Nedelko, Z. (2020). *Society 5.0: balancing of Industry 4.0, economic advancement, and social problems*. Kybernetes.
4. Permatasari, D., Iqbal, M. (2019). Strengthening entrepreneurship with a grit system and transformational technology to face society 5.0 in the Asian community. In *International Conference of One Asia Community*, pp. 104–109. Salgues, B. (2018). *Society 5.0: Industry of the Future, Technologies, Methods and Tools*. John Wiley & Sons.
5. Jawahir, I., Badurdeen, F., Rouch, K. (2015). *Innovation in Sustainable Manufacturing Education*. Universitätsverlags der TU Berlin.
6. United States Department of Commerce Website (2009). Available at: http://web.archive.org/web/20170127160912/http://trade.gov/competitiveness/sustainablemanufacturing/how_doc_defines_SM.asp. Accessed on October 5th, 2012.
7. National Council for Advanced Manufacturing (NACFAM) (2009). Sustainable manufacturing. Available at: http://web.archive.org/web/20170127160912/http://trade.gov/competitiveness/sustainablemanufacturing/how_doc_defines_SM.asp. Accessed on November 11th, 2013.
8. Guide to industry 4.0 to 5.0, Global electronics services, i-scoop, Industry 4.0: fourth industrial revolution guide to Industrie 4.0 (i-scoop.eu).
9. Industry5.0.whatthisapproachon,howitwillbeachievedandhowitisalreadybeing implemented, https://www.theinternetofthings.eu/industry-50-what-approach-focused-how-it-will-be-achieved-and-how-it-already-being-implemented.
10. Industry 5.0 the bridge between capitalism and sustainability.
11. Jawahir, I., Dillon Jr, O. (2007). Sustainable manufacturing processes: New challenges for developing predictive models and optimization techniques. In Paper presented at the *Proceedings of the first International Conference on Sustainable Manufacturing*, Montreal, Canada.
12. Jawahir, I.S., Badurdeen, F., Hapuwatte, B.M., Aydin, R. (2020). 6R principles of sustainable manufacturing for achieving circular economy. Unpublished results.
13. Enyoghasi, C., Badurdeen, F. (2021). Industry 4.0 for sustainable manufacturing: Opportunities at the product, process, and system levels. *Resources, Conservation and Recycling*, 166, p. 105362. https://doi.org/10.1016/j.resconrec.2020.105362
14. Blasco, J.L., King, A. (2017). The road ahead: The KPMG survey of corporate responsibility reporting 2017. KPMG International. https://assets.kpmg/content/dam/kpmg/be/pdf/2017/kpmg-survey-of-corporate-responsibility-reporting2017.pdf. (Accessed November 20th 2019).
15. Calabrese, A., Costa, R., Ghiron, N.L., Menichini, T. (2019). Materiality analysis in sustainability reporting: a tool for directing corporate sustainability towards emerging economic, environmental, and social opportunities. *Technological and Economic Development of Economy*, 25(5), pp. 1016–1038.

16. Shuaib, M., Seevers, D. (2014). A metrics based framework to evaluate the total life cycle sustainability of manufactured products, research and analysis, doi:10.1111/ijec.12179.

17. Geissdoerfer, M., Savaget, P., Bocken, N.M., Hultink, E.J. (2017). The circular economy – A new sustainability paradigm? *Journal of Cleaner Production*, 143, pp. 757–768.

18. Raworth, K. (2017). *Doughnut Economics: Seven Ways to Think Like 21st Century*. Chelsea Green Publishing.

19. Nahavandi S.(2019). Industry 5.0-A human-centric solution, https://www.fastcompany.com/90410693/capitalism-is-dead-live-capitalism

20. https://www.fastcompany.com/90410693/capitalism-is-dead-long-live-capitalism

21. Rundle, E. (2017). The 5th industrial revolution, when it will happen and how, https://devops.com/5th-industrial-revolution-will-happen

22. Shelzer, R. (2017). What is industry 50-and How will it affects manufactures, https://blog.gesrepair.com/industry-5-0-will-affect-manufatures

23. Rada, M. (2018). Industry 5.0 defination, https://medium.com/@michael.rada/industry-5-0-defination-6a2f9922dc48

24. Jayal, A.D., Badurdeen, F., Dillon Jr, O.W., Jawahir, I.S. (2010). Sustainable manufacturing: modeling and optimization challenges at the product, process and system levels. *CIRP Journal of Manufacturing Science and Technology*, 2(3), pp. 144–152.

25. Haapala, K.R., Zhao, F., Camelio, J., Sutherland, J.W., Skerlos, S.J., Dornfeld, D.A., Rickli, J.L. (2013). A review of engineering research in sustainable manufacturing. *Journal of Manufacturing Science and Engineering*, 135(4).

26. Ghadimi, P., Yusof, N.M., Saman, M.Z.M., Asadi, M. (2013). Methodologies for measuring sustainbillity of product/process: A review. *Pertan. J. Sit Technol.*, 21(2), pp. 303–326.

27. Hasan, M.F., Saman, M.Z.M., Mahmood, S., Nor, N.H.M., Rahman, M.H.A. (2017). Sustainbillity assessment methodology in product design:a review and directions for future research. *Jurnal Teknologi*, 1, p. 79.

28. Alkahla, I., Pervaiz, S. (2017). Sustainability assessment of shielded metal arc welding (SMAW) process. *IOP Conference Series: Materials Science and Engineering* 244(1).

29. Linke, B.S. 2013. Sustainability indicators for grinding applied to dressing strategies *Journal of Manufacturing Science and Engineering*135(5).

30. Wee, D., Kelly, R., Cattel, J., Breunig, M. (2015). *Industry 4.0-How to Navigate Digitization of Manufacturing Sector*. McKinsey Campany, p. 58.

31. Miller, C.C. (n.d.). Evidence that robots are winning the race for American jobs. Available at: https://www.nytimes.com/2017/03/28/upshot/evidence-that-robots-are-winning-the-race-for-american-jobs.html (Accessed on March 15th, 2019)

32. Nahavandi, S. (2019). Industry 5.0—A human-centric solution. *Sustainability*, 11, p. 4371, doi: 10.3390/su11164371

33. Thoben, K. D., Wiesner, S., Wuest, T. (2017). 'Industrie 4.0' and smart manufacturing–A review of research issue and application examples. *International Journal of Automation Technology*, 11, pp. 4–16, doi: 10.20965/ijat.2017.p0004.

34. Chand, S., Davis, J. F. (2010). What is smart manufacturing. *Time Magazine Wrapper*, pp. 28–33.
35. Choi, S., Kim, B. H., Do Noh, S. (2015). A diagnosis and evaluation method for strategic planning and systematic design of a virtual factory in smart manufacturing systems. *International Journal of Precision Engineering and Manufacturing*, 16(6), pp. 1107–1115.
36. Davis, J., Edgar, T., Graybill, R., Korambath, P., Schott, B., Swink, D., Wang, J., Wetzel, J. (2015). Smart manufacturing. *Annual Review of Chemical and Biomolecular Engineering*, 6, pp. 141–160.
37. Fedorov, A., Goloschchapov, E., Ipatov, O., Potekhin, V., Shkodyrev, V., Zobnin, S. (2015). Aspects of smart manufacturing via agent-based approach. *Procedia Engineering*, 100, pp. 1572–1581.
38. Mittal, S., Kahn, M., Davis, J., Wuest, T. (2016). *Smart Manufacturing: Characteristics and Technologies*. Columbia, SC.
39. Ed Burns (n.d.). Smart manufacturing, https://internetofthingsagenda.techtarget.com/definition/smart-manufacturing-SM

Chapter 15

Surface modification of titanium for drug-eluting stents

Monalisha Mohanta and A. Thirugnanam

National Institute of Technology Rourkela, Rourkela, Odisha, India

CONTENTS

15.1 INTRODUCTION

Cardiovascular diseases are emerging as the vital genesis of death globally, and 90% of coronary intercessions comprise stenting techniques [1]. Two types of stents developed for treating cardiovascular diseases are bare metal and drug-eluting stents (DESs). The bare metallic stent has some disadvantages such as restenosis and thrombus formation. The restenosis process makes the luminal narrow without the growth of the plaque mass, whereas

DOI: 10.1201/9781003319375-15

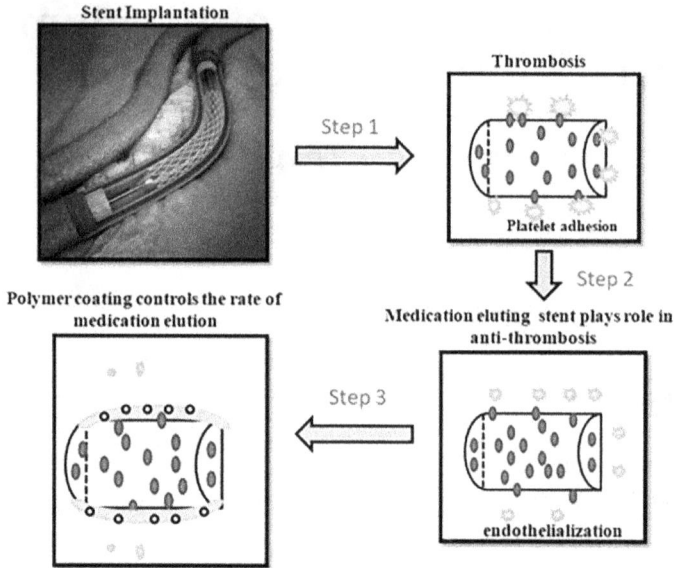

Figure 15.1 Schematic representation of drug-eluting stent.

thrombosis formation is due to the development of plaque tissue in the luminal. So, to overcome these problems, DES has been implemented. The DES has been designed to deliver the drug for a specific period at the location of injury to reduce plaque formation [2, 3]. It is mainly manufactured using metal and metal alloys. But the limitation with the metallic DES is the insufficient quantity of drug elution within the stipulated time. Thus, polymeric coatings have been developed to better-stack drug particles on the stent surface and improve designed supremacy over drug discharge [4]. Polymer layers on the stent surface should restrain the drug from being eroded from the surface and provides an engineering structure for controlling the drug elution. Further, creating the layer-by-layer polymer coating makes the stent platform more biocompatible after the drug has been eluted out [5]. Besides this, an ideal stent should possess high radial strength, good radiopacity, low electrical conductivity, and excellent biocompatibility. Therefore, developing a drug-eluting substrate with good, optical, and antithrombotic properties is necessary, which are the basic requirements to overcome prime clinical challenges. Figure 15.1 shows the schematic representation of a DES.

15.1.1 Metal-based drug-eluting stent

An ideal mechanism of action (MOA) followed by the drugs incorporated in the DES is represented in Figure 15.2. Ideally, the drug would act early in the G1 phase and prevent it from entering the S phase [6]. Limus family drugs such as paclitaxel, sirolimus, biolimus A9, zotarolimus, everolimus,

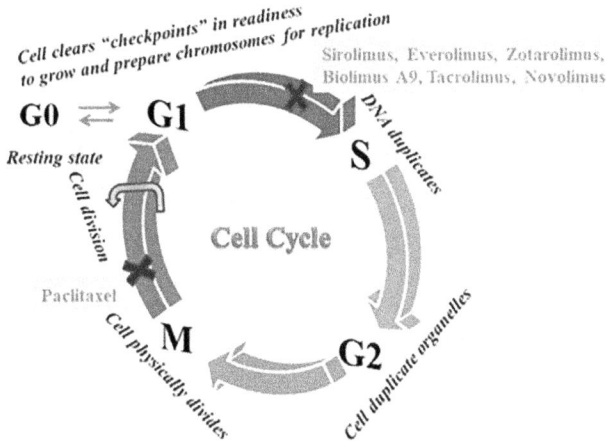

Figure 15.2 Cyclic representation of the ideal mechanism of action in drugs.

tacrolimus, and novolimus are incorporated in stents that selectively inhibit smooth muscle cell (SMC) proliferation and migration, which thereby enhances endothelization of cells. These drugs are structurally related and execute in the same stage of the cell cycle. Cell cycles are arrested before entering the dividing cycle and is regarded as cytostatic agents.

Drugs are divided into two groups, cytostatic and cytotoxic drugs, which are represented in Figure 15.3. Limus drugs are cytostatic, which can stop cell proliferation before the cell is committed to division. But, paclitaxel is a cytotoxic drug that contributes to cell death. Therefore, the MOA should be cytostatic and non-inflammatory [7]. Different types of commercially used metallic stent and their properties are represented in Table 15.1.

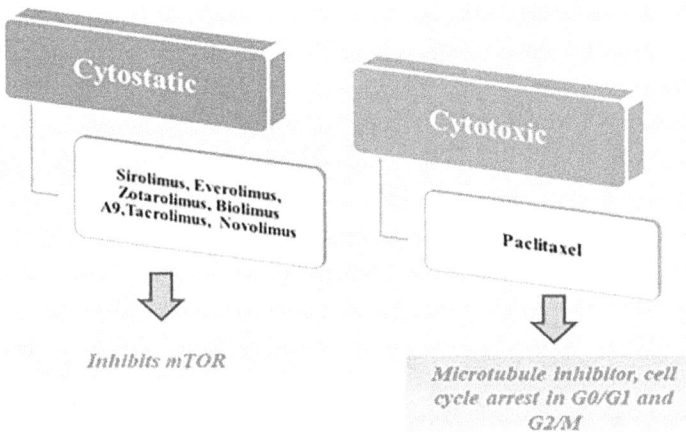

Figure 15.3 Block diagram of properties of drugs.

Table 15.1 Different types of commercially used drug-eluting stents and their properties.

Sl. No.	Types of stent	Metal used	Strut thickness (μm)	Magnetic property	Polymer coating	Characteristic	Duration of elution	Ref.
1	**Paclitaxel-eluting**							
	TAXUS Express	Stainless steel	132	Paramagnetic	SIBS	Durable	>180 days	[9, 10]
	TAXUS Liberté	Stainless steel	81	Paramagnetic	SIBS	Durable	30 days	[11]
	TAXUS Element	PtCr	81	antiferromagnetic-like coupling	SIBS	Durable	30 days	[12]
	Conor	Stainless steel	140	Paramagnetic	PLGA	Biodegradable	6 months to 1 year	[13]
	eucaTAX	Stainless steel	85	Paramagnetic	PLGA	Biodegradable	6 months to 1 year	[15]
	Infinnium	Stainless steel	135	Paramagnetic	PDLLA-co-PGA, PNVP, PLLA-co-PL	Biodegradable	9 months	[16]
	Luc-chopin	Stainless steel	160	Paramagnetic	PLGA	Biodegradable	9 months	[17]
	JACTAX	Stainless steel	97	Paramagnetic	PLA	Biodegradable	9 months	
2	**Sirolimus-eluting**							
	Cypher	Stainless steel	120	Paramagnetic	PEVA/PBMA	Durable	90 days	[20]
	Orsiro	CoCr	60–80	Ferromagnetic to spin glass	PLLA	Biodegradable	3–15 months	[21]
	Ultimaster	CoCr	80	Paramagnetic	PDLLA-PCL	Biodegradable	3–4 months	[22]
	Combo	Stainless steel	100	Antiferromagnetic	PDLLA-PLGA	Biodegradable	30–40 days	[23]
	Drug-filled stent	CONi with Ta	81	Paramagnetic	Polymer free	Polymer free	180 days	[24]
	Meres	Pt-Ir markers	100	Non-ferromagnetic	PDLLA	Biodegradable	12 months	[25]
	Magmaris	Mg,Ta markers	150	Paramagnetic	PLLA	Biodegradable	12 months	[26]
	Sparrow	Nitinol	67	Non-ferromagnetic	PLLA, PLGA, PLC	Biodegradable	12 months	[27]
	Supralimus	Stainless steel	60	Paramagnetic	PLLA-PLGA	Biodegradable	8 months	[28]
	Biomime	CoCr	65	Non-ferromagnetic	PLA	Biodegradable	30 days	[29]
	Excel	Stainless steel	81–140	Non-ferromagnetic	PDLLA	Biodegradable	18 months	[30]
	NOYA	CoCr	81	Paramagnetic	PLLA-PDLLGA	Biodegradable	9 months	[31]
	Inspiron	CoCr	75	Non-ferromagnetic	PLGA	Biodegradable	6–9 months	[32]
	TIVOLI	CoCr	80	Paramagnetic	PDLLA	Biodegradable	8 months	[33]
	BuMa	Stainless steel	100	Paramagnetic	PLA-PLGA	Biodegradable	3 months	[34]
	Firehawk	CoCr	86–96		PDLLA	Biodegradable	9 months	[35]
	Cardiomind	Nitinol	61		SynBiosys	Biodegradable	3 months	[36]
	Yucon choice PC	Stainless steel	87		PLA	Bioresorbable	6–9 months	

	Crystalline-sirolimus eluting MiStent	CoCr	64	Non-ferromagnetic	PLGA	Bioresorbable	9 months	[37]
4	**Everolimus-eluting** Synergy Champion Absorb Absorb-GTI	CoCr Stainless steel Pt markers Pt markers	74 81	Non-ferromagnetic Paramagnetic Paramagnetic Paramagnetic	PLGA PLA PLLA PLLA	Biodegradable Biodegradable Biodegradable Biodegradable	3–4 months 4–6 months 9 months 12 months	[40] [41]
5	**Zotarilimus-eluting** Resolute Integrity Resolute Onyx	CoNi CoNi with Pt-IR	91 81	Ferromagnetic Paramagnetic	Biolinx Biolinx	Biodegradable Biodegradable	12 months 12 months	[42] [43]
6	**Biolimus A9-eluting** Axxess Nobori Biomatrix Biomatrix flex BioFreedom	NiTi Stainless steel Stainless steel 316L SS 316L SS	180 125 114–120 84–88 112	Paramagnetic	PLA PLA PLA PDLLA Polymer free	Biodegradable Biodegradable Biodegradable Biodegradable	9 months 6–9 months 6–9 months 12 months 12 months	[44] [45] [46] [47] [48]
7	**Tacrolimus-eluting** MAHOROBA	CoCr	75	Non-ferromagnetic	PLGA	Biodegradable	6 months	[49]
8	**Novolimus-eluting** DESyne BD	CoCr	81	Non-ferromagnetic	PLLA	Bioresorbable	3–9 months	[50]

15.1.1.1 Paclitaxel-eluting stent

Paclitaxel is exceptionally lipophilic, which works with fast intimal cell response while limiting scattering in the plasma. Paclitaxel has hepatic utilization, with 16% of a fundamental portion being disposed of through the urination process. The structural organization of paclitaxel contains a protein-bounded drug (88–98%). The enormous metabolite of paclitaxel, 6α hydroxypaclitaxel, is, similar to the most active metabolite, 3′-p-hydroxypaclitaxel. Paclitaxel (from the family of the Taxus) is an antineoplastic agent that's generally utilized to deal with several coronary illnesses and ovarian cancer. The antiproliferative movement of paclitaxel is an effect of fixation reliant and reversible which restricts to microtubules, i.e., β-subunit of tubulin on the N-terminal place [8].

Both NIR® and Express™ were utilized in clinical preliminaries of the paclitaxel-eluting stent (PES). These stents are manufactured from stainless steel (SS) and covered with a non-degradable polyolefin (Translute™). It contains paclitaxel about 1 µg/mm². The polymeric coating was found to be compatible with the vascular tissue of porcine which shows no coagulation over the surface. It means the polymer coating was biocompatible with the vascular intimal tissue of porcine. Sterilization assessments have proven that no delamination of the polymeric coating takes place within the *in-vivo* circumstance. Besides, the Express™ stent is adequately adaptable to avoid injury due to its structural construction. It is coated with a preventive polymer (a substance compound) called Translute™, which was grown explicitly for the TAXUS Express™ Stent. The Translute™ polymer is also called as SIBS [poly (styrene-b-isobutylene-b-styrene)]. The role of the polymer is to convey and ensure the medication at the specific site during the methodology. When the stent is embedded in the coronary artery, the polymer assists in control in drug discharge inside the blood vessel. It also ensures the even and steady circulation of the drug from the stent. The paclitaxel–polymer covering has been intended to consider a steady and controlled arrival of drug from the stent surface into the lumen wall while limiting its discharge into the circulation system. Both the medication and discharge rates are important during the cycle to limit the restenosis process, thereby reducing the unnecessary requirement for additional treatment in the stented region [9, 10]. The TAXUS Liberté™ paclitaxel-eluting coronary stent system comprises an inflatable expandable VeriFLEX™ stent covered with triblock copolymer framework, which ensures slow delivery of paclitaxel (8.8 %) [11]. Further, a platinum–chromium (PtCr) alloy replaces the 316L SS, which is utilized more for second-generation TAXUS stents in the TAXUS™ Element stent framework. The PtCr alloy consists of Fe (37%), Pt (33%), Cr (18%), Ni (9%), Mo (2.6%), and Mn (0.05%). The growth within the platinum (Pt) content inside the PtCr alloy pretty much increases its electricity. Besides, keep appropriate radial power and recoiling effect in spite of an 81 µm strut thickness. The natural idle polymer framework of the

TAXUS™ stent framework is covered with SIBS. It is generally an elastomeric hydrophobic copolymer. The paclitaxel mixed with SIBS polymer matrix and incorporated over the stent deprived of a top layer. This polymer structure shields the whole such as the luminal and albuminal stent surfaces [12]. Further, Conor™ stents are comprised of a cobalt–chromium paclitaxel-eluting coronary stent, which has been explicitly intended for vascular medication conveyance. Conor™ stent surface fuses with many pores, and each is considered a repository into which drug-polymer syntheses can be stacked. Also, these stents utilize biodegradable polymers that are engrossed by the body after the medication is delivered, thereby leaving no long-lasting remaining polymers or medication at the objective site [13]. Also, eucaTAX™ configuration utilizes 316L SS as a stent stage which is covered with a long-lasting Camouflage® as the first layer. The second and outer stent layer presents a paclitaxel drug mix biodegradable polyethylene glycolic acid (PLGA) polymer network. The total debasement of the polymer takes place around 2 months. So, it helps decrease neointimal hyperplasia by counteracting its proliferative medication arrangement on the stent surface. Generally, it is coated for a medication content of about $0.25\mu m/mm^2$, which discharges within 8–10 weeks after being controlled release. Also, it uses a PLGA polymer with a low portion of medication, not causing infection, and maintains carbon dioxide and water within proper limits inside the body [14]. Further, the Infinnium™ – CE mark supported, is manufactured by Sahajanand Medical Technologies. This stent framework uses paclitaxel, which has been demonstrated to diminish the narrowing of coronary vessels after stent implantation. This stent has a thinner strut (0.0032″) with high deliverable efficiency [15]. Another one is Luc-Chopin™ – coronary DES with 316L SS paclitaxel and biodegradable polymer. It is developed by the Balton organization, consolidating the most elastic layer of biodegradable polymer, paclitaxel, and a material that has an antiproliferative impact. The polymer layers discharge paclitaxel in a controlled manner by slowly biodegrading within 2 months, thereby hindering neointima development [16]. Also, Jactax™ (Boston Scientific Corporation, Natick, MA), a new-age PES, has been developed with an ultrathin biodegradable albuminal polymer. However, it is obscure whether polyethersulfone with insignificant biodegradable albuminal covering improves strut insertion while forestalling neointimal hyperplasia. Utilizing optical intelligence tomography as the imaging methodology, the authors evaluated the extent of the strut for half a year in PES coated with sturdy ultrathin (<1 µm) biodegradable albuminal polymers [17].

15.1.1.2 Sirolimus-eluting stent

Sirolimus is also called rapamycin. It is a characteristic macrolide anti-toxin with strong immunosuppressive characteristics. It was first endorsed by the US FDA in 1999 as an immunosuppressant for organ transplantation. The most clinical use for sirolimus is as a macrolide anti-infection. It also possesses strong antifungal and antitumor properties. It is also useful for the prevention of certain diseases such as choroid neovascularization and diabetic macular edema [18]. One problem with metal as a stent platform is drug stacking due to the excessively smooth surface. Therefore, a porous surface gives the chance of stacking a massive amount of drug content than the smooth metal surface due to the large surface region and controls its release rate. Medication conveyance through biodegradation is the overall strategy of this stent, and it has been broadly monitored in the exploration of muscular, visual, and cardiovascular applications [19].

The Cypher™ stent, the world's first medication eluting coronary stent, presents another age of coronary stent. The stent is covered with a polymer that continuously delivers the sirolimus into the vessel wall to stop tissue narrowing due to blockage formation. Sirolimus is a cytostatic agent, which means it prevents cells from isolating without obliterating them. It was noticed that sirolimus can repress smooth muscle, endothelial cell multiplication, and T-lymphocyte initiation. Sirolimus mostly ties with the immunophilin, FK binding protein-12 (FKBP-12) present in the cell. The sirolimus/FKBP-12 complex binds and hinders the mammalian target of rapamycin (mTOR) actuation, prompting the restraint of cell cycle movement from the G1 to the S stage. The super-thin Orsiro™ DES has become an FDA endorsement and is currently monetarily available in the United States via gadget producer Biotronik. It is manufactured by utilizing cobalt–chromium metal alloy, which has a bioabsorbable medication transporter polymer coating. It breaks down after the infected vessel wall repair. This arrangement might reduce the requirement for double antiplatelet treatment (DAPT) [20]. Ultimate™ has been specially designed for vascular repairing. This stent has a creative albuminal biodegradable layer of poly (D,L-lactic acid) PDLLA-PCL polymer. Coating of sirolimus diminished the risk of breakage of polymer and delamination when the stent expanded. Also, polymer absorption and sirolimus discharge occur inside the artery within 3–4 months [21]. Another advanced therapeutic DES is COMBO™ Plus. It is made up of customary DES segments with the biological coating as shown in Figure 15.4. The bioabsorbable polymer is covered on the albumin side of the stent, permitting directional medication discharge within 30 days and polymer holding for 90 days. The stent platform is extremely robustly designed using CoCr with enhanced deliverability. Anti-CD34 antibodies grafting persuades the efficient endothelial cells. COMBO's double helix configuration is designed for tremendous radial strength and normal conformability [22].

The drug-filled stent (DFS) is based on the demonstrated establishment of the Resolute Integrity™ DES with continuous sinusoid technology (an interesting Medtronic technique for stent production that molds one strand of wire into a sinusoidal wave empowering a nonstop scope of movement).

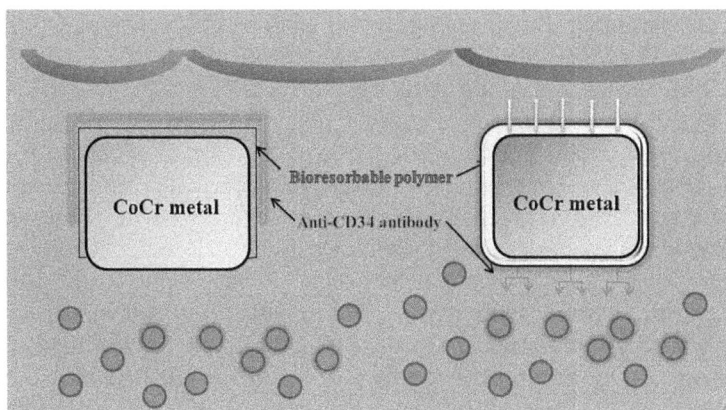

Figure 15.4 Schematic representation of sirolimus release from bioresorbable COMBO stent.

The new era Resolute Onyx™ DES with CoreWire™ technology allows a denser-centered metal coated with a cobalt alloy external layer. The new DFS demonstrates an original tri-layer wire plan, which permits the polymer layer to degrade, so that the bare stent lumen behaves as an interior medication supply. The drug is eluted during the stent implantation through albuminal openings outside of the stent, which is considered a controlled release into the blood vessel [23]. Similarly, the MeRes™, a bioabsorbable stent (BRS), has been manufactured with a decreased strut thickness for better deliverability and quicker debasement, and thrombosis formation is potentially lower. It is a sirolimus-eluting bioresorbable metallic framework [24]. It was demonstrated to work on the paclitaxel-eluting DREAMS™ stage, which is tried in BIOSOLVE-I™ preliminary. The stent platform is made of absorbable magnesium (Mg), which is considered radiolucent. Here, long-lasting tantalum radiopaque with twofold markers is placed at the distal and proximal end of the stent for visualization. These markers are rotated to 90° when worked under radiological signals during imaging. They are silicon-shrouded to protect the Mg compound because Mg has extraordinary synthetic and galvanic affectability [25]. Magmaris™ BRS, in the past known as DREAMS™ 2G, is the first bio-corrodible metallic BRS accessible available, recognized by CE endorsement in Europe in June 2016. It is an inflatable, sirolimus conveying, biodegradable metal framework to control the sirolimus elution [26]. The Cardiomind Sparrow™ stent has an NiTi-based platform which was planned explicitly for injuries inside blood vessel walls (2.00–2.75 mm). This stent has a strut thickness of 61 μm and is pre-stacked on a 0.014-inch guide wire with 2–3 cm of the radiopaque guidewire at the distal end monitoring its location inside the vessel [27]. Further, supralimus Grace™ DES has been developed for the prevention of coronary infection. Fourth-generation supralimus Grace™ stent presents better deliverability, radial strength, and radiopacity. It has about 60 μm strut thickness which prevents re-blockage of the artery [28]. Similarly, the Sparrow™ stent has a 67 μm strut thickness. It is usually covered with a 4-μm-thick layer of sirolimus of about 6 μg and an 8 μm of thick biodegradable poly-lactic acid (PLA)/poly D,L-lactic-co-glycolic acid (PLGA) polymer [29]. Likewise, BioMime™ is a sirolimus-eluting coronary stent system containing a super slim strut thickness of about 65 μm. This future sirolimus-eluting stent (SES) has a novel hybrid design with closed cells (on the two closures of the stent) and open cells (in the center). It has been taken into consideration for morphology-mediated expansion which provides sufficient radiopacity and a lesser recoiling effect. Here, a biodegradable polymer about 2 μm low and of steady thickness was used to coat the SES platform [30]. The pre-clinical examination on an SES covered with PLA polymer (Excel™, JW Medical frameworks, China) exhibited that it requires around a month for sirolimus to be delivered from the covering and around a half year for the polymer to completely corrupt. Their prime objective is to provide security and viability of the Excel™ stent in treating human coronary artery diseases. In-stent late thrombosis formation was observed to be around 0.07 mm in diameter. There was no major antagonistic heart surgery, and no malposition was observed after post-Excel™ implantation. The capacity to decrease the occurrence of major unfavorable heart occasions and the danger of restenosis was comparable with the Excel™, Cypher9 sturdy polymer SES [31]. COMBO™ stent is the main double treatment stent to speed up endothelial. It controls neointimal expansion by discharging a sirolimus drug elution from a biodegradable polymer that accomplishes complete dispersal by 90 days. The following innovation comprises a counteracting agent surface covering that arrests endothelial proliferation cells (EPCs) flowing in the blood by forming an endothelial layer over the stent that gives assurance against thrombosis [32]. INSPIRON™ is one of the third-era DESs intended to make a quick and homogeneous endothelialization. Its configuration, material (CoCr), thin strut (75 μm), and conveyance framework give it great traversability, adaptability, great hybrid profile, moderate radiopacity, and high inflatable crack pressing factor. Its abluminal covering is made out of PLA-PLGA polymers and has low sirolimus measurements, bringing about a moderate medication elution profile, i.e., 60 % in 10 days and 100 % in 45 days. Also, its polymer coating completely degrades within 6–9 months. It provides a quick and homogeneous endothelialization [33]. Moreover, the BuMA™ DES has a latent electrochemical processed and biodegradable PLGA drug transporter. Together, these two layers structure a vigorous polymer with an elution kinetic that adjusts the decrease of hyperplasia with the formation of an active endothelium [34]. Firehawk™ is also a DES with an in-groove stent strut surface and target drug-delivering. Firehawk™ strut is made up of thin hair-like and robust CoCr alloy. The drugs are correctly stacked into the small notches by completely programmed 3D printing to accomplish miniature depression filling.

It joins the benefits of the exposed metal stent and DES, which permits it to accomplish similar clinical adequacy with primarily lower drug stacking [35]. The Yucon choice PCTM is the first DES to furnish the collaboration of biodegradable polymer with microporous surface secured with a thermally insulative protection layer for best features and ideal execution [36].

Similarly, the MiStent™ is a CoCr-based SES with a super-thin strut thickness of 64 μm. Their surface is generally coated with a bioresorbable polymer, facilitating a controlled drug discharge from microcrystalline sirolimus as a repository mounted on the strut surface [37].

15.1.1.3 Everolimus-eluting stent

Everolimus-eluting stent (EES) was a second-era DES supported by the FDA in July 2008. Its CoCr stent plan, high deliverability, and everolimus drug covering were used to avert irregular tissue development. It is a lipophilic compound design that makes it quickly ingest into the blood vessel wall, thereby making it a superior medication conveyer after stent implantation. Everolimus belongs to the sirolimus family (a 2D subordinate of the limus with a solitary insignificant change in 40 sub-atomic layout of hydroxyl group, alkalinized with a 2-hydroxyethyl bunch) and commonly dissolves in natural solvents. The everolimus/FKB12 composite meddles with FKB12/ rapamycin, an administrative protein that directs mobile digestion and multiplication via the p70 S6 kinase and 4E-BP1 phosphorylation process [38]. Subsequently, FKB12/rapamycin arrests the cellular cycle on the past due G1 level to minimize the protein problem, as shown in Figure 15.2. Also, the FUTURE I and FUTURE II preliminaries utilize the EES stage comprised of 316L SS alloy which was manufactured by way of Biosensors (Newport Beach, CA). It is a tube-formed stent with everolimus blanketed over the stent surface with a biodegradable poly-L-lactide (PLLA) polymer which discharges ≈ 70% of medicine in 30 days and ≈ 85% within 90 days. The foundation of the Xience V® (Promus®) stent is the most extensively utilized EES made out of L-605 CoCr. It has 81 μm strut thickness, excessive adaptability, extraordinarily deliverable, true consistency, and fantastic radio-opacity houses. The Xience V® is mainly coated with polymer with the aid of layers – an emaciated bond layer of poly (n-butyl methacrylate) and poly (vinylidene fluoride-co-hexafluoropropylene) [39].

The SYNERGY™ stent was the primary FDA-supported DES with bioabsorbable polymer coating. It was intended to overcome the difficulties related to super-durable polymer stents like atherosclerosis and late stent thrombosis [40]. The CHAMPION™ everolimus-eluting coronary stent system is made up of 316L SS. But, the 316L SS stent stage didn't meet the Guidant's interior execution or quality guidelines. Therefore, after additional testing and assessment for 6 months, the organization has recognized a blend of material preparation and stent plan changes that further develops the stainless steel stent stage and addresses execution and quality concerns.

ABSORB II and ABSORB III, manufactured by Abbotts technology, are specifically designed to reduce stent thrombosis. Abbott's Absorb stent utilized a PLLA polymer covered with everolimus, with a thin strut thickness ($\leq$ 120 μm versus 150 μm) [41].

15.1.1.4 Zotarolimus-eluting stent

Zotarolimus is an immunosuppressant. It is a semi-engineered subsidiary of sirolimus (rapamycin). It was intended for use in stents with phosphorylcholine as a transporter. Zotarolimus, or ABT-578, was initially utilized on Abbott's coronary stent to diminish thrombosis and restenosis [42].

Resolute integrity™ is a CoCr-based zotarolimus-eluting stent. It is manufactured to meet common clinical challenges by offering tremendous radial strength, great deliverability, conformability, security, and effectiveness [43]. Similarly, Resolute Onyx™ is the only DES with platinum–iridium (PtIr) core within the CoNi, thereby enhancing the radiopacity for précising the stent location. It consists of a biodegradable polymer (Biolinx) and a drug combination layer as shown in Figure 15.5. It is specially designed to reduce bleeding risk and provide better stent adaptability, deliverability, flexibility, large dimension matrix, and visibility. It also has excellent thrombosis resistance properties [44].

15.1.1.5 Biolimus A9-eluting stent

Biolimus-eluting stents (BESs) are a new-era DES. It is mainly coated with biodegradable polymer for drug discharge. Biolimus A9™ (BA9™) is highly lipophilic among the limus drugs, due to which the tissue quickly engrosses it. It is resorbed by the tissue within 6–9 months. Also, it limits the fundamental exposure and decreases the number of drugs flowing into the circulatory system. A rapid decrease in-stent thrombosis was observed with BES compared to a robust polymer-based early-age SES in an individual's trial during long-haul follow-up [45].

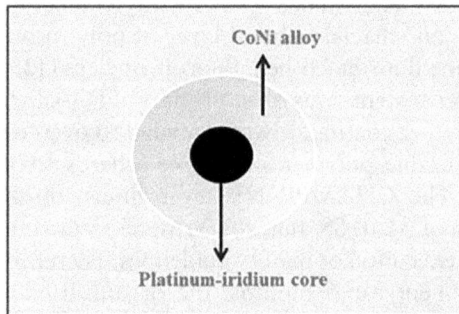

Figure 15.5 Cross section of resolute Onyx strut form.

The Axxess™ bifurcation DES comprises a self-extending narrowly formed by the NiTi stent stage, explicitly intended to adjust to the state of the bifurcation life systems. It upholds the carina structure while safeguarding the side branch. The Axxess™ stent is abluminal covered with a biodegradable PLA polymer that discharges BA9, an anti-restenotic drug licensed by Biosensors [46]. Also, The Nobori™ stent framework uses biolimus A9™, simple sirolimus which is relied upon to decrease tissue contraction through elution from PLA (a biodegradable polymer). The stent conveyance framework is having a hydrophilic covering, which improves deliverability and diminishes blood vessel wall impairment [47]. Similarly, the BioMatrix™ stent is coated with a bioabsorbable PLA polymer at albuminal strut surface for conveying and delivering BA9. The advantages of this new second-era DES had been exhibited in the limus eluted from a durable as compared to the disintegrated polymer coating over the stent surface, which directed a randomized correlation of the clinical outcomes of the BioMatrix™ stent with the CYPHER™. Biosensors' additional innovation is BioFreedom™ which is having an ultra-thin strut thickness of 84μm. It is based on a CoCr-polymer composite and transporter-free biolimus A9™ drug-coated stent. Biolimus A9 is Biosensors' exclusive profoundly lipophilic anti-restenosis drug, grown explicitly for usage in coronary vascular applications [48].

15.1.1.6 Tacrolimus-eluting stent

Like sirolimus, tacrolimus has been utilized for immunosuppression following kidney transplantation and acts by stifling incendiary cytokine emission or T cell actuation. The neointimal multiplication is repressed by working specifically on SMCs. It was observed that tacrolimus-eluting stent (TES) smothered neointimal hyperplasia during the stenting procedure in juvenile swine. However, TES showed lesser clinical outcomes contrasted with sirolimus. Tacrolimus has the property to restrain the vascular smooth muscle cells (VSMCs). The MAHOROBA™ is a TES, which constitutes a thin, adaptable CoCr alloy metal stent stage covered with an incorporated drug-biodegradable polymer. It is planned and created to repress the development of neointimal hyperplasia and keep away from obstruction with the vascular recuperating reaction [49].

15.1.1.7 Novolimus-eluting stent

Novolimus consists of a group of macrocyclic lactones with immuno-suppressive and antiproliferative properties. The macrocyclic lactone novo-limus ties and restrains the beginning of mTOR, a critical administrative kinase to create an immunosuppressive complex. This restraint stifles cyto-kine-driven cell multiplication, repressing the movement from the G1 to the S phase of the cell cycle.

Novolimus is a metabolite form of sirolimus. It is developed to provide efficacy similar to currently available low-dose drugs. Consequently, it requires a lesser amount of polymer coating. The DESyne BD™ stent was manufactured by the Elixir Medical Corporation, which utilizes an exclusive innovation to allow an ultrathin (<3 μm) polymer covering without the requirement for a basic preliminary layer. The biodegradable, polylactide-based polymer allows the sustainable release of Elixir's curative compound (novolimus) to the coronary vessel wall for 6–9 months, leaving behind the bare CoCr metal stent surface to accomplish intense clinical results [50].

15.1.2 Disadvantages of metals used in drug-eluting stents

Metals used for manufacturing DESs are stainless steel, cobalt–chromium, nickel–titanium, and tantalum (as shown in Table 15.1), which provide excellent biocompatibility and radiopacity. Also, other biocompatible metals such as gold and silver are utilized as radiopaque markers to visualize stents better [51]. But all these materials also possess excellent electrical conductivity properties. Thus, stents manufactured from these materials are capable of generating a "Faraday cage" effect. It means that stents functioning as Faraday cages can evade exterior magnetic fields to penetrate the interior bulk of the stent. This phenomenon has a disadvantage in the magnetic resonance imaging (MRI) technique, as demonstrated in Figure 15.6. This technique is used for imaging soft tissue morphologies in the human body. It utilizes a sturdy, time-independent exterior magnetic field (B0 field) and merges with the time-dependent magnetic field (B$_1$ field known as a radiofrequency field). It coincides with the nuclei of the H$^+$ proton of soft tissue configuration. Therefore, a metallic stent working as a Faraday

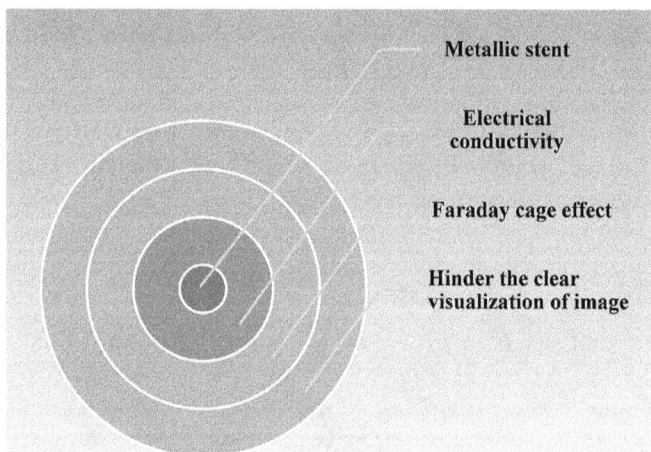

Figure 15.6 Disadvantages of metals used in a drug-eluting stent.

cage covers the lumen from the B_1 radiofrequency field. As a result, the stent lumen would not render clear visualization in an MRI imaging technique.

In an MRI device, radiofrequency (B_1 field) prompts an eddy current in the closed-loop of electrically conductive material whose plane doesn't align parallel to the direction of the imposed time-dependent magnetic field. Consequently, it causes a Faraday cage effect, preventing the MRT device from producing a clear image of the stent lumen. So, it is necessary to reduce the material's electrical conductivity by coating its surface with a polymer and using non-magnetic or paramagnetic metals for manufacturing stents. The magnetic property of metal is necessary because electricity and magnetic field are closely related [52].

15.1.3 Drawbacks of titanium

Titanium (Ti) has been widely used for biomedical applications due to its lightweight and high strength. Also, it has the unique property of developing a titanium oxide (TiO_2) passive oxide layer on its surface, giving it excellent corrosion resistance, and biocompatibility, and further, lowering its electrical conductivity. This inactive protective film is detained within the pH = 7.4 of the human body due to its oxide isoelectric factor. Similarly, in fluid conditions, Ti and its oxides have low particle improvement and are less susceptible to macromolecules. The Ti and its alloys are applied in biomedical embedded units that supplant impaired, tough tissue. It is generally utilized in biomedical applications to manufacturing such as dental and muscular inserts, counterfeit hearts, pacemakers, artificial hip and knee joints, bone plates, cardiovascular valve prostheses, cornea backplates, etc. [53, 54]. Besides, it is having exceptionally alluring characteristics like a higher melting point and higher thermal resistance. It is generally utilized in airplane parts, biomedical embed applications, hand apparatuses, and sports supplies. Titanium can be found in different structures like sheets, strips, plates, foil, adjusts, cylinders, lines, and wires [55]. Generally, Ti6Al4V and commercially pure titanium (cpTi) are used for the production of scientific medical implants. cpTi has higher protection from corrosion. It is widely recognized as a biocompatible metal due to a non-stop dormant oxide layer that suddenly shapes when it oxidizes in the oxidizing media. cpTi and Ti6Al4V produced through the conventional ways are demonstrated by the American Society for Testing and Materials (ASTM) as Grades 1–5. Grade 2 Ti is the precept pure Ti widely utilized in dental embed packages. It has a yield strength of 275 MPa, equal to the yield strength of warm-treated austenitic SS. Grade 5 Ti6Al4V is the most broadly utilized Ti combination in biomedical inserts. The cpTi with ASTM F67 Grade 1: tensile strength is about 240 MPa, elongation is 24% and elastic modulus is between 103 and 107 GPa. ASTM F67 Grade 2: tensile strength is 345 MPa, elongation is 20%, and elastic modulus is between 103 and 107 GPa. ASTM F67 Grade 3: tensile strength is 450 MPa, elongation is 18%, and elastic modulus is

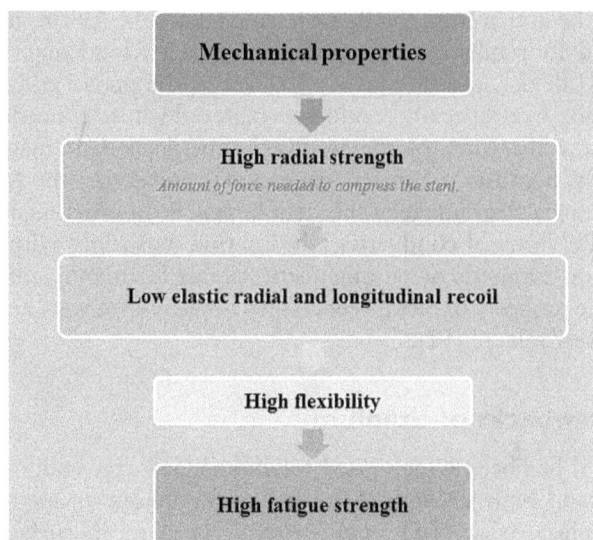

Figure 15.7 Flow chart of the mechanical properties required for coronary stents.

between 103 and 107 GPa. ASTM F67 Grade 4: tensile strength is 550 MPa, elongation is 15%, and elastic modulus is between 103 and 107 GPa. Ti6Al4V with ASTMF136 Grade 5: tensile strength is 860 MPa, elongation is 10 %, and elastic modulus is 114–120 GPa. But, vanadium in the Ti6Al4V combination is cytotoxic, implying that this compound is restricted to certain applications. In some studies, *in-vitro* examination with cpTi showed prolonged osteogenic separation with little cytotoxicity. Therefore, cpTi is considered to be less cytotoxic. Commercial pure titanium has an elastic modulus of around 110 GPa, yield strength of about 485 MPa, and tensile strength of 760 MPa. Titanium is comparable with CoCr for its strength properties, but its high yield strength can lead to material rupture when subjected to an expansion of stents [56–58]. Mechanical properties required for making an ideal stent are represented in Figure 15.7.

These significant constraints restrict Ti to fundamentally one of the surface coatings to give a steady surface oxide, further developing biocompatibility. This alloy has been tried as an independent stent (Titan Stent, Hexacath, France) with promising starting clinical outcomes [59].

15.1.4 Relationship between radiopacity and optical density

The radiopacity estimation in simple film-based radiography can be characterized by ISO 4049 and ISO 9917 global standards. It states that the direct relapse of the logarithm of optical density and aluminium width is equivalent to the aluminium wedge (mm). The "Beer-Lambert law" applies to X-ray passing through the film as presented in Equation (15.1) [60]:

$$T = \frac{I}{I_0} = e^{-\mu x} \qquad (15.1)$$

where T = transmission, I = intensity of the X-ray passes via film, I_0 = incident X-ray intensity, μ = diminution constant, and x = film thickness.

$$\ln(D - f) = \ln(atI_0) - \mu x \qquad (15.2)$$

where D = film optical thickness, f = starting position of optical thickness of the covered film, a = scale aspect of the proficiency of the picture, and t = period to coverage.

From Equation (15.2), the linear logarithm of basis (uncovered) modified optical thickness $(\ln(D\text{-}f))$ and the thickness (x) can give the μ of a film. Also, the dark scale (G) can be predictable with the $D \approx k$ (255-G), where k is a scale and Equation (15.3) can describe as follows [61]:

$$\ln(g - G) = \ln(atI_0) - \mu x \qquad (15.3)$$

The radiodensity of a material can be calculated by the processed film's optical density, which should be in the Beer–Lambert law region. So, the relationship between the absorbance and transmittance represented by Beer's law is shown in Equation (15.4) [62].

According to Beer's law,

$$\%T = \text{antilog}(2 - \text{Absorbance}) \qquad (15.4)$$

where % T is transmittance percentage, which measures the amount of radiation absorbed by the surface. Transmittance is inversely proportional to opacity. The opacity is a proportion of light occurring on the coating separated by the measure of light transmitted. It is described as the absorbance of the coatings on the surface represented in Equation (15.5), where Absorbance is opacity stated as a logarithm to base 10. This estimation requires a white light source and a locator [63].

$$\text{Absorbance} = \log_{10}(\text{opacity}) = \log_{10}\left(\frac{1}{T}\right) \qquad (15.5)$$

15.2 SURFACE ENGINEERING TECHNIQUES

15.2.1 Surface modification with chemical treatment

TiO_2 was first found in the form of ilmenite in 179. The oxide forms are classified into three: brookite, anatase, and rutile. TiO_2 is broadly applied in synthetic, microelectronics, and optics advances because of its high compound reliability, a wide range of optic transmission covering all the apparent and close IR range, high dielectric penetrability, and high oxides refraction index. The optic action of TiO_2, as the most promising photocatalyst, is somewhat effectively concentrated in the responses of the decay of natural compounds.

The photoactivity phenomenon of TiO_2 was comprehended in 1929 due to its white colors in structures. Also, the first logical report regarding this was distributed around 1932–1934. It revealed that the TiO_2 surface treated with ammonia (NH_3) and its salts had been oxidized to nitrites when exposed to ultraviolet (UV) light. Again, the report on photobleaching, which is the phenomenon of the photocatalytic effect of TiO_2, was published in 1938. It was revealed that photobleaching of color happens when UV mildly interacts with TiO_2 surfaces by utilizing formed active oxygen species. The TiO_2 nanostructures have more importance as compared to TiO_2 bulk substances. The TiO_2 nanostructures provide elevated surface regions because of their excessive area at which photo-prompted events take place, thereby enhancing mild assimilation charge, expanding surface photo-instigated current density, upgrading the photons decay rate, and facilitating better surface photoactivity. Simultaneously, the surface-to-volume fraction of the nanoparticles improves the surface retention of -OH and H_2O and also enhances the photocatalytic reaction [64, 65].

The surface of TiO_2 absorbed the UV radiation within a frequency range of 280–400 nm, which facilitates the electron to jump from the valence band to the conduction band of TiO_2. Thus, it leaves a hole inside the valence band (represented in Figure 15.8).

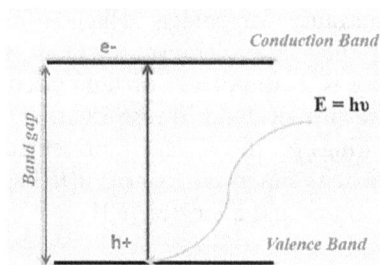

Figure 15.8 Schematic representation of the photocatalytic activity of TiO_2.

In the photo-enacted process, photon activity gets prompted by absorbing responsive species on TiO_2 surfaces to upgrade photon prompted activity. The produced electrons and holes typically recombine on their surfaces, which brings down the photo-initiated acts. Therefore, this process might impede the recombination of electrons and holes to formulate dynamic radicals such as O_2, OH, etc. [66].

In a few research, titanium (Ti) samples have been first inundated in a 10 M NaOH answer for 24 h, accompanied by washing with distilled water. After that, ultrasonic cleansing should be accomplished for 5 min [67]. The pattern is then left for drying and in the end, heated to around 600°C for 1 h as Ti has a sturdy tendency to get oxidized. The alkali treatment forms a nanoporous network over the titanium surface. Therefore, alkali treatment reveals that the TiO_2 layer has a strong tendency to dissolve moderately in the alkaline solution due to the formation of a hydroxyl group [68].

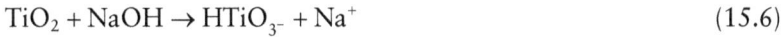

$$TiO_2 + NaOH \rightarrow HTiO_{3^-} + Na^+ \tag{15.6}$$

The above reaction is assumed to continue concurrently with the hydration of titanium.

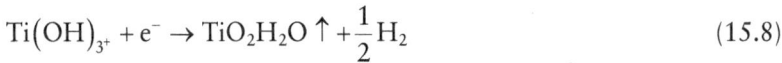

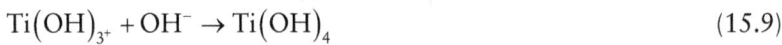

$$Ti + 3OH^- \rightarrow Ti(OH)3^+ + 4e^- \tag{15.7}$$

$$Ti(OH)_{3^+} + e^- \rightarrow TiO_2H_2O \uparrow + \frac{1}{2}H_2 \tag{15.8}$$

$$Ti(OH)_{3^+} + OH^- \rightarrow Ti(OH)_4 \tag{15.9}$$

So, under UV illumination, electrons from the valence band transfer to the conduction band, bordering electrons (e–) with elevated motion. Simultaneously, holes are generated in the valence band. These photocatalytic activities take place on Ti particle surfaces. To accentuate the area of the photocatalytic activities, the synthetic arrangement of the dynamic radicals at Ti surfaces is represented as follows:

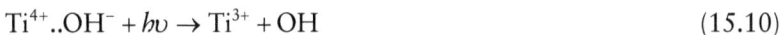

$$Ti^{4+}..OH^- + h\upsilon \rightarrow Ti^{3+} + OH \tag{15.10}$$

These hydroxyl radicals produced in photocatalytic activities are responsible for the oxidation and degradation of organic and inorganic materials. Further, rutile TiO_2 imitates electromagnetic emission which is higher than the anatase phase. So, rutile shades are particularly extreme than anatase

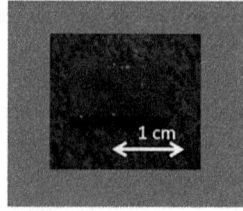

Figure 15.9 Blue color emission of alkali-treated Ti at 365 nm under UV illumination lamp.

colors. In evaluation, anatase phase colorations are pretty bluer than rutile phase colors as shown in Figure 15.9 [69].

The energy bandgap can be calculated for alkali-treated Ti by utilizing the accompanying conditions given in Equation (15.11):

$$(\alpha h\upsilon)^2 = \beta (h\upsilon - E_g) \qquad (15.11)$$

where E_g is the optical energy bandgap and n is the power aspect of the transition forms, which depends upon the behavior of the sample [70].

15.2.2 Surface coating with polymers

The optical and electrical properties of the polymer are important in DES application.

15.2.2.1 Model of optical parameters

To evaluate the light fraction through a material, Beer–Lambert set up the following relationship [71]:

$$\frac{I_t}{I_0} = e^{-\alpha t} \qquad (15.12)$$

where I_0 = intensity of incident light, I_t = intensity of transmitted light

The amount of I_0 to I_t light is determined by the way length (d) of light and the absorption coefficient of polymer constituents at I_0 frequency. The total I_0 light at the subsequent plane is equal to the summation of I_T, I_R, and I_A, individually. The accompanying arithmetical condition is represented as follows:

$$I_o = I_R + I_T + I_A \qquad (15.13)$$

where I_R = reflected light intensity, I_T = transmitted light intensity, and I_A = absorption light intensity.

The refractive index must be changed to develop an effective optical dynamic material. The relationship between the refractive index, reflectance, and extinction coefficient is represented as follows:

$$n = \left(\frac{1+R}{1-R}\right) + \sqrt{\frac{4*R}{(1-R)^2} - K^2} \tag{15.14}$$

where K (extinction coefficient) directly proportional to the frequency and retention coefficient while being inversely proportional to the thickness (t) of the sample. A high refractive index is needed for a polymer framework to be lightweight, high adaptible, and formable. It was observed from the literature that UV–vis spectroscopy was utilized to evaluate the optical and electronic properties of polymer electrolytes. It was also observed that polymer electrolytes soaked with alkali-treated metal salts showed better optical absorption properties. Furthermore, it was stated that the polymer electrolyte film of poly-vinyl chloride incapacitated with $MnSO_4$ shifted the absorbance spectra to the high-frequency level.

One of the studies has demonstrated the effect of UV illumination on the optical and ionic conductivity of PVDF-HFP/$LiClO_4$ samples. It was observed that its visibility enhanced as the absorption spectra shifted towards a higher wavelength. This change in retention has been associated with a sizeable lower of debris to steel debris. In polymethyl methacrylate, a small portion of the absorption peak lies in the UV range. Pure PMMA has a peak at 280 nm because of the $\pi = \pi^*$ in C=O functional group. Also, natural PVA has a confirmed band at 198 nm frequency due to the π–π^* transition. Correspondingly, the absorption level at 281 nm is due to the n–π^* transition. Moreover, one extra peak is observed typically at 208 nm, demonstrating the presence of unsaturated carbonyl (C=O) bonds and, moreover alkene essentially in the pure PVA molecular chain. The huge absorption of TiO_2 is revealed for the TiO_2/PVA mixtures, and the depth relies upon the quantity of TiO_2 load in the mixtures [72].

Also, the polymer coating is utilized for holding and for controlled drug release inside the blood vessel. The techniques for controlled delivery of drugs from polymer matrixes are zero-order, first-order, the Korsmeyer–Peppas model, the Ritger–Peppas model, and the Higuchi model [73].

15.2.2.1.1 Zero-order

Medication disintegration from dose shapes that don't disaggregate and deliver the medication gradually is given by Equation (15.15):

$$Q_t = Q_0 + K_0 t \tag{15.15}$$

where Q_t = medication disintegration at t, Q_0 = the underlying measure of medication in the arrangement at $Q_0 = 0$, and K_0 = the zero request discharge steady constant in units of fixation/time.

This relationship can be utilized to describe the medication disintegration of changed delivery drug measurement structures, as in the instance of a few transdermal frameworks, just as network capsules with low solvent medicines in blanketed systems, osmotic frameworks.

15.2.2.1.2 First-order

The following equation represents the arrival of the medication which followed first-order kinetic:

$$\frac{dc}{dt} = -Kc \qquad (15.16)$$

where K = first-order kinetic rate constant expressed by t^{-1}.

This relationship can be utilized to reveal the medication disintegration in drug measurement structures, for example, those containing water-solvent medications in permeable frameworks. At first it was imagined for planar frameworks and it then reached out to various geometrics and permeable frameworks.

15.2.2.1.3 Higuchi model

The Higuchi version depends on the following speculation that

(i) Preliminary drug attention in the network is better than drug solvency
(ii) Drug dispersion takes place simply in one measurement (area of impact needs to be insignificant)
(iii) Drug debris are a lot more modest than polymer thickness
(iv) Lattice increasing and disintegration are immaterial
(v) Drug diffusion is constant
(vi) deal sink conditions are continuously completed within the discharge condition

The expression for this model is represented by the following:

$$F_t = Q_t = A\sqrt{D(2C - C_s)C_s t} \qquad (15.17)$$

where Q_t is the measure of medication delivered in time t per unit region A (Area), C is the initial medication fixation, C_s is the medication dissolvability in the grid media, and D is the diffusivity of the medication particles (dissemination coefficient) in the grid substance.

This relationship can be utilized to reveal the medication disintegration from changed delivery drug dose structures and lattice tablets with water solvent medications.

15.2.2.1.4 Korsmeyer–Peppas model

Korsmeyer et al. showed a mechanism of drug elution from a polymeric structure. The drug elution mechanism was determined by using 60% of initial drug release data and fitted in the Korsmeyer–Peppas model:

$$Ft = \frac{M_t}{M_\infty} = kt^n \tag{15.18}$$

where F_t = fraction of medication delivered at t, k = the discharge rate constant, and n = the delivery exponential power.

The n esteem is utilized to describe unusual discharge for round and hollow formed polymer lattices. For the instance of tube-shaped tablets, $n \geq$ 0.45 is equivalent to a Fickian dispersion and $0.89 > n > 0.45$ to non-Fickian transport; similarly, $n = 0.89$ is equivalent to the case II mechanism, and $n >$ 0.89 corresponds to super case II (relaxational) mechanism.

15.2.2.1.5 Ritger–Peppas model

Ritger and Peppas et al. presented an exponential correlation $Mt/M = kt^n$ that might be utilized to show the Fickian and non-Fickian discharge conduct of expanding controlled delivery frameworks that swell to a moderate equilibrium level enlarging and are arranged by joining of a medication in a hydrophilic system.

$$Ft = \frac{M_t}{M_\infty} = kt^n \tag{15.19}$$

For a hydrogel film, $n = 0.5$ (Fickian dissemination) and $n = 1.0$ (Case-II transport), prompting a zero-request discharge profile. $0.5 < n < 1.0$ describes non-Fickian transport. The model is appropriate just for $M_t/M_\infty \leq 60\%$ [74].

15.3 POST-CLINICAL ADVANTAGES

15.3.1 Optical and electrical conductivity property

Optical property is the most significant property, which includes absorbance, reflectance, and opacity. The optical properties of a material characterize how it cooperates with light. The refractive index is the most crucial parameter of optical property. Visual image is an apparent proliferation of items that are generated by the reflected and refracted light waves. Optical imaging is especially advantageous for estimating the numerous properties of delicate tissue. It can act as an early marker detection of metabolic changes in organs and tissues due to its tendency to absorb and dissipate light through it. In the case of metal, the refractive index and absorbance coefficient are two important optical constants.

Similarly, electrical conductivity is also one of the essential properties in metals generated due to the electrically charged particles. It is a "free electron" that permits metals to allow electric flow through it. If the imaging signal comes in contact with the conductive metal, it starts producing noise and ultimately leads to image distortion. Consequently, it is necessary for surface modification and metal coating to be made electrically insulative when utilized as a stent. So, optical and electrical conductivity properties are essential for image resolution under X-ray, MRI, CT scan, and CT fluoroscopy devices, which can be achieved through the surface.

15.3.2 Drug elution property

Coronary stents have been developed to elute the drug gradually in *in-vivo* to impede cell growth. This phenomenon averts fibrosis along with thrombosis (clotting of mass). The FDA supported the current clinical drug-eluting drugs after clinical preliminaries. It was shown that they were better than bare metallic stents for the treatment of narrowed coronary arteries and facilitate lower paces of primary antagonistic cardiovascular therapy. DESs are generally constructed to improve the diameter of a damaged artery so that blood can normally flow inside the artery. Therefore, for holding a good quantity of drugs and controlling the rate of drug elution, surface medication and one to at least three layers of polymer coatings over drug encapsulated surface are needed. The elution of medicines from the stent surface with good physicochemical properties was demonstrated in 2001. The aim is to elute a high concentration of medicine to the lesion site with insignificant incidental effects. The post-clinical advantages after surface modification and polymer coating of Ti are represented in Figure 15.10.

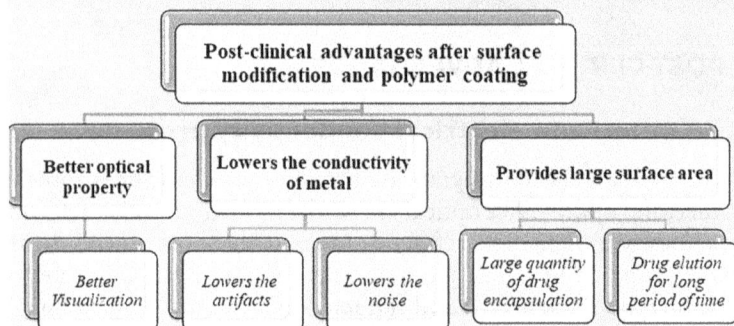

Figure 15.10 Block representation of post-clinical advantages after surface modification and polymer coating.

15.4 SUMMARY

Even though commercial pure titanium has lower mechanical strength, it possesses a few predominant properties such as excellent biocompatibility, inertness, paramagnetic, and corrosion resistance. The recoiling effect can be minimized by controlling its spring-back development. Also, it was observed that the optical density of metal is directly proportional to radiodensity. So, surface modification of cpTi with alkali (NaOH) can improve its photocatalytic activity, enhancing its optical property. In addition, surface modification provides a large surface area by forming nanoporous structures, enabling the encapsulation of drugs. Further, alkali treatment improves its semi-conductive nature by lowering its conductivity. Additionally, the conductivity of cpTi can be diminished by polymer coatings which upgrade its optical property because of the immense shift of band toward the semi-conductive area. Also, the polymer layer improves the antithrombotic property. Therefore, cpTi can be a potent metal to be utilized for cardiovascular stent application.

REFERENCES

1. G. D. Flora, M. K. Nayak. A brief review of cardiovascular diseases associated risk factors and current treatment regimes. *Curr Pharm Des* 25 (2019) 4063–4084. https://doi.org/10.2174/1381612825666190925163827
2. J. Iqbal, J. Gunn, P. W. Serruys. Coronary stents: Historical development, current status and future directions. *Br Med Bull*106 (2013) 193–211. https://doi.org/10.1093/bmb/ldt009
3. D. Buccheri, D. Piraino, G. Andolina, B. Cortese. Understanding and managing in-stent restenosis: A review of clinical data, from pathogenesis to treatment. *J Thorac Dis* 8 (2016) E1150–E1162. https://doi.org/10.21037/jtd.2016.10.93
4. M. Y. Ho, C. C. Chen, S. H. Chang, M. J. Hsieh, C. H. Lee, V. C. C. Wu, I. C. Hsieh. The development of coronary artery stents: From bare-metal to bio-resorbable types. *Metals* 6 (2016) 168. https://doi.org/10.3390/met6070168
5. M. Livingston, A. Tan. Coating techniques and release kinetics of drug-eluting stents. *J Med Device.* 10(1) (2016) 1–13. https://doi.org/10.1115/1.4031718
6. S. Bagga, M. J. Bouchard. Cell cycle regulation during viral infection. *Cell Cycle Control.* 1170 (2014) 165–227. https://doi.org/10.1007%2F978-1-4939-0888-2_10
7. S. R. Bartz, M. E. Rogel, M. Emerman. Human immunodeficiency virus type 1 cell cycle control: Vpr is cytostatic and mediates G2 accumulation by a mechanism which differs from DNA damage checkpoint control. *J Virol* 70 (1996) 2324–2331. https://www.ncbi.nlm.nih.gov/pmc/articles/PMC190074/pdf/702324.pdf
8. P. Ma, R. J. Mumper. Paclitaxel nano-delivery systems: A comprehensive review. *J Nanomed Nanotechnol* 4 (2013) 1000164. https://doi.org/10.4172/2157-7439.1000164
9. O. Waugh, A. J. Wagstaff. The paclitaxel (TAXUS)-eluting stent: A review of its use in the management of de novo coronary artery lesions. *Am J Cardiovasc Drugs* 4 (2004) 257–68. https://doi.org/10.2165/00129784-200404040-00006

10. S. Tanimoto, J. Daemen, P. W. Serruys. Update on stents: Recent studies on the TAXUS® stent system in small vessels. *Vasc Health Risk Manag* 3 (2007) 481–490.

11. W. H. Ahmed. Review of the TAXUS Liberté SR paclitaxel-eluting coronary stent. *Expert Rev Med Devices* 4 (2007) 117–20. https://doi.org/10.1586/17434440.4.2.117

12. D. J. Allocco, M. V. Jacoski, B. Huibregtse, T. Mickley, K. D. Dawkins. Platinum chromium stent series – The TAXUS™ Element™ (ION™), PROMUS Element™ and OMEGA™ stents. *Interv Cardiol* 6 (2011) 134–41. https://doi.org/10.15420/icr.2011.6.2.134

13. P. W. Erruys, G. Sianos, A. Abizaid, J. Aoki, P. den Heijer, H Bonnier. The effect of variable dose and release kinetics on neointimal hyperplasia using a novel paclitaxel–eluting stent platform: The paclitaxel In–Stent Controlled Elution Study (PISCES). *J Am Coll Cardiol* 46 (2005) 253–60. https://doi.org/10.1016/j.jacc.2005.03.069

14. A. Ostovan, M. Rezaie, J Kojuri. The outcome of six-month clinical trial with eucatax paclitaxel-eluting-stent compared with eucastsflex bare metal stent in patients undergoing de novo coronary stenting. *Iran Cardiovasc Res J* 4 (2010) 70–72. https://applications.emro.who.int/imemrf/Iran_Cardiovasc_Res_J/Iran_Cardiovasc_Res_J_2010_4_2_70_73.pdf

15. P. Vranckx, P. W. Serruys, S. Gambhir, E. Sousa, A. Abizaid, P. Lemos, E. Ribeiro, S. I. Dani, J. J. Dalal, V. Mehan, A. Dhar, A. L. Dutta, K. N. Reddy, R. Chand, A. Ray, J. Symons. Biodegradable-polymer-based, paclitaxel-eluting Infinnium stent: 9-Month clinical and angiographic follow-up results from the SIMPLE II prospective multi-centre registry study. *EuroIntervention* 2 (2006) 310–7. PMID: 19755306.

16. P. Buszman, K. Milewski, A. Żurakowski, J. Pająk, Ł. Liszka, P. Buszman, E. Musioł, M. AbuSamra, S. Trznadel, G. Kałuża. Novel biodegradable polymer–coated, paclitaxel–eluting stent inhibits neointimal formation in porcine coronary arteries. *Kardiologia Polska (Polish Heart Journal)* 68 (2010) 503–509.

17. E. Grube, J. Schofer, K. E. Hauptmann, G. Nickenig, N. Curzen, D. J. Allocco, K. D. Dawkins. A novel paclitaxel-eluting stent with an ultrathin abluminal biodegradable polymer 9-month outcomes with the JACTAX HD stent. *JACC Cardiovasc Interv* 3 (2010) 431–8. https://doi.org/10.1016/j.jcin.2009.12.015

18. C. G. Groth, L. Bäckman, J. M. Morales, R. Calne, H. Kreis, P. Lang, J. L. Touraine, K. Claesson, J. M. Campistol, D. Durand, L. Wramner, C. Brattström, B. Charpentier. Sirolimus (rapamycin)-based therapy in human renal transplantation: Similar efficacy and different toxicity compared with cyclosporine: 1, 2. *Transplantation* 67 (1999) 1036–42. https://doi.org/10.1097/00007890-199904150-00017

19. E. H. Seo, K. Na. Polyurethane membrane with porous surface for controlled drug release in drug eluting stent. *Biomater Res* 18 (2014). https://doi.org/10.1186/2055-7124-18-15

20. K. W. Park, J. H. Yoon, J. S. Kim, J. Y. Hahn, Y. S. Cho, I. H. Chae, H. C. Gwon, T. Ahn, B. H. Oh, J. E. Park, W. H. Shim, E. K. Shin, Y. S. Jang, H. S. Kim. Efficacy of Xience/promus versus Cypher in rEducing Late Loss after stENTing (EXCELLENT) trial: Study design and rationale of a Korean multicenter prospective randomized trial. *Am Heart J* 157 (2009) 811–817. https://doi.org/10.1016/j.ahj.2009.02.008

21. J. Zhang, X. Gao, J. Kan, Z. Ge, L. Han, S. Lu, N. Tian, S. Lin, Q. Lu, X. Wu, Q. Li. Intravascular ultrasound versus angiography-guided drug-eluting stent implantation: The ULTIMATE trial. *J Am Coll Cardiol* 72 (2018) 3126–3137. https://doi.org/10.1016/j.jacc.2018.09.013. Epub 2018 Sep 24.

22. A Huseynov, M. Behnes, U. Ansari, S. Baumann, D. Lossnitzer, I. El-Battrawy, C. Fastner, M. Borggrefe, I. Akin. Optimal duration for dual antiplatelet therapy with COMBO dual therapy stent. *J Geriatr Cardiol* 16 (2019) 840–843. https://doi.org/10.11909/j.issn.1671-5411.2019.11.001

23. S. G. Worthely, A. Abizaid, A. J. Kirtane, D. L. Simon, S. Windecker, S. Brar, I. T. Meredith, S. Shetty, A. Sinhal, A. P. Almonacid, D. Chamie, A. Maehara, G. W. Stone. First-in-human evaluation of a novel polymer-free drug-filled stent: Angiographic, IVUS, OCT, and clinical outcomes from the revaluation study. *ACC Cardiovasc Interv* 10 (2017)147–156.

24. P. Chandra, A. U. Mahajan, V. D. Bulani, A. S. Thakkar. Pharmacokinetic study of sirolimus-eluting bioresorbable vascular scaffold system for treatment of *De Novo* native coronary lesions: A sub-study of MeRes-1 trial. *Cardiol Res* 9 (2018): 364–369. https://doi.org/10.14740/cr799

25. M. Haude, R. Erbel, P. Erne, S. Verheye, H. Degen, D. Böse, P. Vermeersch, I. Wijnbergen, N. Weissman, F. Prati, R. Waksman, J. Koolen. Safety and performance of the drug-eluting absorbable metal scaffold (DREAMS) in patients with de-novo coronary lesions: 12 month results of the prospective, multicentre, first-in-man BIOSOLVE-I trial. *Lancet* 381(2013) 836–44. https://doi.org/10.1016/S0140-6736(12)61765-6

26. C. Rapett, M. Leoncini. Magmaris: A new generation metallic sirolimus-eluting fully bioresorbable scaffold: Present status and future perspectives. *J Thorac Dis* 9 (2017) S903–S913. https://doi.org/10.21037/jtd.2017.06.34

27. P. Agostoni, S. Verhey. Step-by-step StentBoost-guided small vessel stenting using the self-expandable Sparrow stent-in-wire. *Catheter Cardiovasc Interv* 73 (2009) 78–83. https://doi.org/10.1002/ccd.21817

28. S. K. Sinha. Ultra-thin (60 μm), ultra-long (≥40 mm) sirolimus eluting stent – Study of clinical and safety profile among real-world patients. *Anatol J Cardiol* 25 (2020). https://doi.org/10.14744/AnatolJCardiol.2020.40909

29. F. R. A. Oliveira, L. A. P. E. Mattos, A. Abizaid, A. S. Abizaid, J. R. Costa, R. Costa, R. Staico, R. Botelho, J. E. Sousa, A. Sousa. Miniaturized self-expanding drug-eluting stent in small coronary arteries: Late effectiveness. *Arq Bras Cardiol* 101(2013) 379–87. https://doi.org/10.5935/abc.20130199

30. K. Upendra, B. Sanjeev. Advantages of novel BioMimeTM sirolimus eluting coronary stent system. Moving towards biomimicry. *Minerva Cardioangiol* 60 (2012) 23–31. PMID: 22322571

31. S. Hiremath, P. Chandra, D. Desai, R. Sivakumar, S. Selvamani, A. Srinivasan, M. Paulose, S. Jose, B. C. Kalmath, V. P. Magarkar, A. Pathak, T. Mehtre. A prospective, multi-centric, observational registry to evaluate performance of Excel™ DES in 'real world, all comers' patient population. *Indian Heart J* 66 (2014) 691–695.

32. R. Zarpak, O. D. Sanchez, M. Joner, L. G. Guy, G. Leclerc, R. Virmani. A novel 'pro-healing' approach: The COMBO™ dual therapy stent from a pathological view. *Minerva Cardioangiol* 63 (2015) 31–43. PMID: 25502187.

33. F. Falcao, F. Cantarelli, R. Cantarelli, F. Mota, M. Navarro, H. Mota, M. Santos, D. Cruz, A. Sansonio, M. Parents, F. Oliveira. One-year clinical outcome of

inspiron stent in all-comers population (analysis from 790 consecutive patients). *J Interv Cardiol* 2020 (2020) 6340716. https://doi.org/10.1155/2020/6340716

34. T. Asano, Y. Katagiri, C. Collet, E. Tenekecioglu, Y. Miyazaki, Y. Sotomi, G. Amoroso, A. Aminian, S. Brugaletta, M. Vrolix, R. Hernandez-Antolin, P. V. Harst, A. I-Romo, L. Janssens, P. C. Smits, J. J. Wykrzykowska, V. G. Ribeiro, H. Pereira, P. C. Silva, J. J. Piek, J. H. C. Reiber, C. N. Birgelen, M. Sabate, Y. Onuma, P. W. Serruys. Functional comparison between the BuMA Supreme biodegradable polymer sirolimus-eluting stent and a durable polymer zotaro-limus-eluting coronary stent using quantitative flow ratio: PIONEER QFR substudy. *EuroIntervention*. 14 (2018) e570–e579. https://doi.org/10.4244/EIJ-D-17-00461

35. Y. Saito, D. Grubman, E. Cristea, A. Lansky. The Firehawk Stent: A review of a novel abluminal groove-filled biodegradable polymer sirolimus-eluting stent. *Cardiol Rev* 28 (2020) 208–212. https://doi.org/10.1097/CRD.0000000000000298

36. E. Xhepa, T. Tada, S. Cassese, L. King, I. Ott, M. Fusaro, A. Kastrati, R. A. Byrne. Safety and efficacy of the Yukon Choice Flex sirolimus-eluting coronary stent in an all-comers population cohort. *Indian Heart J* 66 (2014) 345–349. https://doi.org/10.1097/10.1016/j.ihj.2014.05.003

37. Y. G. R. Tijssen, R. P. Kraak, H. Lu, J. G. Mifek, J. J. Wykrzykowska. Evaluation of the MiStent sustained sirolimus eluting biodegradable polymer coated stent for the treatment of coronary artery disease: Does uniform sustained ablumi-nal drug release result in earlier strut coverage and better safety profile? *Expert Rev Med Devices*, 14 (2017) 325–334. https://doi.org/10.1080/17434440.2017.1318057

38. M. Meng, B. Gao, X. Wang, Z. Bai, R. Sa, B. Ge. Long-term clinical outcomes of everolimus-eluting stent versus paclitaxel-eluting stent in patients undergoing percutaneous coronary interventions: A meta-analysis. *BMC Cardiovasc Disord* 16 (2016) 34. https://doi.org/10.1186/s12872-016-0206-6

39. J. E. Sousa, P. W. Serruys, M. A. Costa. Drug-eluting stents: Part II. *Circulation* 107 (2003) 2383–2389. https://doi.org/10.1161/01.CIR.0000069331.67148.2F

40. G. J. Wilson, A. Marks, K. J. Berg, M. Eppihimer, N. Sushkova, S. P. Hawley, K. A. Robertson, D. Knapp, D. E. Pennington, Y. L. Chen, A. Foss, B. Huibregtse, K. D. Dawkins. The SYNERGY biodegradable polymer everolimus eluting coronary stent: Porcine vascular compatibility and polymer safety study. *Catheter Cardiovasc Interv* 86 (2015) E247–57. https://doi.org/10.1002/ccd.25993

41. J Ge. Limus-eluting stents with Poly-L-lactic acid coating. *Asia Pac Cardiol* 1 (2007) 42–3. https://doi.org/10.15420/apc.2007:1:1:42

42. Y. W. Chen, M. L. Smith, M. Sheets, S. Ballaron, J. M. Trevillyan, S. E. Burke, T. Rosenberg, C. Henry, R. Wagner, J. Bauch, K. Marsh, T. A. Fey, G. Hsieh, D. Gauvin, K. W. Mollison, G. W. Carter, S. W. Djuric. Zotarolimus, a novel sirolimus analogue with potent anti-proliferative activity on coronary smooth muscle cells and reduced potential for systemic immunosuppression. *J Cardiovasc Pharmacol* 49 (2007) 228–35. https://doi.org/10.1097/FJC.0b013e3180325b0a

43. J. Burgos, S. Farrag, D. Mukherjee. Resolute Integrity drug eluting stent safety and efficacy for the treatment of coronary artery disease. *Res Rep Clin Cardiol* 4 (2013) 23–37. https://doi.org/10.2147/RRCC.S31012

44. M. J. Price, S. Saito, R. A. Shlofmitz, D. J. Spriggs, M. Attubato, B. M. A. P. Almonacid, S. Brar, M. Liu, E. Moe, R. Mehran. First report of the resolute onyx 2.0-mm zotarolimus-eluting stent for the treatment of coronary lesions

with very small reference vessel diameter. *JACC Cardiovasc Interv* 10 (2017) 1381–1388. https://doi.org/10.1016/j.jcin.2017.05.004

45. G. J. Vlachojannis, P. C. Smits, S. H. Hofma, M. Togni, N. Vazquez, M. Valdes, V. Voudris, T. Slagboom, J. J. Goy, P. D. Heijer, M. V. D. Ent. Biodegradable polymer biolimus-eluting stents versus durable polymer everolimus-eluting stents in patients with coronary artery disease: Final 5-year report from the COMPARE II trial (abluminal biodegradable polymer biolimus-eluting stent versus durable polymer everolimus-eluting stent). *JACC Cardiovasc Interv* 10 (2017) 1215–1221. https://doi.org/10.1016/j.jcin.2017.02.029

46. J. Rawlins, J. Din, S. Talwar, P. O. Kane. AXXESS™ stent: Delivery indications and outcomes. *Interv Cardiol* 10 (2015) 85–89. https://doi.org/10.15420/icr.2015.10.2.85

47. G. B. Danzi, B. Chevalier, M. Ostoji, M Hamilos, W. Wijns. Nobori™ drug eluting stent system: Clinical evidence update. *Minerva Cardioangiol* 58 (2010) 599–610.

48. D. Capodanno, F. Dipasqua, C. Tamburino. Novel drug-eluting stents in the treatment of de novo coronary lesions. *Vasc Health Risk Manag* 7 (2011) 103–118. https://doi.org/10.2147/VHRM.S11444

49. S. Tanimoto, W. V. Giessen, M. V. Beusekom, O. Sorop, N. Kukreja, K. Fukaya, T. Nishide, R. Nakano, H. Maeda, P. W. Serruys. 'MAHOROBA: Tacrolimus eluting coronary stent.' Drug eluting stent implantation for high risk patients and novel technologies in percutaneous coronary intervention. *EuroInterv* 3 (2007) 149–153.

50. J. Iqbal, S. Verheye, A. Abizaid, J. Ormiston, T. D. Vries, L. Morrison, S. Toyloy, P. Fitzgerald, S. Windecker, P. W. Serruys. DESyne novolimus-eluting coronary stent is superior to Endeavor zotarolimus-eluting coronary stent at five-year follow-up: Final results of the multicentre EXCELLA II randomized controlled trial. *EuroInterv*12 (2016) e1336–e1342. https://doi.org/10.4244/EIJY15M10_04

51. E. R. Edelman, P. Seifert, A. Groothuis, A. Morss, D. Bornstein, C. Rogers. Gold-coated NIR stents in porcine coronary arteries. *Circulation* 103 (2001) 429–434. https://doi.org/10.1161/01.CIR.103.3.42

52. M. Lenhart, M. Volk,C. Manke, W. R. Nitz,M. Strotzer, S. Feuerbach, J. Link. Stent appearance at contrast-enhanced MR angiography: In vitro examination with 14 stents. *Radiology* 217 (2000) 173–8.

53. K. Prasad, O. Bazaka, M. Chua, M. Rochford, L. Fedrick, J. Spoor, R. Symes, M. Tieppo, C. Collins, A. Cao, D. Markwell, K. Ostrikov, K. Bazaka. Metallic biomaterials: Current challenges and opportunities. *Materials (Basel)* 10 (2017) 884. https://doi.org/10.3390/ma10080884

54. M. Niinomi, M. Nakai. Titanium-based biomaterials for preventing stress shielding between implant devices and bone. *Int J Biomater* 2011 (2011) 836587. https://doi.org/10.10.1155/2011/836587

55. R. A. Antunes, C. A. F. Salvador, M. C. L. Oliveira. Materials selection of optimized titanium alloys for aircraft applications. *Mat Res* 21 (2018) e20170979. https://doi.org/10.1590/1980-5373-MR-2017-0979

56. Standard Specification for Unalloyed Titanium for Surgical Implant Applications (UNS R50250, UNS R50400, UNS R50550, UNS R50700) American Society for Testing Materials; West Conshohocken, PA, USA: 2013. ASTM F67–13.

57. Standard Specification for Wrought Titanium-6 Aluminum-4 Vanadium ELI (Extra Low Interstitial) Alloy for Surgical Implant Applications (UNS R56401)

American Society for Testing Materials; West Conshohocken, PA, USA: 2013. ASTM F136–13.

58. A. T. Sidambe. Biocompatibility of advanced manufactured titanium implants—A review. *Materials (Basel)* 7 (2014) 8168–8188.

59. H. Eltchaninoff, R. Pillière, G. Traisnel, J. Berland, S. Cattan, J. Monségu, A. Cribier, Acute and six-month clinical outcome after Helistent stent implantation in coronary arteries: Results of the French Helistent multicenter registry. *Catheter Cardiovasc Interv* 56 (2002) 295–9. https://doi.org/10.1002/ccd.10227

60. S. Kelkar, C. J. Boushey, M. Okos. A method to determine the density of foods using X-ray imaging. *J Food Eng* 159 (2015) 36–41. https://doi.org/10.1016/j.jfoodeng.2015.03.012

61. B. J. Heismann, J. Leppert, K. Stierstorfer. Density and atomic number measurements with spectral x-ray attenuation method. *J Appl Phys* 94 (2003) 2073–9.

62. T. G. Mayerhofer, A. V. Pipa, J. Popp. Beer's law-why integrated absorbance depends linearly on concentration. *Chemphyschem* 20 (2019) 2748–2753. https://doi.org/10.1002/cphc.201900787

63. J. Chou, T. J. Robinson, H. Doan. Rapid comparison of UVB absorption effectiveness of various sunscreens by UV-Vis spectroscopy. *J Anal Bioanal Tech* 8 (2017) 355. https://doi.org/10.4172/2155-9872.1000355

64. S. C. Shei. Optical and structural properties of titanium dioxide films from and starting materials annealed at various temperatures, *Adv Mater Sci Eng* 2013 (2013) 7. https://doi.org/10.1155/2013/545076

65. K. H. Rahman, S. Biswas, A. Kar. Optical properties of titanium-di-oxide (TiO_2) prepared by hydrothermal method. *AIP Conf Proc* 1953 (2018) 030022. https://doi.org/10.1063/1.5032357

66. K. Ozawa, M. Emori, S. Yamamoto, R. Yukawa, R. Hobara, K. Fujikawa, H. Sakama, I. Matsuda. Electron–hole recombination time at TiO_2 single-crystal surfaces: Influence of surface band bending. *J Phys Chem Lett* 5 (2014) 1953–1957. https://doi.org/10.1021/jz500770c

67. M. Mohanta, A. Thirugnanam. Drug release studies of titanium-based polyethylene glycol coating as a multifunctional substrate. *Mater Today: Proc* 47 (2021): 257–260. https://doi.org/10.1016/j.matpr.2021.04.242

68. A. Thirugnanam, T. S. Sampath, U. Chakkingal. Bioactivity enhancement of commercial pure titanium by chemical treatments. *Trends Biomater Artif Organs* 22 (2009) 202–210.

69. D. Ravelli, D. Dondi, M. Fagnoni, A. Albini. Photocatalysis. A multi-faceted concept for green chemistry. *Chem Soc Rev* 38 (2009) 1999–2011. https://doi.org/10.1039/b714786b. Epub 2009 Apr 21.

70. A. King, R. Singh, R. Anand, S. Behera, B. B. Nayak. Phase and luminescence behaviour of Ce-doped zirconia nanopowders for latent fingerprint visualization. *Optik.* 242 (6) 167087. https://doi.org/10.1016/j.ijleo.2021.167087

71. S. B. Aziz, M. A. Brza, M. M. Nofal, R. T. Abdulwahid, S. A. Hussen, A. M. Hussein, W. O. Karim. A comprehensive review on optical properties of polymer electrolytes and composites materials. *Materials.* 13 (2020) 3675. https://doi.org/10.3390/ma13173675

72. L. Yesappa, M. Niranjana, S. P. Ashokkumar, H. Vijeth, M. Basappa, J. Dwivedi, V. C. Petwal, S. Ganesh, H. Devendrappa. Optical properties and ionic conductivity studies of an 8 MeV electron beam irradiated poly(vinylidene

fluoride-*co*-hexafluoropropylene)/LiClO$_4$ electrolyte film for opto-electronic applications. *RSC Adv* 8 (2018) 15297–15309. https://doi.org/10.1039/C8RA00970H

73. R. Gouda, H. Baishya, Z. Qing. Application of mathematical models in drug release kinetics of carbidopa and levodopa ER tablets. *J Develop Drugs* 6 (2017) 2–8. https://doi.org/10.4172/2329-6631.1000171

74. D. Li, P. Lv, L. Fan, Y. Huang, F. Yang, X. Mei, D. Wu. The immobilization of antibiotic-loaded polymeric coatings on osteoarticular Ti implants for the prevention of bone infections. *Biomater Sci* 5 (2017) 2337–2346. https://doi.org/10.1039/c7bm00693d

75. P. L. Ritger, N. A. Peppas. A simple equation for description of solute release I. Fickian and non-fickian release from non-swellable devices in the form of slabs, spheres, cylinders or discs. *J Control Release* 5 (1987) 23–26.

Chapter 16

Dry sliding wear behaviour of HVOF-sprayed cermet coatings (CrC-NiCr, WC-Co and WC-Co-Cr) using statistical analysis and ANN models

R. Suresh
M S Ramaiah University of Applied Sciences, Bangalore, Karnataka, India

Ajith G. Joshi
Canara Engineering College, Benjanapadavu, Bantwal,
Mangalore, Karnataka, India

M. Manjaiah
National Institute of Technology Warangal, Warangal, Telangana, India

CONTENTS

16.1 INTRODUCTION

Thermal spray coatings deposited onto various metallic substrates have been extensively applied in numerous engineering arenas including automotive, aerospace and petrochemical fields. Thermal-sprayed cermet coatings are

DOI: 10.1201/9781003319375-16

commonly used in abrasive and erosive wear resistance, corrosion resistance and applications providing surface protection [1–5]. Especially, Tungsten Carbide (WC) and Chromium Carbide (CrC) coated surfaces find various engineering applications due to their enhanced mechanical and tribological properties. It also helps to repair and restore used parts at a reasonable cost. Currently, spray coatings techniques such as High-Velocity Oxy Fuel (HVOF), plasma spray, wire arc, flame and detonation techniques are widely used in engineering industries. However, cermet coatings sprayed by HVOF systems show low porosity, greater hardness and higher adhesive strength relative to plasma spray coated surfaces. The major advantage is the opportunity for large surface coating, which is reflected in the low-cost production.

Normally, the protection of metallic materials against wear and corrosion is by surface treatment, in addition to enhancing surface quality by HVOF, plasma spray, gas-tungsten arc welding (TIG) and electro spark techniques [6–8]. Thermal spraying processes include fine particles like metals (Mo, Ti, Ni, Cr, Ta, Co, W) and metal alloys NiCoCr and AlY alloy and transition metal carbides (WC, TiC, TaC, Cr_3C_2, NbC, Mo_2) and hard oxides (Cr_2O_3, Cr_7C_3, Al_2O_3, TiO_2) on the base material in a semi-molten or molten state with high temperature and pressure conditions. Generally, CrC, WC and a combination of both materials coatings were extensively used in various industrial applications such as metal cutting, machining, construction and mining, due to their resistance to wear, toughness, strength and hardness [9–11].

Several researchers have explored the effect of process parameters, coating materials and mechanical characteristics of thermal spray ceramic coatings on their elevated temperature tribological performance. Some commonly conducted tests on cermet coated substrates are abrasive wear, erosive wear, corrosion properties, hardness, porosity, adhesive strength and many more. Marques et al. [5] studied the wear behaviour of hot forged punches coated with Cr_3C_2-NiCr and WC-CoCr using the HVOF process. The results reveal that the WC-CoCr coated punches showed high abrasive wear resistance compared to Cr3C2-NiCr coated punches. Liu et al. [7] examined the tribological behaviours of WC-Fe3Al and WC-Co coating under dry tribological conditions. They observed that WC-Fe3Al coatings possess high resistance to wear and oxidizing phenomenon relative to WC-Co coated surface. Fayyazi et al. [10] modelled and optimized the wear process parameters of the WC-Co coated surface using Taguchi experimental design. The optimized parameters show that the type of coating materials, grit sizes and substrate material affected the wear resistance of the coated surface.

From the above literatures, it is revealed that the wear behaviour of coatings is to be measured for understanding the quality of thermal-sprayed cermet coatings and it is also essential to study the rate of wear and wear mechanisms. A majority of the studies emphasized abrasive wear tests, due to having hard matrix phase coating with high hardness. The abrasion wear is one of the major damages caused to the coated surface; due to normal load conditions, surfaces suffer local plastic deformation and adhesion. Thus, optimization of tribological parameters such as sliding speed, load and distance is necessary to produce better tribological characteristics. The present work studied three different types of cermet coating combinations viz. CrC-NiCr, WC-Co and WC-Co-Cr on base material through the HVOF coating method.

16.2 MATERIALS AND METHOD

16.2.1 Substrate material

In the present study, AISI 4340 steel was chosen as a base substrate. AISI 4340 steel is widely used in agriculture, marine, chemical, aerospace and automotive industries and it has very good physical and mechanical properties. The AISI 4340 steel is an amalgamation of nickel, chromium and molybdenum. It has corrosion resistance, ductility and good wear characteristics. The chemical composition of the base substrate of AISI 4340 steel is given in Table 16.1. The AISI 4340 alloy steel samples of a size of 10 × 50 × 150 mm were selected. Samples were pre-machined by CNC milling and finished through a grinding machine with a surface finish of 0.8 μm.

16.2.2 Feedstock materials for coating

Normally, Tungsten Carbide (WC) is available in grey powder form. It is further processed and used in various applications in industries such as industrial machinery, cutting tools, jewellery and measuring instruments. The WC is a very hard material ranked ninth on Mohr's scale and has a Young's modulus of 530–700 MPa. Chrome Carbide is a combination of Chrome-Carbide (Cr_3C_2) with a binder of Nickel-Chrome (NiCr). Chromium Carbide-Nickel Chromium composite powders exhibit excellent tribological characteristics at elevated temperatures, and hence find applications in thermal spray coatings, plasma, welding processes and so on.

The selection of coating material is an essential criterion during coating with various combinations of ceramic/cermet powders and the correct

Table 16.1 Chemical composition of the base substrate (AISI 4340 steel)

Element	C	Ni	S	Si	Mo	Mn	Cr	p	Fe
Chemical Composition (%)	0.36	1.5	0.05	0.15	0.25	0.45	1.5	0.05	Bal.

Table 16.2 Cermet powder compositions

Material	Powder combination
M1	88WC-12CO
M2	86WC-10CO-4Cr
M3	75Cr3C2-25NiCr

proportion needs to be used. The effect of grain size, method of coating and percentage of powder composition affect microstructure, adhesive strength and mechanical and wear behaviour. In the present investigation, the powder combinations for the three samples of particle sizes of 30 μm were used for HVOF spraying. The cermet coating compositions are shown in Table 16.2.

16.2.3 High-velocity oxy fuel

HVOF spray coating is extensively used in many industrial sectors due to its erosive wear resistance when compared to other coating methods like plasma spraying, wire arc, flame and detonation techniques. It comprises the high rate of powder deposition on the base substrate surface through a high-velocity flame and produces a thick deposition on the base substrate material. The HVOF coating device delivers the powder, gases and fuel at the entry level. The HVOF coating device is surrounded by a water cooling jacket for cooling the device and controlling the temperature. In the combustion chamber, a high stream of coating particles accelerate toward the base substrate at 3000°C temperature and 1200 m/s super-sonic velocity and causes the melting of coating material, and faster solidification takes place during the deposition time. Table 16.3 shows the HVOF coating parameters. Figure 16.1 shows the HVOF coated samples.

Table 16.3 Parameters employed for HVOF thermal coating

Process parameter	Value
Supply of kerosene and pressure	1.1657 MPa and 0.828 MPa
Flow rate of kerosene	5.556×10^{-6} m³/s
Combustion pressure	0.71 MPa
O_2 supply and flow pressure	1.42 MPa and 0.967 MPa
Flow rate of N_2	4.480×10^{-4} m³/s
Flow rate of O_2	1.54×10^{-2} m³/s
Pressure of N_2	0.50 MPa
Spray distance	0.38 m

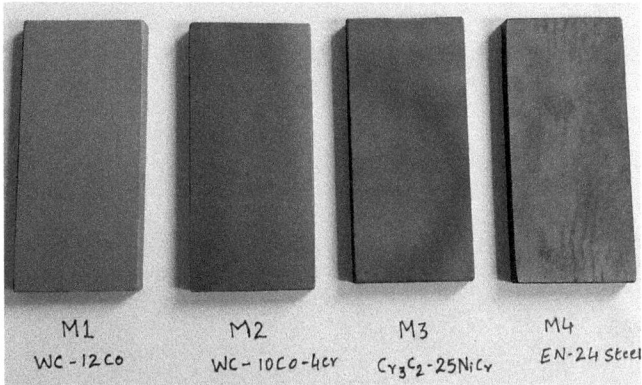

Figure 16.1 HVOF coated samples.

16.2.4 Microstructure analysis of coating

The microstructure of HVOF-sprayed Cr3C2NiCr, WC-Co and WC-Co-Cr coated samples are shown in Figure 16.2. The WC particles embedded into a matrix of W, C and Co (Figure 16.2(a and b)) reveal round edges, owing to amorphization and decarburization at elevated temperatures. Figure 16.2(c) illustrates the better adherence of coating on the substrate. EDX analysis (Figure 16.3) shows the exact composition of WC-Co coating distribution along with binder phase WC-Co-Cr coating is distributed uniformly. The

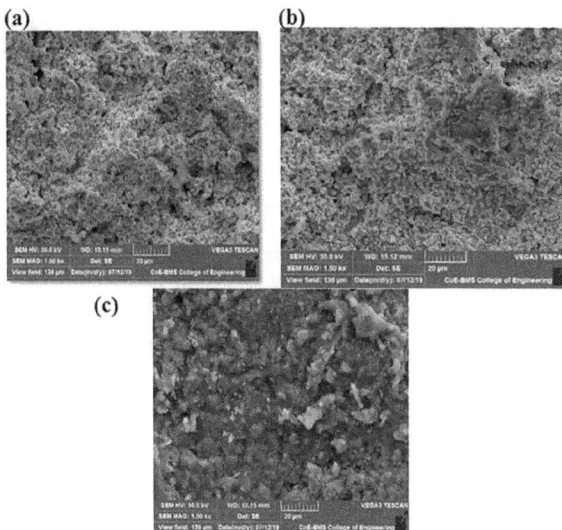

Figure 16.2 Microstructure of coated specimens: (a) WC-Co, (b) WC-Co-Cr and (c) Cr3C2NiCr.

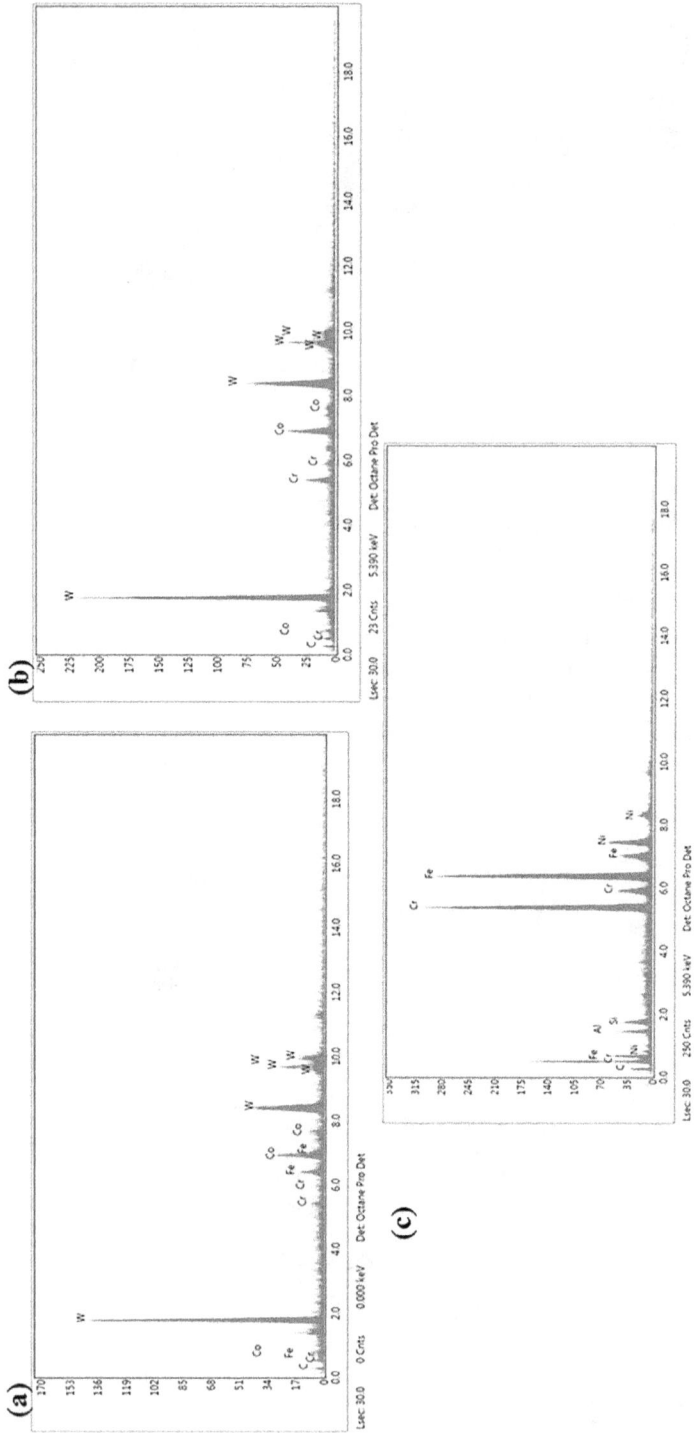

Figure 16.3 EDAX analysis of (a) coated WC–Co, (b) coated WC–Co-Cr and (c) coated Cr3C2NiCr.

higher percentage of Cr is more pronounced in Figure 16.3(c). The porosity inside the coating is limited to a small percentage. The porosity is the crack nucleation site when it is exposed to wear impact.

16.2.5 Dry sliding wear test

HVOF-sprayed samples were employed on pin-on-disc (Figure 16.4) wear test to study dry sliding wear behaviour as per ASTM G99 standard. Hardened AISI D2 steel (65 HRc) disc was used as the counter body. The parameters, along with their levels considered in the study, are illustrated in Table 16.4. Various dry sliding wear tests were conducted under a controlled room temperature of 26°C. To create the consistency of wear of coated samples, each coating condition was repeated and the average was recorded. Mass loss of each sample at various experimental conditions was recorded based on mass before and after the test with a 0.1 mg accuracy. The acquired mass loss was used to compute the wear rate. The wear behaviour was studied as a function of wear rate.

Figure 16.4 Experimental setup (pin-on-disc apparatus).

Table 16.4 Experimental parameters and their levels

Sl. No.	Descriptions	L_1	L_2	L_3
1	Sliding speed (m/s)	1.0	2.0	3.0
2	Sliding distance (m)	1000	2000	3000
3	Normal load (N)	20	40	60

Table 16.5 Taguchi L_{27} orthogonal array and obtained results

Trial. No.	Sliding distance (m)	Sliding speed (m/s)	Normal load (N)	Wear rate of WC-Co (mm³/min)	Wear rate of WC-Co-Cr (mm³/min)	Wear rate of Cr3C2NiCr (mm³/min)
1	1000	1	20	0.87	0.60	0.93
2	1000	1	40	1.23	0.82	1.87
3	1000	1	60	2.19	2.08	3.09
4	1000	2	20	1.19	0.75	1.62
5	1000	2	40	1.81	1.63	2.44
6	1000	2	60	2.10	1.98	2.70
7	1000	3	20	1.30	1.05	1.88
8	1000	3	40	2.12	1.93	3.00
9	1000	3	60	2.63	2.52	3.94
10	2000	1	20	1.50	1.26	2.25
11	2000	1	40	2.06	1.85	2.63
12	2000	1	60	2.44	2.33	3.47
13	2000	2	20	1.80	1.60	2.44
14	2000	2	40	2.37	2.22	3.45
15	2000	2	60	2.72	2.63	4.16
16	2000	3	20	1.82	1.57	2.60
17	2000	3	40	2.52	2.42	3.60
18	2000	3	60	3.19	2.98	4.96
19	3000	1	20	2.06	1.92	2.63
20	3000	1	40	2.25	2.13	3.37
21	3000	1	60	2.75	2.67	4.33
22	3000	2	20	2.23	2.11	3.13
23	3000	2	40	2.64	2.55	4.06
24	3000	2	60	3.00	2.88	4.66
25	3000	3	20	2.44	2.35	3.56
26	3000	3	40	2.92	2.83	4.63
27	3000	3	60	3.76	3.71	5.38

16.2.6 Experimental plan

The Taguchi design of experiments method was adopted to study the dry sliding wear of HVOF coated samples. Taguchi's orthogonal array (L_{27}) was selected to plan the experimental run as per Table 16.5 (columns 2, 3 and 4). Each row indicates the combination of parameters employed for each trial of experiments. The wear results (columns 5, 6 and 7 of Table 16.5) were assessed with the aid of the MINITAB-16 software package. The significant contribution of a single parameter was studied using the main effect plot. The regression models were established using acquired results and wear (output) data was predicted using the Artificial Neural Network (ANN) approach in accordance with inputs parameters like load (N), sliding distance (m) and sliding speed (m/s).

16.3 RESULTS AND DISCUSSION

16.3.1 Statistical analysis

The ANOVA results for wear rates of Cr3C2NiCr, WC-Co and WC-Co-Cr coated substrate material are shown in Tables 16.6–16.8. The ANOVA analysis was carried out with a confidence interval of $\alpha = 0.05$. The p-values obtained for individual parameters were less than 0.05, while for interaction parameters, the p-values were greater than 0.05. Thus, the effect of individual parameters was statistically significant and the interaction of parameters effect was insignificant. Applied load and sliding distance were the most

Table 16.6 Wear rate of WC-Co

Parameters	SS	DOF	MS	F-value	P-value	Remarks
Applied load (L)	5.1422	2	2.57111	152.95	0.00	Significant
Sliding distance (D)	4.2142	2	2.10711	125.34	0.00	Significant
Sliding speed (S)	1.6289	2	0.81445	48.45	0.00	Significant
L × D	0.0787	4	0.01968	1.17	0.392	Insignificant
L × S	0.2499	4	0.06248	3.72	0.054	Insignificant
S × D	0.0449	4	0.01123	0.67	0.632	Insignificant
Error	0.1345	8	0.01681			
Total	11.4934	26				

Table 16.7 Wear rate of WC-Co-Cr

Parameters	SS	DOF	MS	F-value	P-value	Remarks
Applied load (L)	6.1787	2	3.08933	152.95	0.00	Significant
Sliding distance (D)	5.3594	2	2.67971	125.34	0.00	Significant
Sliding speed (S)	1.8078	2	0.90392	48.45	0.00	Significant
L × D	0.1836	4	0.04591	1.98	0.190	Insignificant
L × S	0.2622	4	0.06554	2.83	0.098	Insignificant
S × D	0.1852	4	0.01667	0.72	0.602	Insignificant
Error	0.1345	8	0.02315			
Total	14.0436	26				

Table 16.8 Wear rate of Cr3C2NiCr

Parameters	SS	DOF	MS	F-value	P-value	Remarks
Applied load (L)	13.6313	2	6.81567	129.30	0.00	Significant
Sliding distance (D)	11.3868	2	5.69342	108.01	0.00	Significant
Sliding speed (S)	4.4862	2	2.24308	42.55	0.00	Significant
L × D	0.0425	4	0.01061	0.20	0.931	Insignificant
L × S	0.4307	4	0.10768	2.04	0.181	Insignificant
S × D	0.1053	4	0.02633	0.50	0.738	Insignificant
Error	0.4217	8	0.05271			
Total	30.4045	26				

Table 16.9 ANN MSE data of different studied samples

	Mean-squared error		
	Training	Validation	Test
WC-Co	1.307×10^{-27}	0.0158	0.0087
WC-Co-Cr	1.4322×10^{-10}	0.0318	0.0277
Cr3C2NiCr	1.546×10^{-26}	0.0322	0.0676

significant parameters affecting the wear rate, while sliding speed was found to be the least influencing parameter within the range of studied parameters.

The main effect plots (Figures 16.5–16.7) indicate increased wear rate with increase in studied parameters. However, a rapid increase in wear rate was recorded with the increased sliding distance and load, compared to speed confirming the results of ANOVA. The coating of ceramics on the steel substrate has drastically increased its wear resistance. The hardness of the surface has a major contribution to enhancing the wear resistance of the specimen. Hardness and wear rate are inversely proportional as per the Archard law. The wear rate was comparatively greater for Cr3C2-NiCr followed by WC-Co and WC-Co-Cr. Sidhu et al. [12] have illustrated that WC-Co coating possesses high hardness compared to Cr3C2-NiCr coating. Ozkavak et al. [13] showed that WC-Co-Cr has high wear resistance compared to WC-Co coating due to greater hardness. Therefore, it can be deduced that the hardness of coatings follows the order of WC-Co-Cr > WC-Co > Cr3C2-NiCr. Owing to a decrease in wear rate of the different types of studied coatings, at low load conditions, the increased applied load rises the contact stress at the pin-disc contact zone. Further, the wear rate of coated substrates is highly dependent on several factors like wear conditions, coating material, bonding strength, adhesion behaviour, hardness, the interface of coating and homogeneity of coating.

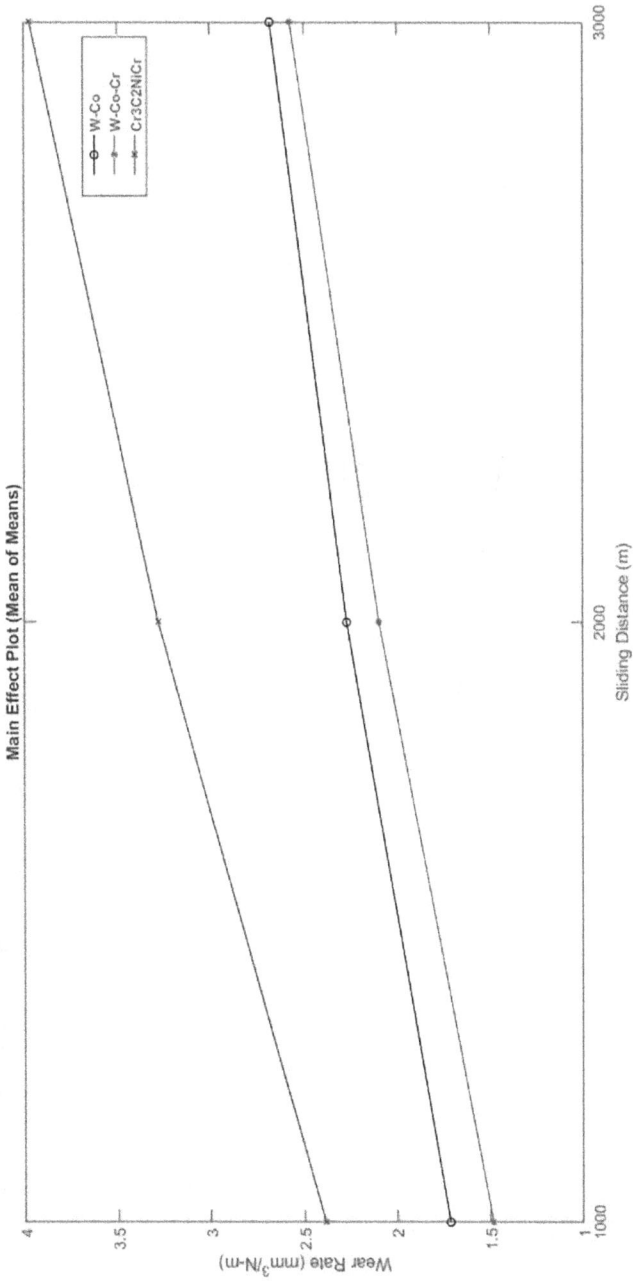

Figure 16.5 Effects of plot of sliding distance versus wear rate.

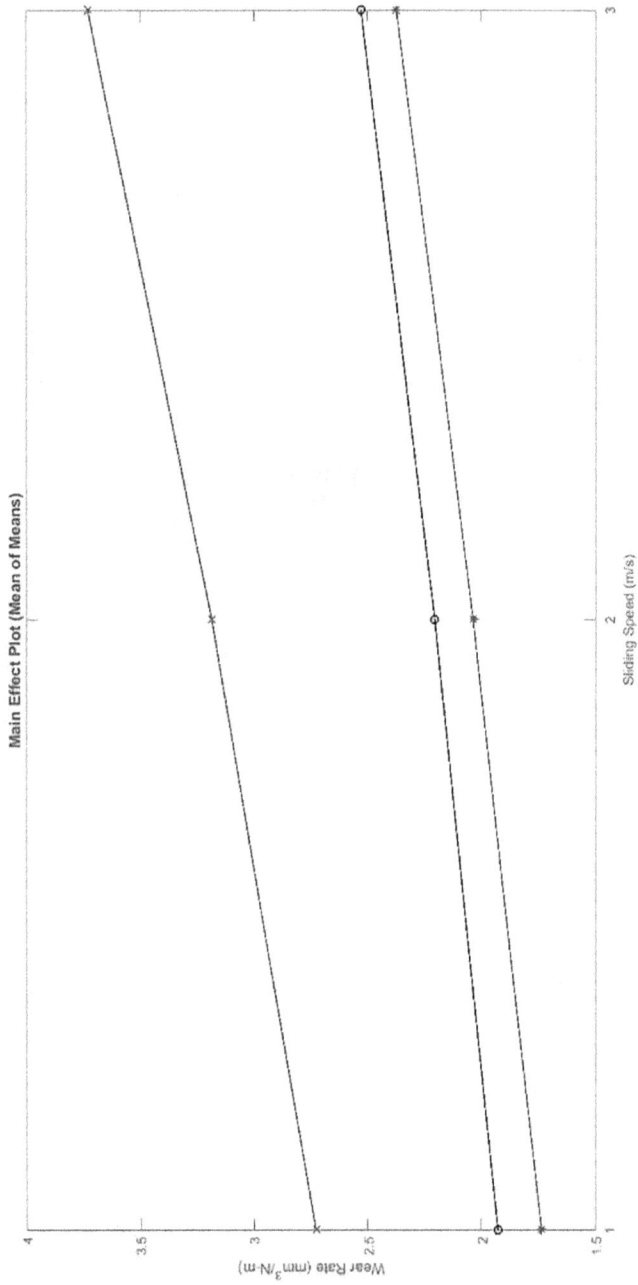

Figure 16.6 Effects of plot of sliding speed versus wear rate.

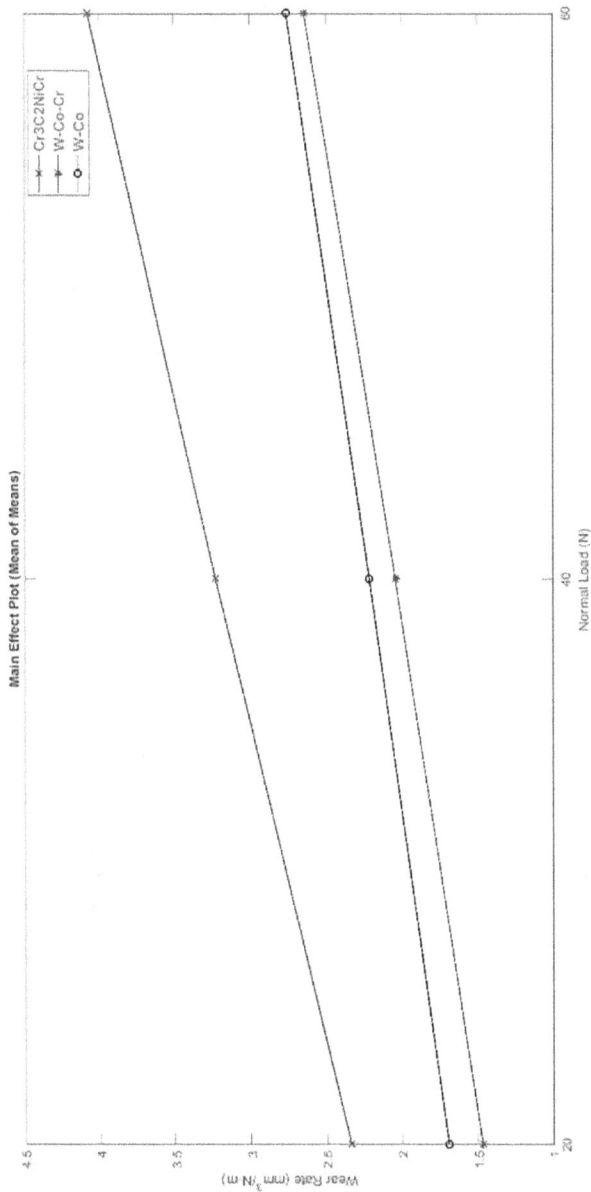

Figure 16.7 Effects of plot of normal load versus wear rate.

The correlation between wear rate and wear parameters was established through regression analysis. The established models were presented as Equations (1), (2) and (3) indicating equations for W-Co, W-Co-Cr and Cr3C2NiCr coated substrates, respectively. The coefficients of interaction terms were comparatively very small and negligible, confirming the absence of interaction effects of parameters on wear rate. Further, the R-Sq. (Adj.), R-Sq and R-Sq. (Pred.) of presented models are greater than 90% and the difference among them is less than 20% depicting their feasibility and adequacy.

$$W_{W-Co} = -0.153 + 0.000543D + 0.051S + 0.02318L + 0.00003 \times S \times D$$
$$- 0.000003 \times D \times L + 0.00476 \times S \times L \tag{1}$$
$$R-Sq.(Adj.)\ 95.7\%,\ R-Sq.\ 96.69\%\ R-Sq.(Pred.)\ 93.03\%$$

$$W_{W-Co-Cr} = -0.899 + 0.000730D + 0.0128S + 0.03217L + 0.000015 \times S$$
$$\times D - 0.000005 \times D \times L + 0.00369 \times S \times L \tag{2}$$
$$R-Sq.(Adj.)\ 95.3\%,\ R-Sq.\ 96.37\%\ R-Sq.(Pred.)\ 92.64\%$$

$$W_{Cr3C2NiCr} = -0.702 + 0.000781D + 0.253S + 0.03601L + 0.000027 \times S$$
$$\times D - 0.000001 * D * L + 0.00479 \times S \times L$$
$$R-Sq.(Adj.)\ 95.93\%,\ R-Sq.\ 96.87\%\ R-Sq.(Pred.)\ 93.74\%$$
$$\tag{3}$$

16.3.2 ANN modelling

ANN is one of the soft computing techniques that can represent the complex relationship between process parameters and the responses. The multilayer perceptron (MLP)-based ANN model was established in the current study. MLP is associated with interconnected neurons named perceptrons. The perceptron calculates a sole outcome from several real value contributions by creating a linear grouping corresponding to weight input. Later, perceptron sets and adjusts the outcome with the help of a few nonlinear activation functions. Mathematically, denoted by Equation (4).

$$y = \varnothing \left(\sum_{i=1}^{n} \omega_i x_i + b \right) \tag{4}$$

where, ω, x, b and $\varnothing$ represent the weight vector, input vector, bias and activation function, respectively. Generally, MLP-ANN includes one or more hidden layers, an input layer and an output layer. Each hidden layer corresponds to a set of source nodes, computing nodes and output nodes. The input signal is propagated layer by layer through the network. The MLP-ANN developed in the study with ten neurons in a hidden layer is shown in Figure 16.8. Table 16.8 shows the ANN mean-squared error (MSE) data of different studied samples.

Furthermore, to simplify the comparisons between ANN predicted value and the target value, an error is calculated using MSE. The experimental data were divided into training data sets of 70%, *validation data* set of 15% and *test* data set of 15%. The model was produced using only training data and validated through validation data. The model is subjected to a feasibility test using test data. It enables understanding the model performance with formerly unobserved data. The training process involves finding the best nonlinear connection between input and output data. The trained MLP-ANN model can be utilized for rapid computation of output data for any given input data sets. ANN model was trained using the Levenberg–Marquardt (LM) method with three input parameters and one output.

In the current study, a regression mathematical model was developed and compared with ANN models. The input independent variables are normal load (L), sliding distance (SD) and sliding speed (SV). The dependent variable is the wear rate (WR) in (mm^3/min). Table 16.8 shows the validation data sets. The correlation of the ANN model with the experimental data for these sets for wear behaviour of coated samples Cr_3C_2NiCr, WC-Co and WC-Co-Cr results is as in Figure 16.9(a–c).

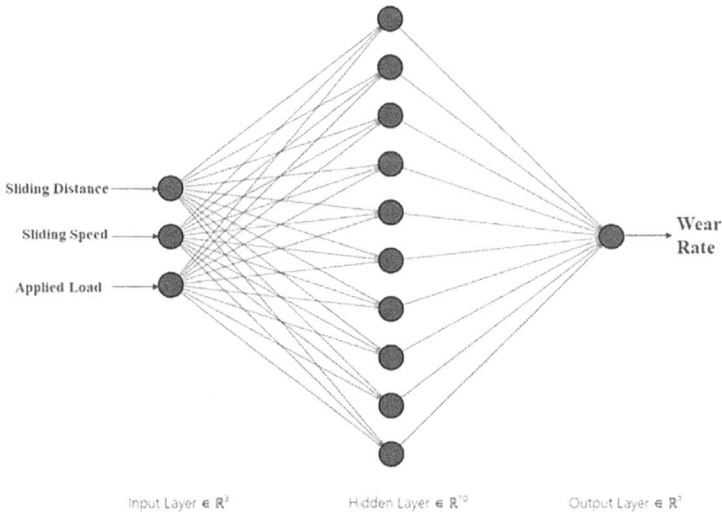

Figure 16.8 MLP network graph.

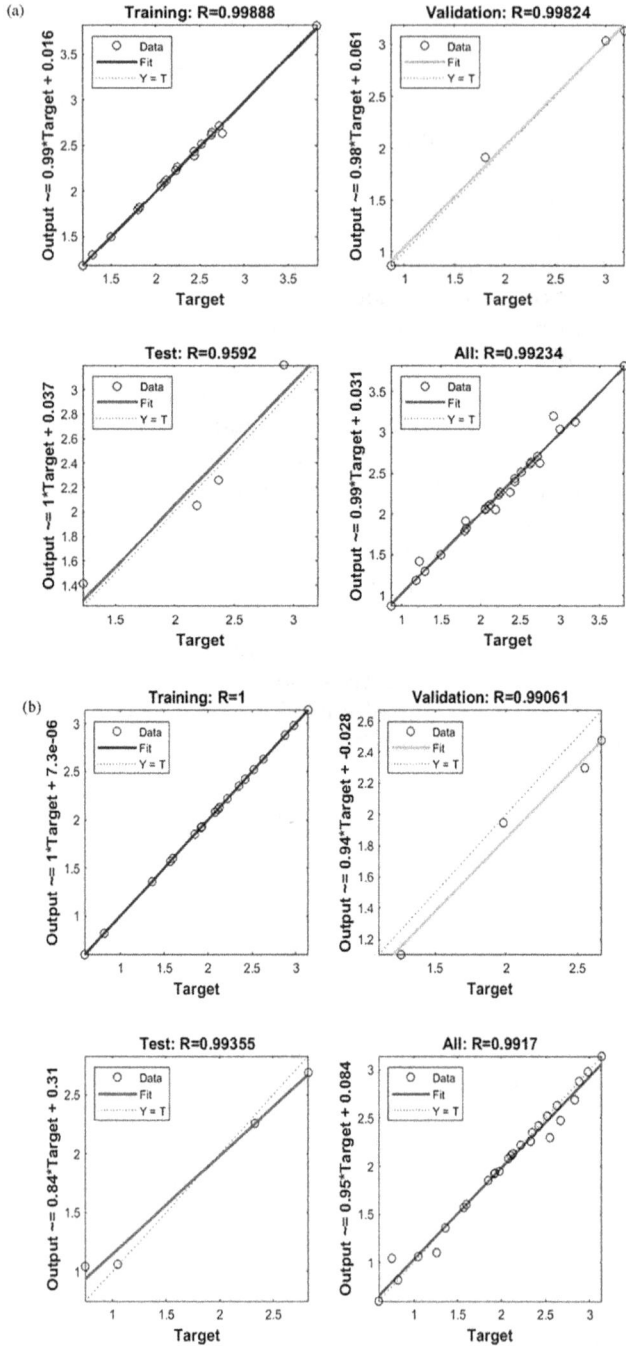

Figure 16.9 Correlation of neural network and experimental training, testing, validation data of (a) WC-Co, (b) WC-Co-Cr and (c) Cr3C2NiCr coated samples.

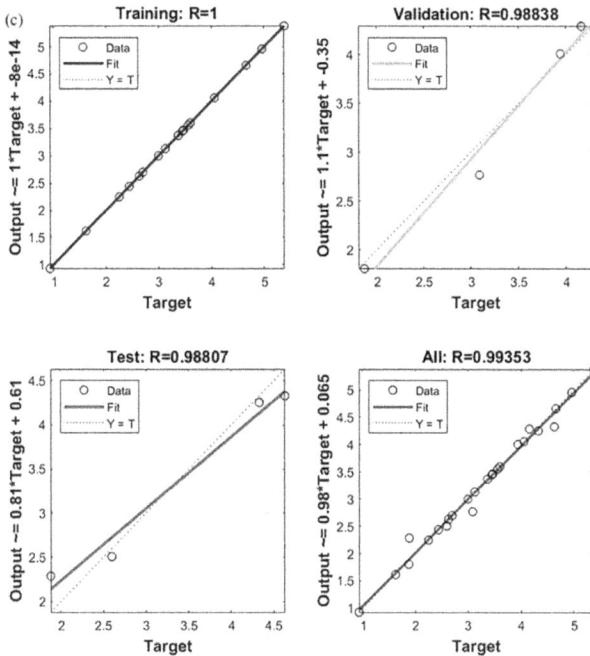

Figure 16.9 (Continued) Correlation of neural network and experimental training, testing, validation data of (a) WC-Co, (b) WC-Co-Cr and (c) Cr3C2NiCr coated samples.

The correlation coefficients of training, validation and test are greater than 0.9, which depicts that the characteristic of the network model is satisfactory as well as adequate. MSEs of wear rate are negligible for the training set and relatively small for validation data set and test data set (tending towards zero). If the MSE tends to zero, the model characteristic is considered excellent [14]. The experimental and predicted values of regression and ANN were compared for three different coated samples (Figure 16.10). It reveals the existence of a good agreement between predicted values with experimental values. Also, ANN predicted values are close to experimental values compared to regression predicted values with error lies in the acceptable range. Hence, the developed ANN model can be considered feasible for wear rate prediction of coated samples considered in the work within the range of studied parameters.

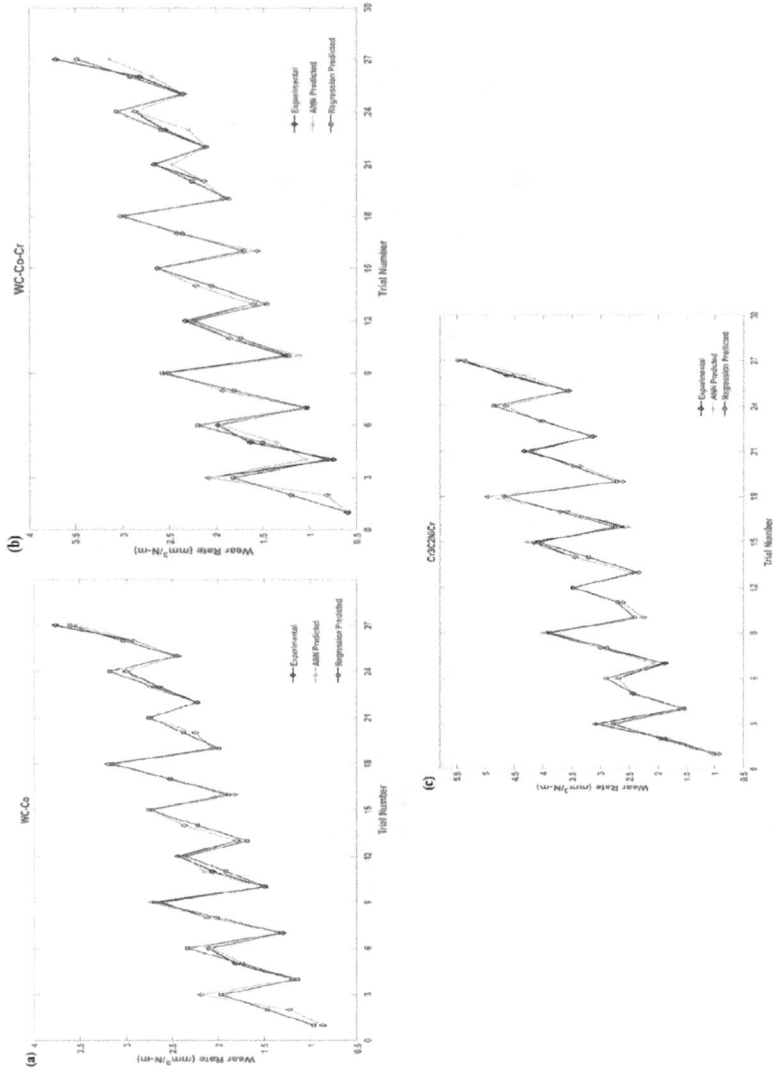

Figure 16.10 Comparison of experimental and predicted results of (a) WC-Co, (b) WC-Co-Cr and (c) Cr3C2NiCr coated samples.

16.4 CONCLUSION

The current research is focussed on the wear of Cr3C2NiCr, WC-Co and WC-Co-Cr HVOF coated AISI 4340 steel substrate. The effect of process parameters on the wear of coated samples was investigated. The current investigation has led to the following conclusions.

- The microstructural characterization reveals the presence of porous coated surfaces. Besides, a greater amount of porosity was found in Cr3C2NiCr coated substrate.
- The main effect plot of experimental results showed that wear rate increased with increased wear parameters. However, a drastic increase in the rate of wear was found under the increase in normal load.
- The minimum wear rate of coated sample was recorded for WC-Co-Cr, followed by WC-Co and Cr3C2NiCr owing to their decrease in hardness. It depicts that WC-Co-Cr possesses better wear resistance compared to the other two coatings.
- ANOVA of experimental results for different studied samples revealed that individual parameters effect was statistically and physically significant, while interaction effect of wear parameters was not found. In addition, the normal load was the most significant parameter followed by sliding distance and speed, respectively.
- Regression models were presented with a correlation of wear parameters with wear rate. The developed models were presented in the chapter. The R-Sq. (Adj.), R-Sq. and R-Sq. (Pred.) of established models were greater than 90% illustrating the feasibility adequacy of models for the prediction of wear rate.
- ANN model was developed with 3-10-1 network architecture and trained with the LM algorithm. The mean-squared error values of ANN models were much low and tended towards zero indicating their better model characteristics. The R-values (>0.9) of ANN models confirms their feasibility. The model exhibited better adequacy compared to regression models in predicting wear rate.

REFERENCES

1. G. Bolelli, L. Lusvarghi, M. Barletta, HVOF-sprayed WC-CoCr coatings on Al alloy: Effect of the coating thickness on the tribological properties, *Wear* 267 (2009), No. 5, pp. 944–953 DOI:10.1016/j.wear.2008.12.066.
2. A. K. Maiti, N. Mukhopadhyay, R. Raman, Improving the wear behavior of WC-CoCr-based HVOF coating by surface grinding, *Journal of Materials Engineering and Performance* 18 (2009), No. 8, pp. 1060–1066 DOI:10.1007/s11665-009-9354-5.

3. X. Ding, X. D. Cheng, C. Li, X. Yu, Z. X. Ding, C. Q. Yuan, Microstructure and performance of multi-dimensional WC-CoCr coating sprayed by HVOF, *International Journal of Advanced Manufacturing Technology* 96 (2018), No. 5, pp. 1625–1633 DOI:10.1007/s00170-017-0837-5.

4. R. S. C. Paredes, W. Nikkel, G. B. Sucharski, P. P. B. Costa, Optimization of arc-sprayed iron-based tungsten carbide hard coatings on harvester blades, *Journal of the Brazilian Society of Mechanical Sciences and Engineering* 41 (2019), No. 5, pp. 1–9 DOI:10.1007/s40430-019-1719-6.

5. A. S. Marques, L. D. L. de Costa, G. R. dos Santos, A. da Silva Rocha, Wear study of hot forging punches coated with WC-CoCr and Cr_3C_2-NiCr through high-velocity oxygen fuel (HVOF) process, *International Journal of Advanced Manufacturing Technology* 100 (2019), No. 1, pp. 3–11 DOI:10.1007/s00170-018-2693-3.

6. V. Bolleddu, V. Racherla, P. P. Bandyopadhyay, Comparative study of air plasma sprayed and high velocity oxy-fuel sprayed nanostructured WC-17wt%Co coatings, *International Journal of Advanced Manufacturing Technology* 84 (2016), No. 5, pp. 1601–1613 DOI:10.1007/s00170-015-7824-5.

7. Y. Liu, J. Cheng, B. Yin, S. Zhu, Z. Qiao, J. Yang, Study of the tribological behaviors and wear mechanisms of WC-Co and WC-Fe3 Al hard materials under dry sliding condition, *Tribiology International* 109 (2017), pp. 19–25 DOI:10.1016/j.triboint.2016.12.023.

8. S. Hui, S. Tianyuan, K. Dejun, Surface and cross-section characteristics and friction–wear properties of high velocity oxy fuel sprayed WC-12Co coating, *International Journal of Applied Ceramic Technology* 15 (2018), No. 5, pp. 1229–1239 DOI:10.1111/ijac.12884.

9. T. Gong, P. Yao, X. Zuo et al., Influence of WC carbide particle size on the microstructure and abrasive wear behavior of WC-10Co-4Cr coatings for aircraft landing gear, *Wear* 362–363 (2016), pp. 135–145 DOI:10.1016/j.wear.2016.05.022.

10. S. Fayyazi, M. Kasraei, M. E. Bahrololoom, Improving impact resistance of high-velocity oxygen fuel-sprayed WC-17Co coating using Taguchi experimental design, *Journal of Thermal Spray Technology*, 28 (2019), No. 4, pp. 706–716 DOI: 10.1007/s11666-019-00844-6.

11. R. Ahmed, O. Li, N. H. Faisal, N. M. Al-Anazi, S. Al-Mutairi, F.-L. Toma, L.-M. Berger, A. Potthoff, M. F. A. Goosen, Sliding wear investigation of suspension sprayed WC-Co nanocomposite coatings, *Wear* 322–323 (2015), pp. 133–150 DOI:10.1016/j.wear.2014.10.021.

12. H. S. Sidhu, B. S. Sidhu, S. Prakash, Wear characteristics of Cr3C2–NiCr and WC–Co coatings deposited by LPG fueled HVOF, *Tribology International* 43 (2010) 887–890 DOI:10.1016/j.triboint.2009.12.016.

13. H. Ozkavak, S. Sahin, M. Sarac, Z. Alkan, Wear properties of WC–Co and WC–CoCr coatings applied by HVOF technique on different steel substrates, *Materials Testing*, 62 (2020), No. 12), pp. 1235–1242 DOI:10.3139/120.111609.

14. Y. Sun, W. D. Zeng, X. M. Zhang, Y. Q. Zhao, X. Ma, Y. F. Han, Prediction of tensile property of hydrogenated Ti600 titanium alloy using artificial neural network, *Journal of Materials Engineering and Performance*, 20 (2011), No. 3, pp. 335–340.

Index

For Product Safety Concerns and Information please contact our EU
representative GPSR@taylorandfrancis.com
Taylor & Francis Verlag GmbH, Kaufingerstraße 24, 80331 München, Germany